UCLA Symposia on Molecular and Cellular Biology, New Series

Series Editor, C. Fred Fox

RECENT TITLES

Volume 108
Acute Lymphoblastic Leukemia, Robert Peter Gale and Dieter Hoelzer, *Editors*

Volume 109
Frontiers of NMR in Molecular Biology, David Live, Ian M. Armitage, and Dinshaw J. Patel, *Editors*

Volume 110
Protein and Pharmaceutical Engineering, Charles S. Craik, Robert J. Fletterick, C. Robert Matthews, and James A. Wells, *Editors*

Volume 111
Glycobiology, Ernest G. Jaworski and Joseph R. Welply, *Editors*

Volume 112
New Directions in Biological Control: Alternatives for Suppressing Agricultural Pests and Diseases, Ralph R. Baker and Peter E. Dunn, *Editors*

Volume 113
Immunogenicity, Charles A. Janeway, Jr., Jonathan Sprent, and Eli Sercarz, *Editors*

Volume 114
Genetic Mechanisms in Carcinogenesis and Tumor Progression, Curtis Harris and Lance A. Liotta, *Editors*

Volume 115
Growth Regulation of Cancer II, Marc E. Lippman and Robert B. Dickson, *Editors*

Volume 116
Transgenic Models in Medicine and Agriculture, Robert B. Church, *Editor*

Volume 117
Early Embryo Development and Paracrine Relationships, Susan Heyner and Lynn M. Wiley, *Editors*

Volume 118
Cellular and Molecular Biology of Normal and Abnormal Erythroid Membranes, Carl M. Cohen and Jiri Palek, *Editors*

Volume 119
Human Retroviruses, Jerome E. Groopman, Irvin S.Y. Chen, Myron Essex, and Robin A. Weiss, *Editors*

Volume 120
Hematopoiesis, David W. Golde and Steven C. Clark, *Editors*

Volume 121
Defense Molecules, John J. Marchalonis and Carol L. Reinisch, *Editors*

Volume 122
Molecular Evolution, Michael T. Clegg and Stephen J. O'Brien, *Editors*

Volume 123
Molecular Biology of Aging, Caleb E. Finch and Thomas E. Johnson, *Editors*

Volume 124
Papillomaviruses, Peter M. Howley and Thomas R. Broker, *Editors*

Volume 125
Developmental Biology, Eric H. Davidson, Joan V. Ruderman, and James W. Posakony, *Editors*

Volume 126
Biotechnology and Human Genetic Predisposition to Disease, Charles R. Cantor, C. Thomas Caskey, Leroy E. Hood, Daphne Kamely, and Gilbert S. Omenn, *Editors*

Volume 127
Molecular Mechanisms in DNA Replication and Recombination, Charles C. Richardson and I. Robert Lehman, *Editors*

Volume 128
Nucleic Acid Methylation, Gary A. Clawson, Dawn B. Willis, Arthur Weissbach, and Peter A. Jones, *Editors*

Volume 129
Plant Gene Transfer, Christopher J. Lamb and Roger N. Beachy, *Editors*

Volume 130
Parasites: Molecular Biology, Drug and Vaccine Design, Nina M. Agabian and Anthony Cerami, *Editors*

Volume 131
Molecular Biology of the Cardiovascular System, Robert Roberts and Joseph F. Sambrook, *Editors*

Volume 132
Obesity: Towards a Molecular Approach, George A. Bray, Daniel Ricquier, and Bruce M. Spiegelman, *Editors*

Volume 133
Structural and Organizational Aspects of Metabolic Regulation, Paul A. Srere, Mary Ellen Jones, and Christopher K. Mathews, *Editors*

Please contact the publisher for information about previous titles in this series.

UCLA Symposia Board

C. Fred Fox, Ph.D., Director
Professor of Microbiology, University of California, Los Angeles

Charles J. Arntzen, Ph.D.
Director, Plant Science and Microbiology
E.I. du Pont de Nemours and Company

Floyd E. Bloom, M.D.
Director, Preclinical Neurosciences/
Endocrinology
Scripps Clinic and Research Institute

Ralph A. Bradshaw, Ph.D.
Chairman, Department of Biological
Chemistry
University of California, Irvine

Francis J. Bullock, M.D.
Vice President, Research
Schering Corporation

Ronald E. Cape, Ph.D., M.B.A.
Chairman
Cetus Corporation

Ralph E. Christoffersen, Ph.D.
Executive Director of Biotechnology
Upjohn Company

John Cole, Ph.D.
Vice President of Research
and Development
Triton Biosciences

Pedro Cuatrecasas, M.D.
Vice President of Research
Glaxo, Inc.

Mark M. Davis, Ph.D.
Department of Medical Microbiology
Stanford University

J. Eugene Fox, Ph.D.
Vice President, Research
and Development
Miles Laboratories

J. Lawrence Fox, Ph.D.
Vice President, Biotechnology Research
Abbott Laboratories

L. Patrick Gage, Ph.D.
Director of Exploratory Research
Hoffmann-La Roche, Inc.

Gideon Goldstein, M.D., Ph.D.
Vice President, Immunology
Ortho Pharmaceutical Corp.

Ernest G. Jaworski, Ph.D.
Director of Biological Sciences
Monsanto Corp.

Irving S. Johnson, Ph.D.
Vice President of Research
Lilly Research Laboratories

Paul A. Marks, M.D.
President
Sloan-Kettering Memorial Institute

David W. Martin, Jr., M.D.
Vice President of Research
Genentech, Inc.

Hugh O. McDevitt, M.D.
Professor of Medical Microbiology
Stanford University School of Medicine

Dale L. Oxender, Ph.D.
Director, Center for Molecular Genetics
University of Michigan

Mark L. Pearson, Ph.D.
Director of Molecular Biology
E.I. du Pont de Nemours and Company

George Poste, Ph.D.
Vice President and Director of Research and
Development
Smith, Kline and French Laboratories

William Rutter, Ph.D.
Director, Hormone Research Institute
University of California, San Francisco

George A. Somkuti, Ph.D.
Eastern Regional Research Center
USDA-ARS

Donald F. Steiner, M.D.
Professor of Biochemistry
University of Chicago

UCLA Symposia Board membership at the time of the meeting is indicated on the above list.

Nucleic Acid Methylation

Nucleic Acid Methylation

Proceedings of a Hoffman-La Roche-UCLA Colloquium on
Nucleic Acid Methylation Held at
Frisco, Colorado, March 31–April 7, 1989

Editors

Gary A. Clawson
Department of Pathology
George Washington University Medical Center
Washington, DC

Dawn B. Willis
Department of Virology/Molecular Biology
St. Jude's Children's Research Hospital
Memphis, Tennessee

Arthur Weissbach
Roche Institute of Molecular Biology
Nutley, New Jersey

Peter A. Jones
Norris Cancer Research Institute
University of Southern California
Los Angeles, California

A JOHN WILEY & SONS, INC., PUBLICATION
New York • Chichester • Brisbane • Toronto • Singapore

Address all Inquiries to the Publisher
Alan R. Liss, Inc., 41 East 11th Street, New York, NY 10003

Copyright © 1990 Alan R. Liss, Inc.

Printed in United States of America

Under the conditions stated below the owner of copyright for this book hereby grants permission to users to make photocopy reproductions of any part or all of its contents for personal or internal organizational use, or for personal or internal use of specific clients. This consent is given on the condition that the copier pay the stated per-copy fee through the Copyright Clearance Center, Incorporated, 27 Congress Street, Salem, MA 01970, as listed in the most current issue of "Permissions to Photocopy" (Publisher's Fee List, distributed by CCC, Inc.), for copying beyond that permitted by sections 107 or 108 of the US Copyright Law. This consent does not extend to other kinds of copying, such as copying for general distribution, for advertising or promotional purposes, for creating new collective works, or for resale.

The publication of this volume was facilitated by the authors and editors who submitted the text in a form suitable for direct reproduction without subsequent editing or proofreading by the publisher.

Library of Congress Cataloging-in-Publication Data

Nucleic acid methylation : proceedings of a UCLA colloquium held at
  Frisco, Colorado, March 31-April 7, 1989 / editors, Gary A. Clawson
  ... [et al.].
      p.   cm. -- (UCLA symposia on molecular and cellular biology ;
  new ser., v. 128)
    "UCLA Symposia Colloquium on Nucleic Acid Methylation"--Pref.
    The Colloquium was sponsored by Hoffmann-La Roche, inc.
    Includes bibliographical references.
    ISBN 0-471-56727-2
    1. DNA--Methylation--Congresses. 2. RNA--Methylation--Congresses.
  I. Clawson, Gary A. II. UCLA Symposia Colloquium on Nucleic Acid
  Methylation (1989 : Frisco, Colo.) III. Hoffmann-La Roche, inc.
  IV. Series.
    [DNLM: 1. DNA--congresses. 2. Methylation--congresses. 3. RNA-
  -congresses.   W3 U17N new ser. v. 128 / QU 58 N9637 1989]
  QP624.N83 1989
  574.87'3282--dc20
  DNLM/DLC
  for Library of Congress   89-24997
                                                                 CIP

Contents

Contributors xi

Preface
Peter A. Jones, Gary A. Clawson, Dawn B. Willis,
and Arthur Weissbach xix

I. RIBOSOMAL RNA METHYLATION

In Vitro Methylation of *Escherichia coli* 16S RNA, 23S RNA, and 30S Ribosomes by Homologous Cell-Free Extracts
Didier Nègre, Carl Weitzmann, and James Ofengand 1

On the Evolutionary Relationships Between Adenosine Dimethylation in 16S and 23S Ribosomal RNA
Bob van Gemen and Peter H. van Knippenberg 19

Modified Nucleotides in Eukaryotic Ribosomal RNA
B. Edward H. Maden 37

II. CAPPING REACTIONS

Structure and Function of Yeast mRNA Capping Enzyme
Kiyohisa Mizumoto, Yoshio Shibagaki, Naoto Itoh, Hisafumi Yamada, Shigekazu Nagata, and Yoshito Kaziro 45

A Nuclear (Guanine-7-)Methyltransferase Isolated From Ehrlich Ascites Cells
Guojun Bu and Thomas O. Sitz 57

Effects of Hypermethylated and Aberrant 5′-Caps on Translation of β-Globin and Sindbis Virus 26S mRNAs
Stanley M. Tahara, Edward Darzynkiewicz, Janusz Stepinski, Irena Ekiel, Youxin Jin, Chang Hahn, James H. Strauss, Teju Sijuwade, and Dorota Haber 67

III. FUNCTION OF RNA METHYLATION

Internal N^6-Methyladenosine Sites in Rous Sarcoma Virus RNA
Tunde Csepany, Susan E. Kane, Alison Lin, Carl J. Baldick, Jr., and Karen Beemon 83

The Relationship Between Protein Synthesis and Ribosomal RNA Methylation
Gary A. Clawson, Julie Sesno, and Kathy Milam.................... 93

The Role of RNA Methylation in the Differentiation of L5 Myoblast Cell Line
S. Scarpa, L. Di Renzo, and R. Strom........................... 103

IV. DNA METHYLATION IN BACTERIA AND VIRUSES

DNA Methylation and Restriction Processes in *Escherichia coli*: Insights by Use of Bacterial Viruses T3 and T7
Detlev H. Krüger, Thomas A. Bickle, Monika Reuter, Claus-Dietmar Pein, and Cornelia Schroeder....................... 113

***Erm* Methylases in Staphylococci and Their Mode of Spread**
Donald T. Dubin, Lucia E. Tillotson, Leslie Huwyler, and Ellen Murphy.. 125

Transcription of a Methylated DNA Virus
Dawn B. Willis, Karim Essani, Rakesh Goorha, James P. Thompson, and Allan Granoff.. 139

The Role of DNA Methylation in HIV Latency
D.P. Bednarik, J.A. Cook, and P.M. Pitha......................... 153

The Role of DNA Methylation in a Retrovirus-Induced Insertional Mutation of the Murine Alpha 1(I) Collagen Gene
Michael Breindl, Hedy Chan, and Stefan Hartung................... 171

V. DNA METHYLATION IN PLANTS

Role of DNA Methylation in the Regulation of Gene Expression in Plants
Richard M. Amasino, Manorama C. John, Manfred Klass, and Dring N. Crowell... 187

Biosynthesis and Distribution of Methylcytosine in Wheat DNA. How Different Are Plant DNA Methyltransferases?
Hartmut Follmann, Hans-Jörg Balzer, and Roland Schleicher........... 199

VI. INHIBITORS OF DNA METHYLATION

A Novel Strategy for Identifying Potential Targets of Altered DNA Methylation in Neoplastic Transformation
Andrew P. Feinberg and Shirley Rainier.......................... 211

The Activation/Inactivation-Prone Prolactin Gene in GH_3 Rat Pituitary Cells: Silencing by EMS and Reactivating by 5-Azacytidine
Iain K. Farrance, Julie Morris, Todd E. Arnold, Iris S. Hall, and Robert Ivarie... 229

Antiviral Activities of 3-Deaza Nucleosides—Indirect Inhibitors of Methylation
Peter K. Chiang... 247

Contents   ix

VII. REGULATION OF GENE EXPRESSION BY DNA METHYLATION

Changing Methylation Patterns During Development
Aharon Razin, Dale Frank, Michal Lichtenstein, Zeev Paroush,
Yehudit Bergman, Moshe Shani, and Howard Cedar................. 257

Regulation of Transcription by DNA Methylation in SV40 and the HSV TK Gene in Oocytes
F. Götz, H. Wagner, H. Kröger, and D. Simon..................... 275

DNA Methylation in the 5' Region of the Mouse PGK-1 Gene and a Quantitative PCR Assay for Methylation
J. Singer-Sam, T.P. Yang, N. Mori, R.L. Tanguay, J.M. Le Bon,
J.C. Flores, and A.D. Riggs..................................... 285

Studying DNA Methylation and Protein-DNA Interaction Sites *In Vivo*
Hanspeter Saluz, Stefan Wölfl, Antonio Milici, and Jean-Pierre Jost...... 299

Hormone-Dependent DNA Demethylation and Protein DNA Interactions in the Promoter Region and Estradiol Response Element of Avian Vitellogenin Gene
Jean-Pierre Jost, Hanspeter Saluz, Iain McEwan, Melya Hughes,
Hai-Min Liang, and Marylin Vaccaro............................. 313

On the Mechanism of Promoter Inactivation by Sequence-Specific Methylation
Walter Doerfler, Miklos Toth, Ralf Hermann, Ursula Lichtenberg, and
Arnd Hoeveler... 329

Methylated DNA-Binding Protein From Mammalian Cells
Melanie Ehrlich, Xian-Yang Zhang, Clement K. Asiedu, Rana Khan,
and Prakash C. Supakar.. 351

VIII. ROLE OF DNA METHYLATION IN IMPRINTING

Transgene Methylation Imprints Are Established Post-Fertilization
Ross McGowan, Thu-Hang Tran, Jean Paquette, and Carmen Sapienza... 367

Allele-Specific Methylation in Human Cells
Hamid Ghazi, Anne E. Erwin, Robin J. Leach, and Peter A. Jones...... 381

The Genetics of Mosaic Methylation Patterns in Humans
Alcino J. Silva, Kenneth Ward, and Raymond White................ 395

Index... 409

Contributors

Richard M. Amasino, Department of Biochemistry, University of Wisconsin-Madison, Madison, WI 53706 [**187**]

Todd E. Arnold, Department of Genetics, University of Georgia, Athens, GA 30602 [**229**]

Clement K. Asiedu, Department of Biochemistry, Tulane Medical School, New Orleans, LA 70112 [**351**]

Carl J. Baldick, Jr., Department of Biology, Johns Hopkins University, Baltimore, MD 21218 [**83**]

Hans-Jörg Balzer, Fachbereich Chemie (Biochemie) der Philipps-Universität, D-3550 Marburg, Federal Republic of Germany [**199**]

D. P. Bednarick, Oncology Center, The Johns Hopkins University, Baltimore, MD 21205 [**153**]

Karen Beemon, Department of Biology, Johns Hopkins University, Baltimore, MD 21218 [**83**]

Yehudit Bergman, Department of Experimental Medicine, The Hebrew University-Hadassah Medical School, Jerusalem, Israel [**257**]

Thomas A. Bickle, Department of Microbiology, Biocenter, University of Basel, CH-4056 Basel, Switzerland [**113**]

Michael Breindl, Department of Biology and Molecular Biology Institute, San Diego State University, San Diego, CA 92182 [**171**]

Guojun Bu, Department of Biochemistry, Virginia Tech, Blacksburg, VA 24061 [**57**]

Howard Cedar, Department of Cellular Biochemistry, The Hebrew University-Hadassah Medical School, Jerusalem, Israel [**257**]

Hedy Chan, Department of Biology and Molecular Biology Institute, San Diego State University, San Diego, CA 92182 [**171**]

Peter K. Chiang, Department of Applied Biochemistry, Walter Reed Army Institute of Research, Washington, DC 20307-5100 [**247**]

Gary A. Clawson, Department of Pathology, George Washington University, Washington, DC 20037 [**xix, 93**]

The numbers in brackets are the opening page numbers of the contributors' articles.

Contributors

J. A. Cook, Oncology Center, The Johns Hopkins University, Baltimore, MD 21205 [153]

Dring N. Crowell, Department of Biochemistry, University of Wisconsin-Madison, Madison, WI 53706 [187]

Tunde Csepany, Department of Biology, Johns Hopkins University, Baltimore, MD 21218; present address: Department of Neurology and Psychiatry, Medical University of Debrecen, Debrecen, Hadju 4012, Hungary [83]

Edward Darzynkiewicz, Department of Biophysics, Institute of Experimental Physics, University of Warsaw, Warsaw 02-089, Poland [67]

L. Di Renzo, Department of Human Pathology and 1st Institute of Clinical Surgery, Università 'La Sapienza', Rome 00161, Italy [103]

Walter Doerfler, Institute of Genetics, University of Cologne, Cologne, Germany [329]

Donald T. Dubin, Department of Molecular Genetics and Microbiology, UMDNJ-Robert Wood Johnson Medical School, Piscataway, NJ 08854 [125]

Melanie Ehrlich, Department of Biochemistry, Tulane Medical School, New Orleans, LA 70112 [351]

Irena Ekiel, Biotechnology Research Institute, National Research Council of Canada, Montreal, Quebec N4P 2R2, Canada [67]

Anne E. Erwin, Department of Biochemistry, University of Southern California Medical School, Los Angeles, CA 90033 [381]

Karim Essani, Department of Virology and Molecular Biology, St. Jude Children's Research Hospital, Memphis, TN 38101 [139]

Iain K. Farrance, Department of Genetics, University of Georgia, Athens, GA 30602; present address: Department of Anatomy, University of California, San Francisco, CA 94143 [229]

Andrew P. Feinberg, Howard Hughes Medical Institute, Departments of Internal Medicine and Human Genetics, University of Michigan Medical School, Ann Arbor, MI 48109 [211]

J. C. Flores, Biology Division, Beckman Research Institute of the City of Hope, Duarte, CA 91010 [285]

Hartmut Follmann, Fachbereich Chemie (Biochemie) der Philipps-Universität, D-3550 Marburg, Federal Republic of Germany; present address: Fachbereich Biologie-Chemie der Universität, D-3500 Kassel, Federal Republic of Germany [199]

Dale Frank, Department of Cellular Biochemistry, The Hebrew University-Hadassah Medical School, Jerusalem, Israel [257]

Hamid Ghazi, Department of Biochemistry, University of Southern California Medical School, Los Angeles, CA 90033 [381]

Rakesh Goorha, Department of Virology and Molecular Biology, St. Jude Children's Research Hospital, Memphis, TN 38101 [139]

Contributors xiii

F. Götz, Department of Biochemistry, Robert Koch-Institut, D-1000 Berlin 65, Federal Republic of Germany **[275]**

Allan Granoff, Department of Virology and Molecular Biology, St. Jude Children's Research Hospital, Memphis, TN 38101 **[139]**

Dorota Haber, Department of Biophysics, Institute of Experimental Physics, University of Warsaw, Warsaw 02-089, Poland **[67]**

Chang Hahn, Division of Biology, California Institute of Technology, Pasadena, CA 91125 **[67]**

Iris S. Hall, Department of Genetics, University of Georgia, Athens, GA 30602; present address: New England Biolabs, Beverly, MA 01915 **[229]**

Stefan Hartung, Department of Biology and Molecular Biology Institute, San Diego State University, San Diego, CA 92182 **[171]**

Ralf Hermann, Institute of Genetics, University of Cologne, Cologne, Germany **[329]**

Arnd Hoeveler, Institute of Genetics, University of Cologne, Cologne, Germany; present address: Centre d'Immunologie, INSERM-CNRS, Marseille, France **[329]**

Melya Hughes, Friedrich Miescher-Institut, CH-4002 Basel, Switzerland **[313]**

Leslie Huwyler, The Public Health Research Institute, New York, NY 10016 **[125]**

Naoto Itoh, The Institute of Medical Science, The University of Tokyo, Tokyo 108, Japan; present address: DNAX Research Institute, Palo Alto, CA 94304 **[45]**

Robert Ivarie, Department of Genetics, University of Georgia, Athens, GA 30602 **[229]**

Youxin Jin, Department of Microbiology, University of Southern California School of Medicine, Los Angeles, CA 90033-1054; present address: Shanghai Institute of Biochemistry, Academia Sinica, Shanghai, People's Republic of China **[67]**

Manorama C. John, Department of Biochemistry, University of Wisconsin-Madison, Madison, WI 53706 **[187]**

Peter A. Jones, Kenneth Norris Jr. Comprehensive Cancer Center, University of Southern California Medical School, Los Angeles, CA 90033 **[xix, 381]**

Jean-Pierre Jost, Friedrich Miescher-Institut, CH-4002 Basel, Switzerland **[299, 313]**

Susan E. Kane, Department of Biology, Johns Hopkins University, Baltimore, MD 21218; present address: Laboratory of Molecular Biology, National Cancer Institute, Bethesda, MD 20892 **[83]**

Yoshito Kaziro, The Institute of Medical Science, The University of Tokyo, Tokyo 108, Japan **[45]**

Contributors

Rana Khan, Department of Biochemistry, Tulane Medical School, New Orleans, LA 70112 [351]

Manfred Klaas, Department of Biochemistry, University of Wisconsin-Madison, Madison, WI 53706 [187]

H. Kröger, Department of Biochemistry, Robert Koch-Institut, D-1000 Berlin 65, Federal Republic of Germany [275]

Detlev H. Krüger, Institute of Medical Virology, Humboldt University, DDR-1040 Berlin, German Democratic Republic [113]

Robin J. Leach, Kenneth Norris Jr. Comprehensive Cancer Center, University of Southern California Medical School, Los Angeles, CA 90033 [381]

J. M. Le Bon, Biology Division, Beckman Research Institute of the City of Hope, Duarte, CA 91010 [285]

Hai-Min Liang, Friedrich Miescher-Institut, CH-4002 Basel, Switzerland [313]

Ursula Lichtenberg, Institute of Genetics, University of Cologne, Cologne, Germany; present address: The George William Hooper Foundation, University of California, San Francisco, CA 94143 [329]

Michal Lichtenstein, Department of Experimental Medicine, The Hebrew University-Hadassah Medical School, Jerusalem, Israel [257]

Alison Lin, Department of Biology, Johns Hopkins University, Baltimore, MD 21218; present address: Baylor College of Medicine, Houston, TX 77030 [83]

B. Edward H. Maden, Department of Biochemistry, Liverpool University, Liverpool L69 3BX, England [37]

Iain McEwan, Friedrich Miescher-Institut, CH-4002 Basel, Switzerland [313]

Ross McGowan, Department of Developmental Genetics, Ludwig Institute for Cancer Research, Montreal, Quebec, Canada H3A 1A1 [367]

Kathy Milam, Department of Pathology, George Washington University, Washington, DC 20037 [93]

Antonio Milici, M. D. Anderson Tumor Institute, Houston, TX 77030 [299]

Kiyohisa Mizumoto, The Institute of Medical Science, The University of Tokyo, Tokyo 108, Japan [45]

N. Mori, Biology Division, Beckman Research Institute of the City of Hope, Duarte, CA 91010; present address: Division of Biology 216-76, California Institute of Technology, Pasadena, CA 91125 [285]

Julie Morris, Department of Genetics, University of Georgia, Athens, GA 30602 [229]

Ellen Murphy, The Public Health Research Institute, New York, NY 10016 [125]

Contributors xv

Shigekazu Nagata, The Institute of Medical Science, The University of Tokyo, Tokyo 108, Japan; present address: Osaka Bioscience Institute, Suita-shi, Osaka 565, Japan **[45]**

Didier Nègre, Department of Biochemistry, Roche Institute of Molecular Biology, Roche Research Center, Nutley, NJ 07110; present address: Laboratoire de Biologie Moléculaire, Université Claude Bernard, Villeurbanne Cedex, France **[1]**

James Ofengand, Department of Biochemistry, Roche Institute of Molecular Biology, Roche Research Center, Nutley, NJ 07110 **[1]**

Jean Paquette, Department of Developmental Genetics, Ludwig Institute for Cancer Research, Montreal, Quebec, Canada H3A 1A1 **[367]**

Zeev Paroush, Department of Cellular Biochemistry, The Hebrew University-Hadassah Medical School, Jerusalem, Israel **[257]**

Claus-Dietmar Pein, Department of Chemistry, Humboldt University, DDR-1040 Berlin, German Democratic Republic **[113]**

P. M. Pitha, Department of Molecular Biology and Genetics, The Johns Hopkins University, Baltimore, MD 21205 **[153]**

Shirley Rainier, Howard Hughes Medical Institute, Departments of Internal Medicine and Human Genetics, University of Michigan Medical School, Ann Arbor, MI 48109 **[211]**

Aharon Razin, Department of Cellular Biochemistry, The Hebrew University-Hadassah Medical School, Jerusalem, Israel **[257]**

Monika Reuter, Institute of Medical Virology, Humboldt University, DDR-1040 Berlin, German Democratic Republic **[113]**

A. D. Riggs, Biology Division, Beckman Research Institute of the City of Hope, Duarte, CA 91010 **[285]**

Hanspeter Saluz, Friedrich Miescher-Institut, CH-4002 Basel, Switzerland **[299, 313]**

Carmen Sapienza, Department of Developmental Genetics, Ludwig Institute for Cancer Research, Montreal, Quebec, Canada H3A 1A1 **[367]**

S. Scarpa, Department of Human Biopathology and 1st Institute of Clinical Surgery, Università 'La Sapienza', Rome 00161, Italy **[103]**

Roland Schleicher, Fachbereich Chemie (Biochemie) der Philipps-Universität, D-3550 Marburg, Federal Republic of Germany **[199]**

Cornelia Schroeder, Institute of Medical Virology, Humboldt University, DDR-1040 Berlin, German Democratic Republic **[113]**

Julie Sesno, Department of Pathology, George Washington University, Washington, DC 20037 **[93]**

Moshe Shani, Institute of Animal Sciences, ARD, The Volcani Center, Bet Dagon, Israel **[257]**

Yoshio Shibagaki, The Institute of Medical Science, The University of Tokyo, Tokyo 108, Japan **[45]**

Teju Sijuwade, Department of Microbiology, University of Southern California School of Medicine, Los Angeles, CA 90033-1054 [67]

Alcino J. Silva, Department of Human Genetics and Howard Hughes Medical Institute, University of Utah, Salt Lake City, UT 84132; present address: Department of Biology, Massachusetts Institute of Technology, Center for Cancer Research, Cambridge, MA 02139 [395]

D. Simon, Department of Biochemistry, Robert Koch-Institut, D-1000 Berlin 65, Federal Republic of Germany [275]

J. Singer-Sam, Biology Division, Beckman Research Institute of the City of Hope, Duarte, CA 91010 [285]

Thomas O. Sitz, Department of Biochemistry, Virginia Tech, Blacksburg, VA 24061 [57]

Janusz Stepinski, Department of Organic Chemistry, Institute of Chemistry, University of Warsaw, Warsaw 02-093, Poland [67]

James H. Strauss, Division of Biology, California Institute of Technology, Pasadena, CA 91125 [67]

R. Strom, Department of Human Biopathology and 1st Institute of Clinical Surgery, Università 'La Sapienza', Rome 00161, Italy [103]

Prakash C. Supakar, Department of Biochemistry, Tulane Medical School, New Orleans, LA 70112 [351]

Stanley M. Tahara, Department of Microbiology, University of Southern California School of Medicine, Los Angeles, CA 90033-1054 [67]

R. L. Tanguay, Biology Division, Beckman Research Institute of the City of Hope, Duarte, CA 91010 [285]

James P. Thompson, Department of Virology and Molecular Biology, St. Jude Children's Research Hospital, Memphis, TN 38101; present address: Department of Medicine, University of Tennessee, Memphis, TN 38163 [139]

Lucia E. Tillotson, Department of Molecular Genetics and Microbiology, UMDNJ-Robert Wood Johnson Medical School, Piscataway, NJ 08854 [125]

Miklos Toth, Institute of Genetics, University of Cologne, Cologne, Germany; present address: Institute of Biochemistry, Biological Research Center, Hungarian Academy of Sciences, Szeged, Hungary [329]

Thu-Hang Tran, Department of Developmental Genetics, Ludwig Institute for Cancer Research, Montreal, Quebec, Canada H3A 1A1 [367]

Marylin Vaccaro, Friedrich Miescher-Institut, CH-4002 Basel, Switzerland [313]

Bob van Gemem, Department of Biochemistry, State University of Leiden, 2333 AL Leiden, The Netherlands; present address: Department of Biochemistry, Gorleaus Laboratories, University of Leiden, 2300 RA Leiden, The Netherlands [19]

Peter H. van Knippenberg, Department of Biochemistry, State University of Leiden, 2333 AL Leiden, The Netherlands; present address: Department of Biochemistry, Gorleaus Laboratories, University of Leiden, 2300 RA Leiden, The Netherlands [19]

H. Wagner, Department of Biochemistry, Robert Koch-Institut, D-1000 Berlin 65, Federal Republic of Germany [275]

Kenneth Ward, Department of Obstetrics and Gynecology, University of Utah, Salt Lake City, UT 84132 [395]

Arthur Weissbach, Roche Institute of Molecular Biology, Nutley, NJ 07110 [xix]

Carl Weitzmann, Department of Biochemistry, Roche Institute of Molecular Biology, Roche Research Center, Nutley, NJ 07110 [1]

Raymond White, Department of Human Genetics and Howard Hughes Medical Institute, University of Utah, Salt Lake City, UT 84132 [395]

Dawn B. Willis, Department of Virology and Molecular Biology, St. Jude Children's Research Hospital, Memphis, TN 38101; present address: Department of Research, American Cancer Society, Atlanta, GA 30392 [xix, 139]

Stefan Wölfl, Institut für Molekularbiologie und Biochemie, Freie Universität Berlin, D-1000 Berlin 33, Federal Republic of Germany [299]

Hisafumi Yamada, The Institute of Medical Science, The University of Tokyo, Tokyo 108, Japan; present address: Department of Molecular Biophysics and Biochemistry, Yale University, New Haven, CT 06511 [45]

T. P. Yang, Biology Division, Beckman Research Institute of the City of Hope, Duarte, CA 91010; present address: Department of Biochemistry and Molecular Biology, University of Florida, College of Medicine, Gainesville, FL 32610 [285]

Xian-Yang Zhang, Department of Biochemistry, Tulane Medical School, New Orleans, LA 70112 [351]

Preface

This first UCLA meeting on **Nucleic Acid Methylation** was sponsored by Hoffmann-La Roche and attended by approximately 100 participants interested in mechanisms and biological functions of nucleic acid methylation. The meeting was organized along traditional lines and included plenary and poster sessions and two workshops. Sessions on RNA modification and DNA modification were interspersed with each other throughout the meeting. The Rocky Mountains provided more than two feet of snow at the beginning of the meeting, which encouraged early interactions between those interested in the two branches of nucleic acid methylation. Subsequently the sun provided superb spring skiing conditions that were much appreciated by all participants.

The meeting opened with a talk by Aharon Razin that summarized the considerable advances that have been made in the field of DNA methylation over the last ten years. Recent developments in the field of capping and methylation of messenger RNA were then treated. The function and structure of ribosomal RNA methylation and the relevance of this methylation to the structure of the ribosome also formed an interesting session of discussion.

There were exciting reports on DNA methylation in fungi and plants, and the work of Eric Selker on genetic instability in *Neurospora* was particularly fascinating. The role of DNA methylation and the control of gene expression was treated in quite some detail, emphasizing the newer techniques of genomic sequencing now being used with particular precision by Hanspeter Saluz, Jean-Pierre Jost, and Walter Doerfler to define exact changes in promoter methylation. Another advance, which was highlighted by a workshop, was the cloning and sequencing by Timothy Bestor of the mammalian DNA methyltransferase. The cloning of this enzyme has been particularly difficult, and the availability of the sequence and the possibility of altering methyltransferase levels in cells by appropriate antisense technology should provide a direct assessment of the actual role of methylation in controlling cellular processes. The roles of DNA methylation in differentiation and tumorigenesis were also discussed. Art Riggs of the City of Hope expanded

Preface

on the concept that DNA methylation might play a fundamental role in cell memory and the stabilization of transcriptionally incompetent states, particularly as they relate to the silencing of the X chromosome.

The final session of the meeting discussed the relationship between DNA methylation and genomic imprinting. Some of the results that have been obtained with transgenic mice are particularly fascinating and show that methylation patterns can be altered by passage through the paternal and maternal germ lines. While it remains to be seen whether this altered methylation has functional significance for differential gene utilization, the availability of testable hypotheses should now spur further experimentation.

The meeting was enjoyed by all the participants, and Walter Doerfler captured the spirit of the meeting during the banquet by elaborating on the general good camaraderie which has characterized this field.

We wish to thank Hoffmann-La Roche, Inc., for its generous support of this meeting. We gratefully acknowledge additional support from the Council for Tobacco Research, U.S.A. Support was also provided by the Director's Fund established by gifts from: Amoco Technology Company, Biotechnology Division; Bristol-Myers Company; Bio-Rad Laboratories; DNAX Research Institute; Chiron Corporation; CIBA-Geigy Corporation, Pharmaceuticals Division; Eastman Kodak Life Sciences Research Laboratories; Genetics Institute; Life Technologies, Inc.; New England Biolabs; Pfizer Central Research, Pfizer, Inc.; Pharmacia LKB Biotechnology, Inc.; and Qiagen, Inc. We also thank the UCLA Symposia staff for their terrific organziation.

Peter A. Jones
Gary A. Clawson
Dawn B. Willis
Arthur Weissbach

IN VITRO METHYLATION OF ESCHERICHIA COLI 16S RNA, 23S RNA, AND 30S RIBOSOMES BY HOMOLOGOUS CELL-FREE EXTRACTS

Didier Nègre,[1] Carl Weitzmann, and James Ofengand

Roche Institute of Molecular Biology, Roche Research Center
Nutley, New Jersey 07110

ABSTRACT Synthetic 16S RNA transcribed *in vitro* or 30S particles reconstituted from it are substrates for m^5C, m^2G, m^3U, and m^6_2A-forming enzymes in *Escherichia coli* extracts. Most of the measured m^2G-forming activity was ribosome-specific, while all of the detectable m^5C activity was RNA-specific. The m^5C-forming enzyme was purified 39-fold. The native enzyme had a MW of 31-38 kilodaltons on a calibrated sizing column. The m^5C-forming enzyme did not recognize naturally-methylated 16S RNA, methylated or unmethylated 23S RNA, poly(dI-dC), or unmethylated 30S ribosomes. Only one m^5C per RNA molecule was synthesized by this enzyme. C967, not C1407, was the methylated residue. A 30S subunit-specific m^2G-synthesizing activity was also partially purified. The enzyme methylated only G966. The 23S RNA gene was cloned and transcribed *in vitro* by T7 polymerase, and was a substrate for methyltransferases. Activities for the synthesis of m^5C, m^5U, and m^2G were found.

INTRODUCTION

The 16S and 23S ribosomal RNAs of *Escherichia coli* each contain a defined set of methylated nucleotides. In 16S RNA, 10 methylated nucleotides are known (1,2). These modified residues are clustered in the 3'-third of the RNA with three ($m^4Cm1402$, m^5C1407, m^3U1498) in the two highly conserved single-stranded sequences 1393-1408 and 1492-

---

[1]Present address: Laboratoire de Biologie Moleculaire, Université Claude Bernard, Villeurbanne Cedex, FRANCE

1505, three (m^2G1516, m^6_2A1518, m^6_2A1519) in the 3'-terminal stem, two (m^2G966, m^5C967) in a partially conserved small loop, and one (m^2G1207) positioned elsewhere in the 3'-third region. The only methylated base found in the 5'-two-thirds of the RNA, m^7G527, is also located in a highly conserved segment. In the three-dimensional structure proposed by Brimacombe *et al* (3), these seemingly separated bases are brought together to form a narrow band around the neck of the small subunit. Within this band, most of the methylated nucleotides are grouped around the cleft of the subunit in the vicinity of the decoding site (4).

The function of the methylated bases of 16S RNA is unknown. Lack of the two m^6_2A residues makes the ribosome kasugamycin-resistant, which fact provided a phenotype for the isolation of mutants unable to form m^6_2A (5), but did not lead to a loss of normal ribosomal function (6,7). However, their absence does affect translational fidelity (8), and modifies the stability of the adjacent stem structure (9). The other methylated bases also appear non-essential for protein synthesis, since ribosomes constructed from unmethylated 16S RNA are able to carry out all of the partial reactions of *in vitro* protein synthesis, albeit at a somewhat lower efficiency (10-12).

With regard to a role in assembly of the 30S subunit, it is believed that methylation occurs late in this process (13,14), but little or nothing is known about the detailed temporal or sequential nature of the methylation reactions. Nothing is known about the substrate specificity either, except that m^6_2A formation requires a 30S particle as substrate (15).

In 23S RNA, so far 16 to 18 methylated nucleotides have been found: 2 m^5C, 1 Cm, 2 to 3 m^5U, 1 m^3U, 1 Um, 1 m^1G, 3 m^2G, 1 m^7G, 1 Gm, 1 m^2A, 2 to 3 m^6A (2, 16). The positions of 13 are known (1,17). As with 16S RNA, the methylation sites are not randomly distributed. Other than the 3 base cluster of m^1G745-$\Psi746$-m^5U747 and m^6A1618, there are no methylations in the 5'-1910 bases of the RNA. Most of the remaining sites are in the vicinity of the peptidyl transferase center loop in domain V (1). Moreover, crosslinking of the segment 739-748 to the peptidyl transferase region indicates that residues 745-747 are also likely to form part of this domain (18). Intriguing as these non-uniform locations may be, no 23S RNA methylations have yet been directly implicated in any ribosomal function. In contrast to 30S assembly, 23S RNA methylation appears to be an early event in 50S assembly (14), although maturation can occur in the absence of at least some methylated nucleotides such as m^1G745, m^5U747, and m^7G2069 (19). There is no known role for methylation in 50S assembly.

RESULTS

Methylation of Synthetic 16S RNA and 30S Subunits

We previously described a system for the *in vitro* synthesis of functional 30S ribosomes. This involved transcription of full-length 16S RNA *in vitro* with T7 RNA polymerase, followed by reconstitution of this RNA with 30S proteins to form a ribosome. These ribosomes had the morphological features of 30S subunits as determined by electron microscopy, sedimented like 30S, and possessed 35-62% of the functional activity of isolated 30S ribosomes, depending on the assay used (10,12). Since the rRNA was synthesized *in vitro* using a cloned polymerase, no methylation of the rRNA was expected, and none was found by direct analysis (10). The ribosome assembly process also did not result in any rRNA methylation (10).

The synthetic 16S RNA and 30S ribosomes proved to be excellent substrates for homologous methylation by *E. coli* extracts. Table 1 shows that using an unfractionated S100 cell extract, five different methylated nucleotides could be obtained in good yield. The m^6_2A-forming activity has been described before (15). m^6A synthesis presumably is an intermediate stage in m^6_2A formation, since there is no m^6A described in *E. coli* 16S RNA. Three previously undescribed activities were found, namely those for m^5C, m^3U, and m^2G synthesis. Even though no attempt was made to optimize conditions, a good extent of methylation was

TABLE 1
METHYLATION ACTIVITIES IN A CRUDE *E. COLI* EXTRACT

Ribosome	Mg^{++}	Methylated bases (moles/mole)				
		m^5C	m^3U	m^2G	m^6A	m^6_2A
SYN	2	0.04	0.07	0.86	0.14	1.02
NAT	2	<0.02	<0.02	<0.02	<0.02	<0.02
SYN	9	0.09	0.21	0.51	0.21	0.40
NAT	9	<0.02	<0.02	<0.02	<0.02	<0.02

SYN, 30S ribosomes reconstituted with 16S RNA transcribed *in vitro*; NAT, isolated 30S ribosomes. Methylated bases were analyzed by HPLC after digestion to nucleosides (20).

obtained at one or the other Mg++ concentration, except for m^5C formation which remained low. The low activity for m^5C is due to the use of 30S ribosomes as substrate, as described in more detail below. The use of isolated, naturally methylated ribosomes was an important control that showed the lack of methylation at incorrect sites.

In these and subsequent experiments, the nature and amount of methylated nucleotide formed was analyzed by HPLC of the [^{3}H]CH$_3$-nucleosides produced by enzymatic digestion of the labelled RNA (20). This system was capable of resolving, in order of elution, m^5C, Cm, m^7G, m^5U, m^3U, Um, m^4Cm, CmC, Gm, m^1G, m^2G, m^4CmC, m^2A, m^6A, and m^6_2A (see also ref. 16). m^5C could not be separated from m^4C in this system so that assignment as m^5C relies on knowledge of the methylated nucleotide content and location in 16S and 23S RNA.

These results encouraged us to examine other cellular fractions and to use both free RNA as well as 30S subunits as substrate (Table 2). In this experiment the 0.5 M NH4Cl wash of ribosomes and the S100 extract were fractionated by ammonium sulfate. Since the purpose of the analysis was to survey the distribution of methyltransferase activities, only a single enzyme concentration and a fixed reaction time were used. Under these

TABLE 2
DISTRIBUTION OF METHYLTRANSFERASES
IN FRACTIONATED *E. COLI* EXTRACTS

Enzyme	Methylated bases (moles/mole)			
	$m^5C(m^4C)$		m^2G	
	RNA	RIB	RNA	RIB
HSW (35-55)	0.2	<0.1	0.2	0.8
S100 (35-55)	0.6	0.1	0.1	1.1
HSW (55-85)	0.8	0.1	0.1	1.7
S100 (55-85)	0.7	0.1	<0.1	1.4

RNA, 16S RNA transcribed *in vitro*; RIB, 30S ribosomes reconstituted with this RNA; HSW, ribosomal 0.5 M NH4Cl wash; (35-55) and (55-85) refer to the % ammonium sulfate used for fractionation. The Mg++ concentration in the assays was 2 mM. $m^5C(m^4C)$ means that the methylated base could be either one or a mixture.

conditions, non-stoichiometric amounts of methylated products were obtained. The m^3U activity seen in Table 1 was no longer detectable after further fractionation. Perhaps it was lost in the 0-35% ammonium sulfate precipitate, or was inactivated during the purification procedure. The main feature of interest, however, was the substrate specificity of the m^5C and m^2G activities. The m^5C activity preferred RNA to 30S by up to 8-fold, while the m^2G activity was the reverse, preferring ribosomes to RNA by as much as 17-fold. The striking specificity of the m^5C activity for RNA suggested that one or the other (or both) m^5C residue(s) might be involved in the early assembly of the 30S subunit. Consequently, this activity was chosen for further study.

Purification and Properties of the m^5C Methyltransferase

As over half of the total RNA methylating activity, but only 10% of the protein, was found in the ribosomal 0.5 M NH_4Cl wash (Table 3), this

TABLE 3
PURIFICATION OF m^5C METHYLTRANSFERASE

Purification Step	Protein	Total Activity	Specific Activity
	mg	units x 10^{-3}	units/mg
S30 Supernatant	3588	2260	630
[S100 Supernatant]	3264	636	195
Ribosomal 0.5 M NH_4Cl Wash			
[$(NH_4)_2SO_4$, 0 to 50%]	110	85	773
$(NH_4)_2SO_4$, 50% Sup.	255	1116	4376
DEAE Sepharose CL6B	39	901	23,100
P11 Phosphocellulose	13	281	21,620

Values are calculated for 100 g of cell paste. Assays were done in 100 mM Hepes, pH 7.5, 200 mM NH_4Cl, 2 mM $Mg(OAc)_2$, 6 mM 2-mercaptoethanol, 4.3 µM S-adenosylmethionine, 800 units/ml RNasin, and 200 nM 16S RNA. One unit is the amount of enzyme required for the incorporation of one pmole of [^3H]-methyl into TCA precipitable material in 1 h at 37°C. Protein was determined by the Bradford assay (21). Steps in [ ] were not part of the purification procedure.

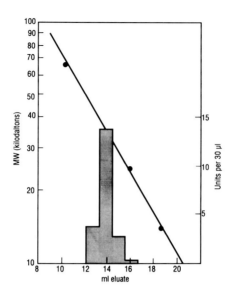

FIGURE 1. Determination of the molecular weight of the native 16S RNA methyltransferase. An aliquot of the phosphocellulose fraction was chromatographed on a Pharmacia Superose 12 10/30 FPLC column. The molecular weight standards were, in decreasing order of size, bovine serum albumin (67 kD), α-chymotrypsinogen A (24.5 kD), and lysozyme (14 kD).

fraction was used for further purification. The ammonium sulfate step served mainly to concentrate the NH_4Cl wash. The DEAE step removed nucleic acids as well as some protein. While the P11 column removed protein, there was also a large loss of activity. The overall degree of purification was modest, reflecting the usual difficulties with purification of nucleic acid methyltransferases. However, the enzyme was reasonably stable (over 35 days) when stored at 0°C at 0.6 mg/ml protein. The fraction from the P11 column was used in all of the following studies.

Chromatography on a calibrated sizing column (Fig. 1) was used to determine the molecular weight of the native enzyme. The molecular weight obtained was 31-38 kD. The substrate specificity is shown in Table 4. The amount of enzyme used gave approximately linear kinetics over the time range chosen. Natural (methylated) RNA isolated from 30S subunits was essentially inactive as already shown in Table 1. In addition, both unmethylated synthetic 23S RNA (see below) and natural 23S RNA isolated from 50S subunits were inactive. Poly (dI-dC), an excellent

TABLE 4
SUBSTRATE SPECIFICITY OF 16S RNA m^5C METHYLTRANSFERASE

Substrate	Activity[a]		Percent	
Synthetic 16S RNA		8.61		100
Natural 16S RNA		0.17		2
Synthetic 23S RNA		0.26		3
Natural 23S RNA		0.02		0.2
Poly(dI-dC), 1 µg/ml	<0.01		<0.1	

[a]pmoles incorporated in 30 min at 37°C using 60 nM RNA, 200 mM NH4Cl, 100 mM Hepes, pH 7.5, 4 µM [^{3}H]S-adenosylmethionine, 4 mM Mg^{++}, 400 units/ml RNasin, 3% glycerol, 5 mM mercaptoethanol, and an aliquot of the phosphocellulose fraction. For description of the substrates, see the text.

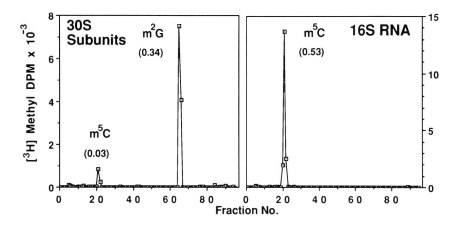

FIGURE 2. Characterization of the product of methylation. Right panel, reaction with synthetic 16S RNA. Left panel, reaction with synthetic 30S ribosomes. Analysis was by HPLC and identification confirmed by the use of internal standards (not shown). Values in brackets are moles of methyl incorporated per mole of substrate under the conditions used.

substrate for DNA methyltransferases (22), was also incapable of being methylated by this enzyme. Clearly the enzyme fraction is highly specific for 16S RNA.

Although the ribosomal salt wash fraction used for the purification had both m^5C activity on RNA and m^2G activity on 30S ribosomes (Table 2), we expected that by this stage of purification activity on 30S subunits would be gone. To our surprise, the phosphocellulose fraction could still methylate ribosomes, as well as RNA.. However, the methylated products were markedly different (Fig. 2). Using an amount of enzyme that gave approximately linear kinetics with free RNA, m^5C but not m^2G was formed with free 16S RNA (right panel), as expected. When 30S subunits were used, very little m^5C was formed, but a significant amount of m^2G was made. Thus, the phosphocellulose fraction contained both an RNA-specific m^5C methyltransferase and a 30S subunit-specific m^2G methyltransferase. When methylated to the plateau level, 1.0 mole of m^5C per mole was incorporated, but only 0.4-0.5 mole of m^2G per mole of ribosomes was obtained (data not shown). As there are two m^5C residues per mole of RNA in fully methylated 30S ribosomes, the stoichiometry for m^5C suggested that only one of the two m^5C residues was being formed. The determination of the actual methylation site is addressed below. The stoichiometry obtained for m^2G is not understood at present, especially since there are three m^2G residues in fully methylated 30S ribosomes. Perhaps it reflects a requirement for functional ribosomes, since the activity of the synthetic 30S is approximately half that of isolated 30S (12), or conversely, perhaps formation of this m^2G may only proceed if the particles are slightly incomplete in some undefined way. This aspect needs further study.

Site of Formation of m^5C and m^2G

We had already shown that even with the HSW(55-85) fraction all of the m^5C formed in free 16S RNA was localized to C967 (20). This result was confirmed using RNA methylated by the phosphocellulose fraction. The technique used for localization was protection of a segment of [^{3}H]CH$_3$-rRNA from RNAse T$_1$ digestion by hybridization with deoxyoligonucleotides of defined sequence. Since each of the two m^5C residues are well separated from each other as are the three m^2G residues, the use of 20mers for hybridization-protection was satisfactory. Fig. 3, panel A, shows that m^5C-containing rRNA was completely protected by the oligomer protecting m^5C967, but was not at all protected by the oligomer covering m^5C1407. This result was expected. The analysis of the m^2G containing rRNA made by methylation of 30S subunits and

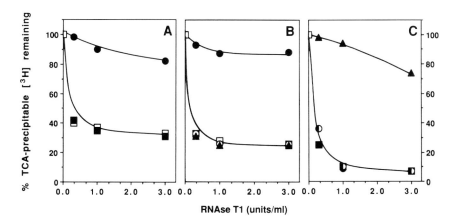

FIGURE 3. Determination of the site of methylation by hybridization-protection. Panel A: m^5C-containing 16S RNA; PANEL B: m^2G-containing RNA from 30S subunits; panel C: m^6_2A-containing RNA from 30S subunits. ●, deoxyoligo 958-977; ○, deoxyoligo 1197-1216; ■, deoxyoligo 1398-1417; ▲, deoxyoligo 1506-1525; ▫, no deoxyoligomer. Oligomer synthesis, preparation of methylated RNAs, hybridization, RNase T_1 digestion, and analysis were done essentially as described by Nègre et al. (20).

extraction of the rRNA is shown in panel B. Surprisingly, not only was all of the m^2G localized to a single site, but the site was G966, just adjacent to the methylated C residue. Neither m^2G1207 nor m^2G1516 was formed. The controls are shown in panel C where the rRNA was labeled at the m^6_2A residues using an enzyme extract specific for these bases (20). The deoxyoligomer 1506-1525 was able to protect the rRNA when labeled at the appropriate site, showing that the failure to protect in panel B is due to the absence of m^2G at position 1516. Conversely, the deoxyoligomer 958-977 was not a general inhibitor of RNAse T_1 since it failed to protect the labeled RNA in panel C.

Cloning, *in vitro* Synthesis, and Methylation of 23S RNA

The 23S rRNA gene, lacking 17 to 31 residues at each end, was excised from the rrnB operon of pKK3535 (23) and ligated into pUC19, together with a synthetic duplex at each end of the 23S rDNA. At the 5'-end, the synthetic duplex supplied the strong class III T7 promoter fused to

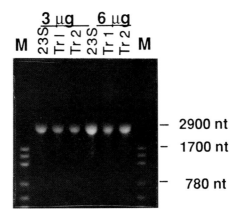

FIGURE 4. Sizing gel analysis of 23S rDNA *in vitro* transcript. The denaturing glyoxal gel was essentially as described by Maniatis *et al* (25). M, RNA standards; 23S, isolated 23S RNA; Tr1 and Tr2, products of 2 separate transcriptions. Three and six mg of each RNA were analyzed.

the mature 23S RNA sequence such that the correct 5'-end of the RNA was produced upon transcription by T7 RNA polymerase. At the 3'-end, the synthetic duplex was used to create a restriction site for linearization in order to terminate the RNA by run-off transcription. The best site that could be obtained was one resulting in the addition of two adenosine residues to the correct 3'-end sequence. The concept and general procedures used follow that of Krzyzosiak et al. (24). The *in vitro* run-off transcript was the same size as isolated 23S RNA (Fig. 4), had the correct 5'-terminal sequence, and the length of the 3' T_1 fragment was also as expected. The RNA sequence was also confirmed through all DNA ligation junctions (data not shown). Typical transcription yields were 500-700 moles RNA per mole of DNA template.

The synthetic 23S RNA was a good substrate for methyl transfer from S-adenosylmethionine catalyzed by *E. coli* cell extracts (Table 5). Natural 23S RNA, which is methylated *in vivo* and served as the control for methylation at incorrect sites, was hardly methylated. In contrast to 16S RNA methylation, where the majority of the activity was found in the ribosomal wash (Table 3), the majority of 23S RNA methylation activity was found in the S100 fraction (Table 5). Nucleoside analysis was done on 23S RNA labeled by both the S100 and the HSW(AS50s) fraction since it was considered possible that the 18% activity in the HSW fraction might be due to a single methyltransferase. However, as shown in Table 6, the

TABLE 5
DISTRIBUTION OF 23S RNA METHYLATION ACTIVITY
IN *E. COLI* EXTRACTS

Fraction	RNA	Activity[a]	%
S30	SYN	17.4	100
	NAT	0.7	
S100	SYN	10.3	59
	NAT	0.3	
HSW(AS50s)	SYN	3.2	18
	NAT	0.1	
HSW(AS50p)	SYN	1.3	7
	NAT	<0.1	

[a]pmoles CH_3 incorporated per hr per mg cells using 97 nM RNA, 200mM NH_4Cl, 100 mM Hepes 7.5, 4 mM [^3H]S-adenosylmethionine, 2 mM Mg^{++}, 800 units/ml RNasin, and 6 mM mercaptoethanol. HSW, ribosomal 0.5 M NH_4Cl wash; AS50s or AS50p, 50% saturated ammonium sulfate supernatant, or precipitate, respectively. SYN, *in vitro* transcript; NAT, isolated from 50S ribosomes.

same methylated bases were produced by both fractions. Nevertheless, the possibility canot be excluded that the activities in the two extracts represent separate sequence-specific enzymes for the same base. Unlike the case with 16S RNA, several methylated nucleotides were produced in free 23S RNA, although only sub-stoichiometric amounts of each were found. As the reaction had reached a kinetic plateau when the samples were taken, it appears that these fractions are still too crude to use in any stoichiometry measurements. The enzymes involved are being purified.

DISCUSSION

In the initial studies of rRNA methyltransferases either heterologous rRNA or homologous rRNA made methyl-deficient by one of several *in vivo* treatments was used as substrate because mature homologous rRNA, being already methylated *in vivo*, could not be further methylated (26-30). Three methyltransferases were purified by this

TABLE 6
NUCLEOSIDE ANALYSIS OF METHYLATED 23S RNA

Nucleoside	Source of Enzyme	
	S100	HSW (AS50s)
	mole per mole RNA	
m^5C	0.3	0.2
m^5U	0.5	0.3
m^2G	0.3	0.2

Synthetic 23S RNA was methylated as described in the text. Analysis was as described in the legend to Table 1.

approach (28, 29). However, none of them were selective for 16S or 23S RNA. In retrospect, part of the problem may have been failure to recognize that the initial breakdown products of 23S RNA are about 16S in size. Thus, it is likely that the m^6A methyltransferase (29, 30) is 23S RNA specific, in agreement with the fact that m^6A is found in 23S RNA (17) but not in mature 16S RNA (1). The apparent formation of m^6_2A in 23S RNA of methyl-deficient *E. coli* rRNA (28) is not so readily explained since m^6_2A has not been detected in 23S RNA. The ability of the purified m^5C methyltransferase to methylate both 16S and 23S RNA (28) is also at variance with what is now known. Although both 23S and 16S RNAs contain m^5C (2, 16), the only m^5C methyltransferase known to recognize 16S RNA is the one described in this work which is not only 16S RNA specific but also position-specific. Moreover, the m^5C-forming activity detected by Bjork and Kjellin-Straby (31) was 23S RNA-specific.

An alternative approach taken to the isolation of rRNA methyltransferases was the use of homologous rRNA isolated from mutagenized cultures of *E. coli*, which was missing a methylated base due to mutation of the cognate methyltransferase (31-33). In this way, enzymes specific for m^5C, m^1G, and m^2G formation in 23S RNA were detected. Two of them, the m^1G and m^2G methyltransferases, were purified (32, 33). Since cells lacking the m^1G methyltransferase could grow, and there is only one m^1G in 23S RNA (2, 16), m^1G formation must be a non-essential event in the life of laboratory-grown *E. coli*.

The ability of these enzymes to react with ribosomes varied. The m^6_2A and m^5C activities did not react with 70S ribosomes (28). The m^6A

activity (29) was not tested. The m^5C (31) and m^2G (32) activities which were 23S RNA-specific appear not to have been tested on ribosomes, whereas the m^1G enzyme (33) was active on 50S as well as 70S particles.

In this work, we capitalized on our ability to transcribe totally unmethylated 16S and 23S RNAs *in vitro* and to reconstitute unmethylated 30S ribosomes to detect, purify, and characterize methyltransferases specific for rRNA. This approach is unique in that it allows systematic study of each methyltransferase using its natural substrate. Three such activities were detected in addition to the previously studied m^6_2A methyltransferase (15). Two of them have been purified. An m^5C methyltransferase specific for C967 of 16S RNA was purified 39-fold. This enzyme did not react with 23S RNA or with 30S ribosomes. In the same enzyme fraction, an m^2G activity was discovered which was specific for G966 of 16S RNA, but only when the RNA was contained in a 30S ribosome. Although it is curious that two enzymes which methylate adjacent bases are found in the same partially purified protein fraction, they are distinct enzymes. A further purification step on a Pharmacia MonoS column clearly separated the two activities (data not shown).

The specificity for C967 in the free RNA state and for G966 only in the 30S ribosome appears to preclude simple exposure or sequestering of these residues as an explanation for the differential activities since the residues involved are adjacent. Moreover, in contrast to the results reported here Moazed et al. (34) found neither G966 nor C967 to be exposed in free native 16S RNA and only a partial exposure of C967 in the 30S subunit.

Some deductions about the sequence of methylation in 30S ribosomes can be made from this work. Clearly, m^5C967 formation occurs at an early stage of ribosome biogenesis and is independent of any other methylation reaction including m^2G966. Moreover m^2G966 synthesis, although it presumably occurs normally after m^5C967 is made, does not require prior m^5C967 synthesis. Also, synthesis of m^6_2A1518 and m^6_2A1519 does not require prior formation of m^2G1516 (Fig. 3C). The formation of m^5C967 at the free RNA level is in agreement with prior studies of Dubin and Günalp (35), who examined the methylation state of 16S precursor RNA in incomplete ribosomal particles. These authors found that methylation was reduced by 81% in these particles, suggesting that most methylation reactions were a late event in ribosome assembly. The remaining 19% methylation was present as m^5C (1.0 mole/mole RNA) and m^2G (0.3 mole/mole RNA). The coincidence of these findings with our independent results implies that the m^5C found in these precursor particles was likely to have been m^5C967. Should this be the case, it would provide further support for the idea that methylation of C967 is an early step in ribosome biogenesis.

Our studies on 23S RNA methylation are less advanced. So far, only free RNA has been examined as 50S assembly from *in vitro* synthesized and unmodified RNA has not yet been accomplished. Our report of an m^5U activity is the first one for this nucleotide in ribosomal RNA. m^5U has been located at residues 747 and 1939 (1,17), and a third one has been claimed (2) but not confirmed (16). The activity for m^5C agrees with the existence of one m^5C at position 2394 (1).and the probable existence of a second one (2,16). The activity detected by us could be the one described previously (31), or it could be an activity for the other m^5C residue. Two or three m^2G residues occur in 23S RNA (2,16).but none of them have been localized. The activity shown in Table 6 could be the same as that purified by Isaksson (32) or it could be specific for a different m^2G residue. Testing of the mutant extracts should indicate whether or not new enzymes for m^5C and m^2G have been detected in our work.

In summary, a system has been developed for the preparation of 16S RNA, 23S RNA, and 30S ribosomes in unmodified form for use as substrates in probing cell-free extracts for the existence of specific rRNA methyltransferases. Several have been detected and the potential exists for finding a number of others. Once these enzymes are available in purified form, it should be possible to examine the role of methylation in assembly and function of the ribosome in a way not possible up to now.

REFERENCES

1. Noller HF (1984). Structure of ribosomal RNA. Ann Rev Biochem 53:119.
2. Hsuchen CC, Dubin DT (1980). Methylation patterns of mycoplasma transfer and ribosomal ribonucleic acid. J Bacteriol 144:991.
3. Brimacombe R, Atmadja J, Stiege W, Schüler D (1988). A detailed model of the three-dimensional structure of *Escherichia coli* 16S ribosomal RNA *in situ* in the 30S subunit. J Mol Biol 199:115.
4. Gornicki P, Nurse K, Hellmann W, Boublik M, Ofengand J (1984). High resolution localization of the tRNA anticodon interaction site on the *Escherichia coli* 30S ribosomal subunit. J Biol Chem 259:10493.
5. Helser T, Davies JE, Dahlberg JE (1972). Mechanism of kasugamycin resistance in *Escherichia coli*. Nature new Biol 235:6.

6. Poldermans B, Van Buul CPJJ, van Knippenberg PH (1979). Studies on the function of two adjacent N^6,N^6-dimethyladenosines near the 3' end of 16S ribosomal RNA of *Escherichia coli*. II. The effect of the absence of the methyl groups on initiation of protein biosynthesis. J Biol Chem 254:9090.
7. Poldermans B, Bakker H, van Knippenberg PH (1980). Studies on the function of two adjacent N^6,N^6-dimethyladenosines near the 3' end of 16S ribosomal RNA of *Escherichia coli*. IV. The effect of the methyl groups on ribosomal subunit interaction.. Nucl Acids Res 8:143.
8. Van Buul CPJJ, Visser W, van Knippenberg PH (1984). Increased translational fidelity caused by the antibiotic kasugamycin and ribosomal ambiguity in mutants harbouring the ksgA gene. FEBS Lett 177:119.
9. Heus HA, van Kimmenade JMA, van Knippenberg PH (1983). High-resolution proton magnetic resonance studies of the 3'-terminal colicin fragment of 16S ribosomal RNA from *Escherichia coli*. Assignment of iminoproton resonances by nuclear Overhauser effect experiments and the influence of adenine dimethylation on the hairpin conformation. J Mol Biol 170:939.
10. Krzyzosiak W, Denman R, Nurse K, Hellmann W, Boublik M, Gehrke CW, Agris PF, Ofengand J (1987). *In vitro* synthesis of 16S ribosomal RNA containing single base changes and assembly into a functional 30S ribosome. Biochemistry 26:2353.
11. Mélançon P, Gravel M, Boileau G, Brakier-Gringras L (1987). Reassembly of active 30S ribosomal subunits with an unmethylated *in vitro* transcribed 16S RNA. Biochem. Cell Biol. 65:1022.
12. Denman, R, Nègre D, Cunningham PR, Nurse K, Colgan J, Weitzmann C, Ofengand J (1989). Effect of point mutations in the decoding site (C1400) region of 16S ribosomal RNA on the ability of ribosomes to carry out individual steps of protein synthesis Biochemistry 28:1012.
13. Feunteun J, Rosset R, Ehresmann C, Stiegler R, Fellner P (1974). Abnormal maturation of precursor 16S RNA in a ribosomal assembly defective mutant of *E. coli*. Nucl Acids Res 1:141.
14. Dahlberg JE, Nikolaev N, Schlessinger D (1975). Post-transcriptional modification of nucleotides in *E. coli* ribosomal RNAs. In Dunn JJ (ed) "Brookhaven Symp. in Biology," Vol. 26, Upton: Brookhaven National Laboratory, p. 194.
15. Poldermans B, Roza L, van Knippenberg PH (1979). Studies on the function of two adjacent N^6,N^6-dimethyladenosines near the 3'end of the 16S ribosomal RNA of *Escherichia coli*. III.

Purification and properties of the methylating enzyme and methylase-30S interactions. J Biol Chem 254:9094.

16. Gehrke CW, Kuo KC (1989). Ribonucleoside analysis by reversed-phase high performance liquid chromatography. J Chromatography 471:3.

17. Branlant C, Krol A, Machatt MA, Pouyet J, Ebel J-P (1981). Primary and secondary structures of *Escherichia coli* MRE 600 23S ribosomal RNA. Comparison with models of secondary structure for maize chloroplast 23S rRNA and for large portions of mouse and human 16S mitochondrial rRNAs. Nucl Acids Res 9:4303.

18. Stiege W, Glotz C, Brimacombe R (1983). Localization of a series of intra-RNA cross-links in the dsecondary and tertiary structure of 23S RNA, induced by ultraviolet irradiation of *Escherichia coli* 50S ribosomal subunits. Nucl Acids Res 11:1687.

19. Chelbi-Alix MK, Expert-Bezançon A, Hayes F, Alix J-H, Branlant C (1981). Properties of ribosomes and ribosomal RNAs synthesized by *Escherichia coli* grown in the presence of ethionine. Eur J Biochem 115:627.

20. Nègre D, Weitzmann C, Ofengand J (1989). *In vitro* methylation of *E. coli* 16S ribosomal RNA. Proc Natl Acad Sci USA (in press).

21. Bradford M (1976). A rapid and sensitive method for the quantitation of microgram quantities of protein utilizing the principle of protein-dye binding. Anal Biochem 72:248

22. Pedrali-Noy G, Weissbach A (1986). Mammalian DNA methyltransferases prefer poly (dI-dC) as substrate. J Biol Chem 261:7600.

23. Brosius J, Dull TJ, Sleeter, DD, Noller HF (1981). Gene organization and primary structure of a ribosomal RNA operon from *Escherichia coli*.. J Mol Biol 148:107

24. Krzyzosiak WJ, Denman R, Cunningham P, Ofengand J (1988). An efficiently mutagenizable recombinant plasmid for *in vitro* transcription of the *E. coli* 16S RNA gene. Anal Biochem 175:373.

25. Maniatis T, Fritsch EF, Sambrook J, (eds) (1982). Molecular Cloning: A Laboratory Manual. Cold Spring Harbor Laboratory, Cold Spring Harbor, NY.

26. Srinivasan PR, Nofal S, Sussman C (1964). Species specificity of ribosomal RNA methylases. Biochem Biophys Res Commun 16:82.

27. Gordon J, Boman HG (1964). Studies on microbial RNA. II. Transfer of methyl groups from methionine to the RNA of a ribonucleoprotein particle. J Mol Biol 9:638.
28. Hurwitz J, Anders M, Gold M, Smith I (1965). The enzymatic methylation of ribonucleic acid and deoxyribonucleic acid. VII. The methylation of ribosomal ribonucleic acid. J Biol Chem 240:1256.
29. Sipe JE, Anderson,Jr WM, Remy CN, Love SH (1972). Characterization of S-adenosylmethionine: Ribosomal ribonucleic acid-adenine (N^6-) methyltransferase of *Escherichia coli* strain B. J Bacteriol 110:81.
30. Anderson,Jr WM, Remy CN, Sipe JE (1973). Ribosomal ribonucleic acid-adenine (N^6-) methylase of *Escherichia coli* strain B: Ionic and substrate site requirements. J Bacteriol 114:988.
31. Björk GR, Kjellin-Straby K (1978). General screening procedure for RNA modificationless mutants: Isolation of *Escherichia coli* strains with specific defects in RNA methylation. J Bacteriol 133:499.
32. Isaksson LA (1973). Partial purification of ribosomal RNA (m^1G)- and rRNA (m^2G)-methylases from *Escherichia coli* and demonstration of some proteins affecting their apparent activity. Biochim Biophys Acta 312:122.
33. Isaksson LA (1973). Formation *in vitro* of 1-methylguanine in 23-S RNA from *Escherichia coli*. The effects of spermidine and two proteins. Biochim Biophys Acta 312:134.
34. Moazed D, Stern S, Noller HF (1986). Rapid chemical probing of conformation in 16S ribosomal RNA and 30S ribosomal subunits using primer extension. J Mol Biol 187:399.
35. Dubin DT, Günalp A (1967). Minor nucleotide composition of ribosomal precursor and ribosomal nucleic acid in *Escherichia coli*. Biochim Biophys Acta 134:106.

ON THE EVOLUTIONARY RELATIONSHIPS BETWEEN ADENOSINE DIMETHYLATION IN 16S AND 23S RIBOSOMAL RNA.

Bob van Gemen and Peter H. van Knippenberg

Department of Biochemistry, State University of Leiden, Wassenaarseweg 64, 2333 AL Leiden, THE NETHERLANDS.

ABSTRACT. A detailed analysis is given of the homologies between *erm* methylases and the *ksgA* methylase, resulting in the support of the idea that the *ksgA* methylase is "ancestral" with respect to the *erm* methylases. The "theoretical" homology between *erm* and *ksgA* methylases is supported by an experiment where it is shown that antibodies against the *ksgA* enzyme cross-react with *ermC* methylase. Next, we describe some differences and similarities in the control of expression of *erm* genes and the *ksgA* gene.
Literature data on the mode of action of erythromycin, and of the mechanism of resistance, are reviewed and a general consensus as to what this antibiotic does is presented. Similarly, the more limited knowledge on kasugamycin action and resistance is summarized. The general conclusion is that both ribosomal subunits are involved in the mode of action of these two antibiotics. It is argued that the adenosines that are affected by methylation (A2058 in 23S rRNA and A1518 A1519 in 16S rRNA) are probably part of one site on the 70S ribosome and that this site is the target for the two antibiotics. There is a (limited) structural homology between these RNA's in the vicinity of the methylatable sites.

INTRODUCTION

Dimethylation of adenosines at the N^6-position in RNA appears to be confined to two cases, both connected with sensitivity for or resistance against antibiotics.
1) In virtually all cases where it has been studied, two neighboring adenosines near the 3' end of 16S(-like) ribosomal RNA are dimethylated (compare 38 for a review).

The sequence $Gm_2^6Am_2^6A$ occurs as a universal feature in the loop of a hairpin structure near the 3' end in the RNA of small ribosomal subunits of almost any source. However, bacterial mutants resistant to the antibiotic kasugamycin (ksgA mutants) typically lack this modification (13). The specific adenosine dimethyltransferase (the "methylase") of E. coli is a protein of 273 amino acids, encoded by the ksgA gene (34). Up till now no studies have been reported on the corresponding enzymes or genes from other organisms.

2) Resistance to erythromycin and other antibiotics of the macrolide-lincosamide-streptogramin (MLS) type in several prokaryotes is brought about by the dimethylation of the adenosine at position 2058 (A2058, E. coli numbering) of 23S ribosomal RNA. The methylases that are responsible for this modification are encoded by erm genes. Several of these genes, which are usually induced by erythromycin and related antibiotics have been studied (compare 7 for a review). Their spread, especially among bacteria of clinical or veterinary importance, has been ascribed to the excessive use of the MLS antibiotics.

The erm methylases can be grouped in three distinct classes on the basis of amino acid similarities (19). When the ksgA gene of E. coli was sequenced it was immediately obvious that the methylase specified by this gene was very similar to the erm methylase (34). This observation led Dubnau and Monod (7) to speculate that the ksgA gene is the ancestral gene from which the erm genes are derived through evolution. We agree with them and this paper presents our view of how the two systems are related. We not only present extensive comparison of the methylases using computer methodology, but also relate the antibiotics and their target sites.

RESULTS

Similarities and differences among the erm methylases and evidence for a ksgA gene origin

Table 1 summarizes erm genes for which the coding sequences are known, and some of the characteristics of their products, and also includes the ksgA methylase of E. coli. Although most of the studied erm genes are plasmid borne, some of them are of chromosomal origin. The inducibility of the erm methylases has been the subject of elaborate studies (6, 11, 14) and will not be discussed here. Instead we intend to compare in this section the various enzymes with the aim to uncover details of evolutionary relationships between them.

Table 1. Genes coding for RNA adenosine dimethyltransferases and their products.

Gene	Origin	Location[a]	Length[b]	Inducible	Ref.
Class I					
ermA	Staphylococcus aureus	tn	243 aa	yes	20
ermAM	Streptococcus sanguis	pl	245 aa	yes	15
ermC	Staphylococcus aureus	pl	244 aa	yes	28
ermC'	Bacillus subtilis (pIM13)[c]	pl	244 aa	no	18
ermG	Bacillus sphaericus	ch	244 aa	yes	19
Class II					
ermD	Bacillus licheniformis	ch	287 aa	yes	11
ermF	Bacteroides fragilis	ch/tn	266 aa	no	30
Class III					
ermE	Saccharopolyspora erythraea[d]	ch	370 aa	no	33
ermAr	Arthrobacter spp.	ch	340 aa	no	25
ermSF	Streptomyces fradiae	ch	320 aa	no	16
Class IV					
ksgA	Escherichia coli	ch	273 aa	no	34

a. tn, transposon; pl, plasmid; ch, chromosome. In case of *ermF* the transposon is stably integrated in the chromosome.
b. Length of the methylases is given in amino acids (aa).
c. The *ermC'* gene of *B. subtilis* is located on the natural plasmid pIM13 and is nearly identical to the *ermC* gene of *S. aureus*.
d. *S. erythraea* was formerly called *Streptomyces erythraeus*.

The length of the proteins[1] varies from 243 (*ermA*) to 370 (*ermE*) amino acids (Table 1). It may be of interest that the shorter versions of the methylases (*erm* A, C, AM, G) are present in bacteria that have been under selection pressure through exposure to MLS antibiotics, while the larger forms (*erm* E, Ar, SF) are found in producers of erythromycin and related antibiotics. As noticed before (7; compare below, however) the extensions of the larger forms of the methylases occur predominantly at the C-

---

[1] footnote: It should be emphasized that except for *ermC* (Shivakumar and Dubnau, 1981) and *ksgA* methylase (Poldermans *et al*, 1979) the molecular weight of the proteins has not been established. Their amino acid lengths and sequences have been deduced from open reading frames in the cloned genes.

terminus and do not appear to affect the overall homology. In previous studies (7, 16, 19, 34) aminoacids of the methylases were lined up by trial-and-error by first searching for identical aminoacids and then extending comparisons by looking for aminoacids with a homologous character (basic, acidic, hydrophilic, hydrophobic). This method was sufficient to establish the homologies, including the resemblance between *erm* methylases and the *ksgA* methylase. Nevertheless, more powerful methods for comparison with the aid of computer programs exist, and we have chosen the COMPARE, DOTPLOT program (5) present in the UWGCG program package to align thesequences. The program selects a preset number (the "window") of amino acids from the first protein, starting from the N-terminus and moving the window one amino acid at a time, and searches the second protein for a homologous sequence. The homology criterium (the "stringency") is preset at a certain number of amino acids that have to be identical or homologous (i.e. basic R, K, H, acidic O, N, E, D, B, Z, hydrophobic, L, I, V, M, aromatic, F, Y, W, neutral, P, A, G, S, T and cross-link forming, C) at exactly corresponding positions. Each time the program scores a homology that meets the criterium, a dot is placed in the X-Y plane. Two identical proteins will display a straight diagonal line where the two proteins show internal homology (or a repetition). Naturally, the patterns depend strongly on the window size and the stringency. To illustrate the method, Fig. 1A displays the pattern with a window of 30 and a stringency of 16, for two proteins which are not identical but very homologous according to the trial-and-error method, i.e. *ermAM* and *ermC* (7).

All *erm* methylase sequences and the *ksgA* methylase sequence were compared using the program with a window setting of 30 and a stringency of 16. In agreement with conclusions drawn from the trial-and-error procedures (7), the *erm* enzymes can be clearly grouped in three classes as indicated in Table I, and the *ksgA* methylase has to be placed in a separate class (IV).

Since it has been suggested that the *ksgA* gene, probably occurring in all cells (38), is the ancestral methylase gene from which *erm* genes evolved (7), it is of special interest to compare the *ksgA* methylase to all others. Fig. 1B-D shows the homology patterns between the *ksgA* enzyme and a representative of each class of *erm* genes: *ermA* (class I, Fig. 1B), *ermD* (class II, Fig. 1C) and *ermE* (class III, Fig. 1D). All the *erm* methylases (including the ones not shown in Fig. 1) share two regions of homology with the *ksgA* gene that survive the scrutiny at the high standard of a stringency of 16 at a window of 30. In Fig. 1B-D these homologous regions of the *ksgA* methylase are indicated by a bar on the Y-axis and Table 2

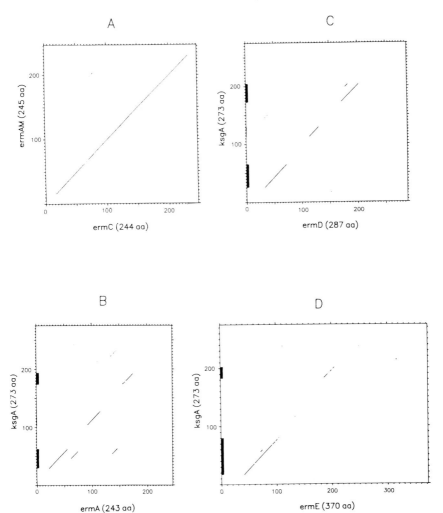

Fig. 1. Comparison of the *ksgA* and *erm* amino acid sequences using the COMPARE and DOTPLOT computer program (Devereux et al., 1984). The amino acid sequences were compared using a window of 30 and a stringency of 16 identical or related amino acids. The six groups of related amino acids used were: Neutral: Pro, Ala, Gly, Ser, Thr; Hydrophilic acid: Gln, Glu, Asn, Asp; Hydrophilic basic: His, Lys, Arg; Hydrophobic: Leu, Ile, Val, Met; Hydrophobic aromatic: Phe, Tyr, Trp; cross-link forming: Cys.
Comparisons made were: A: *ermAM-ermC*; B: *ksgA-ermA*; C: *ksgA-ermD* and D: *ksgA-ermE*.

Table 2. Amino acids of ksgA methylase homologous to erm methylases a derived from DOTPLOT comparisons.

Methylase	N-proximal amino acids	C-proximal amino acids
Class I		
ermA	29 - 63	173 - 193
ermC	26 - 65	173 - 195
ermAM	24 - 61	170 - 193
ermG	24 - 64	173 - 192
Class II		
ermD	26 - 65	172 - 203
ermF	20 - 60	167 - 191
Class III		
ermE	17 - 80	182 - 200
ermAr	15 - 82	178 - 197
ermSF	17 - 65	182 - 200

The DOTPLOTS were made using a window of 30 amino acids and a stringency of 16.

lists these regions for the other *erm* methylases as well. Clearly two regions of the *ksgA* methylase, from approximately 20-60 amino acids and from about 175-200 amino acids from the N-terminus, respectively, find an homologous counterpart in all *erm* methylases at roughly equivalent positions. Size differences between the *erm* methylases and the *ksgA* methylase, but also between the various *erm* methylases, are primarily due to extensions or amputations at the N- or C-terminals (compare also below).

Although the two (or occasionally three) regions of homology are each on a diagonal line, extrapolations of these diagonals do not coincide, as is illustrated in an enlarged diagram of *ksgA* methylase versus *ermE* methylase (Fig. 2A). The diagonal resulting from the extrapolation of the homology near the C-terminus is shifted some 17 amino acids in the direction of this terminus with respect to the diagonal of the N-terminus proximal homology. The *ermE* methylase therefore contains an internal deletion of about 17 amino acids with respect to the *ksgA* enzyme. In

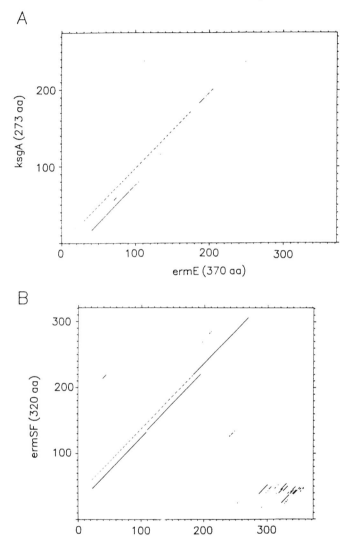

Fig. 2. Comparison of amino acid sequences using the COMPARE and DOTPLOT computer program. C-terminal region of similarities are extrapolated with a dashed line to show the difference in diagonal position, indicating internal deletions in ermE with respect to ksgA (panel A) and ermSF (panel B). For further explanation see legend to Fig. 1 and the text.

fact **all** erm methylases contain, in addition to variations near the N- and C-terminus, deletions with respect to the ksgA methylase, varying in size from approximately 8 to 17 amino acids. Table 3 gives a summary of all deletions and insertions in erm enzymes as compared to the ksgA methylase. The variations in the size of the internal deletions (i.e. 10 for class I, 8 for class II, 17 for ermE, 7-8 for ermAr and ermSF) are also revealed by aligning the erm methylases mutually. As an example Fig. 2B shows a comparison between two methylases of class III (ermE and ermSF) and, although these proteins are nearly identical the ermE enzyme is shorter by 10 amino acids deleted from the ermSF methylase; this is exactly the difference between the deletion in ermE with respect to ksgA (17 amino acids) and the deletions in ermSF as compared to ksgA (7 amino acids) (Table 2).

Table 3. Deletions (-) or insertions (+) in erm methylases with respect to the ksgA methylase.

Methylase	N-terminal[a]	C-terminal[a]	Internal[a]	Total	Length difference[b]
Class I					
ermA	-7	-12	-10	-29	-30
ermC	-6	-11	-10	-27	-29
ermAM	-8	-9	-10	-27	-28
ermG	-7	-11	-10	-28	-29
Class II					
ermD	+7	+13	-8	+12	+14
ermF	-4	+4	-8	-8	-7
Class III					
ermE	+24	+93	-17	+100	+97
ermAr	+19	+55	-8	+66	+66
ermSF	+48	+6	-7	+47	+46

The numbers (1) in the table refer to amino acids.
a. Compare text for explanation of N-terminal, C-terminal and internal domains.
b. Compare table 1; the length difference between the erm methylases and the ksgA methylase.

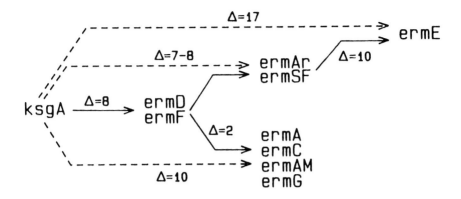

Fig. 3. Schematic evolutionary descendance of *erm* methylases from the *ksgA* methylase. Internal deletion of amino acids are indicated by . Multiple, independent pathways of deletions in the *ksgA* gene are indicated by dashed arrows.

If we, following Dubnau and Monod (7), assume that the *ksgA* gene is the ancestral methylase gene, a scheme of how the other methylase genes evolved from *ksgA* could be drawn as shown in Fig. 3. Since it is less plausible that the *erm* genes evolved from *ksgA* by multiple independent events (dashed arrows in Fig. 3), we favor the evolutionary pathway through *ermD, F* (solid arrows, Fig. 3). It would, however, require a detailed analysis at the nucleotide level to support these pathways. Such an analysis is in progress. It is needless to say that the understanding of the evolutionary relationships would benefit tremendously from the availability of sequences of *ksgA* genes from other sources.

In addition to the homology search described above, other computer programs have been employed to compare the methylases on the basis of hydropathy, secondary structure elements, etc. Some of these comparisons support the above conclusions; however, they are not yet fully worked out and do not add any new information regarding the evolutionary relationships. Clearly more detailed analysis will be very useful.

The hydrophaty plots (not shown) point out that the methylases are rather hydrophilic at the N- and C-termini. Especially the extensions of class III methylases at the termini in comparison to the *ksgA* methylase (Table 3) are extremely hydrophilic.

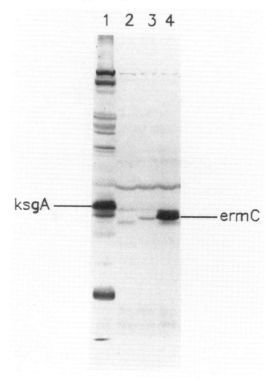

Fig. 4. Cross-reaction of antibodies raised against ksgA methylase with ermC protein. Proteins of cell lysates were separated on a 11% polyacrylamide gel and blotted onto nitrocellulose. Cross-reacting proteins were visualized using anti ksgA methylase antibodies (Van Gemen et al., 1987).
Lane 1: E. coli JM101. Lane 2: B. subtilis BD170. Lane 3: B. subtilis BD170 containing pBD15 harboring the S. aureus ermC gene. Lane 4: B. subtilis BD170 containing pIM13 harboring the B. subtilis ermC gene.

Antibodies against ksgA methylase cross-react with ermC methylase

The homology between the ksgA methylase and one of the erm methylases was tested experimentally by assaying cell extracts that contain one of the latter enzymes. Dr. David Dubnau provided us with B. subtilis carrying either the ermC gene or a close relative of ermC on a plasmid (18). The Western blots of gels after gelelectrophoresis of cell extracts are shown in Fig. 4. No cross-reaction of *ksgA* methylase antibodies is detected in extracts from bacteria harboring no plasmid (lane 2)[2]. The presence of the normal *ermC* gene results in a faint cross-reacting band at the proper position (lane 3). The intensity of this band is strongly increased in cells carrying pIM13 harboring the *ermC* related protein (lane 4). The increase in intensity is due to the fact that this methylase is synthesized constitutively from the latter plasmid (18) resulting in elevated amounts of the protein.

The control of expression of *erm* genes and of the *ksgA* gene

Most of the *erm* genes that have been studied are inducible by erythromycin or a related antibiotic. The mechanism of induction has been described in detail and involves the translation of (an) upstream open reading frame(s) in order to destroy the secondary structure that shields the methylase mRNA from initiating ribosomes (6, 14, 16, 19). However, although *erm* genes of one group (e.g. group I, Table 1) are extremely homologous, the regulating elements in their mRNA may be quite different. This suggests that the coding sequences have either moved through the bacterial kingdom as independent entities, disconnected from their controlling regions, or that the controlling regions, while keeping their function, have rapidly evolved into very different sequences.

As far as we know the *ksgA* methylase is produced constitutively through the activity of a weak promoter (36). Surprisingly, the *ksgA* cistron is not preceded by a standard Shine-Dalgarno signal, and indeed the gene is not expressed when connected to a strong promoter, unless a SD signal is introduced (37), or unless it is translationally coupled to an upstream open reading frame (36). How translation initiation of the *ksgA* mRNA is achieved in the natural situation, i.e. starting from its own promoter, is still subject of investigations in our laboratory, but its mechanism is apparently different from the ones encountered with the inducible *erm* genes.

A property that expression of the *ksgA* gene has in common with one of the *erm* genes, is the autogenous regulation that takes place at the level of translation. This has been described in detail for the *ksgA* gene by Van Gemen *et al.* (37) and by Denoya *et al.* (4) for the *ermC* gene. How the methylases recognize and suppress the

---

[2] Footnote: This does not mean that *B. subtilis* does not contain a *ksgA* methylase. The conditions are such that a chromosomal *ksgA* gene product is present in too low amounts to be detected here (compare Van Gemen *et al.*, 1987).

translation of their mRNA is still unknown, but sequences resembling their substrate site on the ribosomal RNA can be indicated. Future work will have to demonstrate how important this mechanism of regulation is, and how the *ksgA* and *erm* systems correspond.

Interaction of the target sites for erythromycin and kasugamycin on the ribosome

The relatedness of the *ksgA* methylase and the *erm* methylases raises two interesting questions: 1. Is there any resemblance in the target sites of the two types of enzymes, and 2. Is there any similarity in the mode of action of the two antibiotics or do their target sites interact? The first question will be addressed in the next section and here we will dwell upon the antibiotics and their target sites.

The mode of action of MLS antibiotics, especially erythromycin, has been studied by many investigators during more than two decades. A comprehensive review is beyond the scope of this article and the reader is referred to Gale et al. (9) for a summary of older data. We will briefly summarize the essentials and present what we believe is the currently accepted view of how erythromycin acts. The drug binds tightly ($K_d > 10^{-8}$-10^{-7} M) to one site on the 50S ribosomal subunit and, by doing so, interferes with the (proper) binding of peptidyl-tRNA in the P-site of the ribosome. Elongation of polypeptide synthesis is inhibited when the drug is bound to its site on 50S, but the peptidyl-tRNA itself seems to interfere with erythromycin binding. Interference of erythromycin with P-site binding of peptidyl-tRNA appears to depend on the length of the peptide chain and on the nature of the amino acids that have been assembled, especially those that have been polymerized most recently (1, 40). In any event, *in vitro* protein synthesis is most sensitive to erythromycin at an early stage in the polymerization of hydrophilic amino acids (3).

Mehninger and Otto (17) reported that erythromycin enhances the accuracy of protein synthesis *in vivo*, and ascribed this effect to an influence of codon-anticodon interaction between peptidyl-tRNA and mRNA in the P-site: a mismatch after the polymerization of a "wrong" amino acid would lead to a more erythromycin sensitive peptidyl-tRNA binding and result in the abortion of the complex in the presence of the drug.

Mutants of *E. coli*, resistant to high levels of erythromycin, were found to be altered in ribosomal proteins L_4 or L_{22} or in an unidentified component of the 30S subunit specified by the *eryC* gene (22, 23). From the latter observation it was concluded that, although the binding site of erythromycin is located on the 50S subunit, the mode of action of the drug, and indeed its binding to the 70S ribosome, depends on both subunits. This was further substantiated by the finding that introduction of a streptomycin resistance mutations (*strA*, affecting S12) and a spectinomycin resistance mutation (Spc^R, affecting S5) into an erythromycin resistant mutant of *E. coli* (affected in L$_4$) led to a reappearance of the erythromycin sensitive phenotype (27). Ribosomes from such a triple mutant had regained the capacity to bind erythromycin (2). When spontaneous revertants to erythromycin resistance were selected from the erythromycin sensitive triple mutant described above, one of the mutants turned out to be kasugamycin resistant (26). The 30S particles of the quadruple mutant lacked methylation of 16S ribosomal RNA; this was probably due to a mutation in the *ksgA* gene. 70S ribosomes of the mutant had again lost the capacity to bind erythromycin. All these data led to the conclusion that the "functional" binding site of erythromycin on the ribosome is composed of components of both subunits, including the 30S site that is involved in kasugamycin action (26).

Erythromycin resistance based on dimethylation of A2058 in 23S rRNA does not naturally occur in *E. coli*. However, when an *erm* gene from one of the Gram-positive bacteria is introduced in *E. coli*, A2058 becomes (di)methylated and the bacteria become resistant to high levels of erythromycin (31, 32). Similarly, when *E. coli* is transformed with a high-copy-number plasmid carrying a mutation at position 2058 (or nucleotides in the immediate vicinity) in the 23S rRNA, cells become resistant to erythromycin (8, 29, 40). It is therefore generally accepted that A2058, and its immediate vicinity, is directly involved in erythromycin binding. It follows then that other components of the 50S subunit (L$_2$, L$_{22}$) and components of the 30S subunits (S$_5$, S$_{12}$, 16S rRNA), although not directly involved in binding, play a role in shaping the erythromycin binding site.

Much less is known concerning the mode of action of kasugamycin. The available data were reviewed by Van Knippenberg (39) and may be summarized as follows. Kasugamycin (at relatively high concentrations, >+ 100 g/ml) inhibits polypeptide synthesis by interfering with the binding of formyl-methionyl-tRNA to mRNA programmed ribosomes. Although this step is thought to involve only components of the 30S ribosomes and initiation factors, and indeed 30S.mRNA.fMet.tRNA complexes are dissociated by kasugamycin, ribosomes are only resistant to this effect when, in addition to 30S particles from a *ksgA* mutant, also the 50S subunit is present (either from the wild-

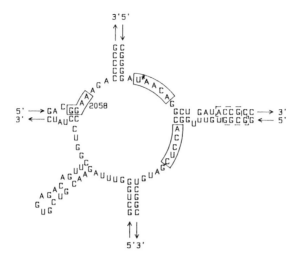

23S rRNA

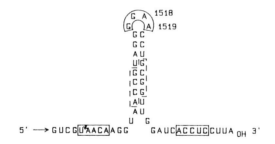

16S rRNA

Fig. 5. Secondary structure models of the 3' end of 16S rRNA and part of 23S rRNA. The dimethylatable adenosines 1518, 1519 in 16S rRNA and 2058 in 23S rRNA are indicated by their number. Homologous regions in both structures are boxed. The solid boxed regions are conserved. The dashed boxed regions are only homologous in *E. coli* and closely related organisms. U* means that uracil moiety is (N3-) methylated in *E. coli*.

type or the *ksgA* mutant). Like in the case of erythromycin this suggests that the "functional" kasugamycin binding site is composed of components of both subunits. Lateron (34), it was found that kasugamycin also affects elongation: it **increases** the accuracy of polypeptide synthesis *in vitro*. It is interesting in this respect that the absence of adenosine dimethylation in 16S rRNA affects *in vivo* fidelity of translation (34) and that the mode of action of erythromycin has also been associated with translational accuracy (compare above).

The interrelationship between erythromycin and kasugamycin was recently dramatically illustrated by our own studies (31). Introduction of an *erm* gene from *Streptococcus pyogenes* into *E. coli* not only makes the cell highly resistant to erythromycin, but at the same time makes them more sensitive to kasugamycin. This effect was not only recorded *in vivo*; ribosomes isolated from such transformed *E. coli* cells displayed kasugamycin resistance *in vitro*, despite the presence of 16S rRNA with the two geminal A's still dimethylated.

Structural homology between (di)methylatable adenosine sites in 23S and 16S rRNA

In view of all the phenomena described above, it is a natural question whether we can detect any homology between the methylatable sites in 23S and 16S rRNA. Fig. 5 shows these regions as they occur in the now well-established secondary structure models of these molecules (21). Except for the presence of the A's in the identical sequence 5' GGAA 3', no structural homology can be detected in the direct vicinity of the dimethylatable adenosines. However, the secondary structure domain in the neighborhood of A2058 in 23S rRNA (occasionally referred to as the "peptidyl transferase center" of the ribosome, e.g. 41) displays a few regions of sequence homology with the 3' end of 16S rRNA. These are boxed in Fig. 5. Especially the sequence UAACA at the 5' side of the hairpin, where the U is methylated in both 23S and 16S rRNA, and the sequence ACCUC at the 3' side of the hairpin stem, might be significant since they are both preserved in a wide range of organisms (12). If these homologous regions are involved in the recognition of the RNAs by the methylases, the juxta-position of the substrate adenosines with respect to these regions might be better described in terms of topology than in terms of (secondary) structure. In any event, an involvement of these regions could now be approached by site-directed mutagenesis of the RNA's.

CONCLUSION

The two systems of adenosine dimethylation in ribosomal RNA are clearly related. The *erm* methylases can be viewed as descendants from an ancestral *ksgA* methylase that have "learnt" to (di)methylate an adenosine residue in the 23S rRNA. Scrutiny of many literature data justifies the conclusion that the target sites of the antibiotics erythromycin and kasugamycin are composed of elements of both ribosomal subunits, despite the fact that mutations that determine resistance affect primarily one of the subunits (50S in the case of erythromycin and 30S in the case of kasugamycin). The results of Suvorov et al. (31) are a very strong indication that A2058 of 23S rRNA is indeed in the direct vicinity of A1518 A1519 of the 16S rRNA in the intact 70S ribosome.

It is our view that the *erm* system became into existence by re-directing an already ancient mechanism of adenosine dimethylation (i.e. the *ksgA* system) in the direction of 23S rRNA.

ACKNOWLEDGEMENTS

The authors gratefully acknowledge the gift of *B. subtilis* strains carrying *ermC* plasmids from dr. D. Dubnau, Public Health Service of New York.

REFERENCES

1. Andersson S and Kurland CG (1987). Biochimie **69**: 901-904.
2. Apirion D and Saltzman L (1974). Mol Gen Genet **135**: 11-18.
3. Chinali G, Nyssen E, Di Giambattista M and Cotico C (1988). Biochim Biophys Acta: in press.
4. Denoya CD, Bechhofer DH and Dubnau D (1986). J Bacteriol **168**: 1133-1141.
5. Devereux J, Haeberli P and Smithies O (1984). Nucl Acids Res **12**: 387-395.
6. Dubnau D (1984) CRC Critical Reviews in Biochemistry **16**: 103-132.
7. Dubnau D and Monod M (1986). Banbury Rep **24**: Antibiotic resistance genes. Cold Spring Harbor Lab, Levy SB & Novick RP (eds), Cold Spring Harbor, N.Y., pp 369-387.
8. Ettayebi M, Prasad SM and Morgan EA (1985). J Bacteriol **162**: 551-557.
9. Gale EF, Cunliffe E, Reynolds PE, Richmond MH and Warning MJ (eds) (1981) In: The molecular basis of antibiotic action (J. Wiley & Sons), 2nd edition, pp 402-547.

10. Gryczan TJ, Grandi G, Hahn J, Grandi R and Dubnau D (1980). Nucl Acids Res **8**: 6081-6097.
11. Gryczan TJ, Israeli-Rechers M, Del Bue M and Dubnau D (1984). Mol Gen Genet **194**: 349-356.
12. Gutell RR and Fox GE (1988). Nucl Acids Res **16** suppl.: 175-270.
13. Helser TL, Davies JE and Dahlberg JE (1971). Nat New Biol **233**: 12-14.
14. Horinouchi S and Weisblum B (1980). Proc Natl Acad Sci USA **77**: 7079-7083.
15. Horinouchi S, Byeon W-H and Weisblum B (1983). J Bacteriol **154**: 1252-1262.
16. Kamimiya S and Weisblum B (1988). J Bacteriol **170**: 1800-1811.
17. Menninger JR and Otto DP (1982). Antimicrob Agents Chemother **21**: 811-818.
18. Monod M, Denoya C and Dubnau D (1986). J Bacteriol **167**: 138-147.
19. Monod M, Mohan S and Dubnau D (1987). J Bacteriol **169**: 340-350.
20. Murphy E (1985). J Bacteriol **162**: 633-640.
21. Noller HF (1984). Ann Rev Biochem **53**: 119-162.
22. Pardo D and Rosset R (1977a). Mol Gen Genet **153**: 199-204.
23. Pardo D and Rosset R (1977b). Mol Gen Genet **156**: 267-271.
24. Poldermans B, Goossen N and Van Knippenberg PH (1979). J Biol Chem **254**: 9094-9100.
25. Roberts AN, Hudson GS and Brenner S (1985). Gene **35**: 259-270.
26. Saltzman L and Apirion D (1976). Mol Gen Genet **143**: 301-306.
27. Saltzman L, Brown M and Apirion D (1974). Mol Gen Genet **133**: 201-207.
28. Shivakumar AG and Dubnau D (1981). Nucl Acids Res **9**: 2549-2562.
29. Sigmund CD, Ettayebi M and Morgan A (1984). Nucl Acids Res **12**: 4653-4663.
30. Smith CJ (1987). J Bacteriol **169**: 4589-4596.
31. Suvorov AN, Van Gemen B and Van Knippenberg PH (1988). Mol Gen Genet **215**: 152-155.
32. Thakker-Varia S, Ranzini AC and Dubin DT (1985). Plasmid **14**: 152-161.
33. Uchiyama H and Weisblum B (1985). Gene **38**: 103-110.
34. Van Buul CPJJ & Van Knippenberg PH (1985). Gene **38**: 65-72.
35. Van Buul CPJJ, Lawson MP, Visser W and Van Knippenberg PH (1984). Febs Lett **177**: 119-124.
36. Van Gemen B, Koets HJ, Plooy CAM, Bodlaender J and Van Knippenberg PH (1987). Biochimie **69**: 841-848.
37. Van Gemen B, Twisk J and Van Knippenberg PH (1988). J Bacteriol : submitted.
38. Van Knippenberg PH, Van Kimmenade JMA and Heus HA (1984).

Nucl Acids Res **12**: 2595-2604.
39. Van Knippenberg PH (1986). In: Structure, Function and Genetics of Ribosomes. Hardesty B and Kramer G (eds), Springer Verlag, New York, N.Y., USA: pp 412-424.
40. Vester B and Garret RA (1987). Biochimie **69**: 891-900.
41. Vester B and Garret RA (1988). EMBO J **7**: 3577-3587.

MODIFIED NUCLEOTIDES IN EUKARYOTIC RIBOSOMAL RNA

B. Edward H. Maden

Department of Biochemistry, Liverpool University,
Liverpool, L69 3BX, UK.

ABSTRACT Ribosomal RNA from most eukaryotes contains large numbers of modified nucleotides, in contrast to ribosomal RNA from prokaryotes, which generally contains relatively few modified nucleotides. Most of the eukaryotic rRNA modifications fall into three classes: 2'-O-methylations (about 110 per human ribosome), base methylations (about 10 per human ribosome) and pseudouridines (about 95 per human ribosome). Most of the methyl groups and many of the pseudouridines have recently been mapped in human rRNA. Almost all of them are in the conserved structural core regions of rRNA. However, they occur in a wide variety of local primary and secondary structures, and there appear to be no simple consensus features to explain the presence of modifications at the large number of specific sites. This lack of obvious recognition features is particularly interesting in view of the rapidity with which most of the modifications occur upon ribosomal RNA. Thus eukaryotic rRNA modification affords a complex and largely unexplored system in which to study the factors governing the recognition of specific features in large RNA molecules.

INTRODUCTION

Although it has been known for many years that rRNA from most eukaryotes contains large numbers of modified nucleotides, only recently has the full complexity of the phenomenon of eukaryotic rRNA modification begun to unfold. This unfolding comes from determining the locations of many of the modified nucleotides in the primary structure and hence the secondary structure of rRNA: specifically the rRNA of humans, some other mammals, Xenopus and yeast. The purpose of this article is to give a descriptive overview of rRNA modification, particularly in vertebrates. The primary data from which the article is written are in references 1-3, and a more detailed review than the present one is in press (ref. 4). In that review the various findings are fully documented by detailed Tables and Figures. In the present article only a limited amount of tabulated information is given.

NUMBERS OF MODIFIED NUCLEOTIDES

Table 1 summarizes information on the numbers of modified nucleotides in human rRNA. There are about 212 modified nucleotides per ribosome, distributed between the small subunit (18S or SSU) rRNA, 5.8S and large subunit (28S or LSU) rRNA.

TABLE 1
NUMBERS OF MODIFIED NUCLEOTIDES IN HUMAN rRNA

	2'O-methyl	Base methyl	Pseudo-uridine	Other	Total modified	Total nucleotides	%Nucleotides
SSU	40	5	36	1	82	1869	4.3
5.8S	2	-	2	-	4	157	2.5
LSU	63-65	5	57	?	126	5025	2.5
Total	105-107	10	95	1-2?	212	7051	3.0

Thus the number of modified nucleotides in a human ribosome is almost three times as many as the total number of nucleotides in a tRNA molecule. However, only about 3% of the nucleotides in rRNA are modified, which is proportionately considerably less than in tRNA, in which as many as 15-20% of the nucleotides may be modified. Moreover, the repertoire of modified nucleotides is qualitatively less complex in rRNA than in tRNA, comprising in human rRNA mainly the following classes: many 2'-O-methyl groups, a few base methylations and many pseudouridines. The numbers and kinds of modified nucleotides in rRNA from other mammals that have been examined are closely similar to those in human rRNA, and the modified nucleotides in Xenopus rRNA are also fairly similar in numbers and kinds to those in human rRNA. Yeast rRNA contains about 112 modified nucleotides, or about 60% of the number in human rRNA, and again 2'-O-methylations and pseudouridines predominate. Available analytical data from other sources indicate that this is the general pattern of nucleotide modification in eukaryotic rRNA.

LOCATIONS IN THE SEQUENCES

During the last several years all of the methyl groups have been located in SSU rRNA of man, some other mammals and the frog Xenopus laevis (1); most of the pseudouridines have been located in mammalian SSU rRNA (2) and most of the methyl groups have been located in human and Xenopus LSU rRNA (3).

The modified nucleotides are distributed throughout large parts of those regions of the sequences known as conserved core regions. Conserved core regions are regions showing major

TABLE 2
MODIFIED NUCLEOTIDES IN CONSERVED CORE SEQUENCES OF HUMAN rRNA

		Core	Expansion
SSU	total nucleotides (approx)	1540	330
	located modified nucleotides(a)	68	6
LSU	total nucleotides (approx.)	3000	2000
	located modified nucleotides(a)	58	2

(a) Not all modified nucleotides have yet been located in the sequences. See the text.

homology at the level of secondary structure between eukaryotes and prokaryotes. The conserved core regions are presumed to be essential to ribosome function. They contrast to eukaryotic expansion segments, which generally vary greatly in size between different eukaryotes, and which are presumed to play a less essential role in ribosome function. (Part of one of the expansion segments in SSU rRNA is rather highly conserved among eukaryotes and this region does contain some modified nucleotides.) Thus in general the modified nucleotides are in conserved regions of the molecule. Many of the 2'-O-methylations are on phylogenetically invariant nucleotides, although the presence or absence of methyl groups at these locations is not phylogenetically invariant (ref. 4).

Within the conserved core regions there are characteristic clusters in the distribution of modifications:

    in the 5' domain of SSU rRNA, a large cluster of 2'-O-methyl groups;

in the central domain of SSU rRNA, fewer 2'-O-methyl groups but many pseudouridines;

in the 3' domain of SSU rRNA, some 2'-O-methyl groups, all five of the base-methylated nucleotides and some pseudouridines;

in domains 5 and 6 of LSU rRNA, about 40 (60%) of located 2'-O-methyl groups in the molecule.

LOCATIONS WITH RESPECT TO LOCAL PRIMARY STRUCTURE

In vertebrate DNA, methylation occurs predominantly in the diad-symmetrical sequence CpG. Are any sequences preferentially modified in rRNA? The following analysis shows that this is not the case. Each 2'-O-methylated nucleotide can be regarded as the central nucleotide of a triplet X-Ym-Z. There are 64 theoretically possible, different, such triplets. Table 3 shows that many of the different theoretically possible triplets are actually methylated in human rRNA.

TABLE 3
FREQUENCIES OF TRIPLETS ENCOMPASSING 2'-O-METHYLATION SITES IN HUMAN rRNA

	Total different 2'-O-methylated triplets
SSU rRNA	29
LSU rRNA	37
SSU + LSU	46(a)

(a) This number is not the sum of the previous two numbers as some of the same triplets are methylated in both SSU and LSU rRNA.

Similarly, the relatively few methylated bases in SSU and LSU rRNA are all in different sequences, as are the located pseudouridines in SSU rRNA. It can be concluded that the specificity for these modifications is not signalled by consensus features at the level of primary structure.

MODIFIED NUCLEOTIDES AND SECONDARY STRUCTURE

Given the lack of any obvious consensus sequences at the level of primary structure it may be asked whether modifications occur at particular types of feature in the secondary structure. There are now reliable models for the secondary structures of vertebrate SSU and LSU rRNA, especially the conserved core regions, based on a large amount of phylogenetic sequence data. Upon following an rRNA molecule from end to end one encounters several types of secondary structure features, namely: single stranded tracts that span regions between helices; long-range helical interactions; local helices; lateral and hairpin loops in local helices. The data in Table 4 show that methyl groups and pseudouridines

TABLE 4
MODIFIED NUCLEOTIDES AND LOCAL CONFORMATIONS
IN HUMAN rRNA

	SSU			LSU	
	2'-O-methyl	base methyl	Ψ	2'-O-methyl	base methyl
Single-stranded link tract	8	2	5	17	3
Long-range interaction	10	-	7	6	-
local helix: 5' limb	7	-	11	12	-
local helix: hairpin	2	3	3	9	2
local helix: 3' limb	12	-	4	7	-
(unlocated)			(8)	(15)	

occur in all of the main types of local secondary structure features. In general, 2'-O-methyl groups that occur in double stranded helical regions are in or near to imperfections in the helices, such as lateral bulges. It should be recalled that nearly all of the modified nucleotides are in the conserved core. However, even within the core there are many examples of the various types of feature listed in Table 4 that do not contain any modified nucleotides. Thus even at the level of secondary structure the basis for nucleotide modification remains enigmatic.

MODIFIED NUCLEOTIDES AND THE PATH OF HUMAN SSU rRNA IN THE RIBOSOME

Recently the path of SSU rRNA in the small ribosomal subunit of Escherichia coli has been deduced in some detail (5,6), and has been related to the overall shape of the subunit previously revealed by electron microscopy (7). Given that the core region of SSU rRNA is largely similar between prokaryotes and eukaryotes it can be inferred that the path of SSU rRNA in the ribosomal subunit is also similar between the prokaryotic and eukaryotic kingdoms. Thus the distribution of modified nucleotides in the overall structure can be inferred in the human small ribosomal subunit. The results of this exercise are summarized in Table 5. As can be seen, modified nucleotides permeate much of the overall structure of the ribosomal subunit. (These results are discussed in more detail in ref. 4.)

TABLE 5
DISTRIBUTION OF MODIFIED NUCLEOTIDES IN THE INFERRED OVERALL STRUCTURE OF HUMAN SSU rRNA IN THE RIBOSOME

Feature	2'-O-methyl	base-methyl	Located pseudouridines
Body	28	0	16
Head	10	2 (amψ,m7G)	8
Cleft	2	1 (m6A)	1
Platform	0	2 (m$_2^6$A x 2)	5

MODIFIED NUCLEOTIDES, RIBOSOMAL PRECURSOR RNA AND RIBOSOME BIOSYNTHESIS

It has been known for many years that the great majority of modifications in human rRNA occur rapidly in the nucleolus upon ribosomal precursor RNA, probably on the nascent transcript (8-10). These long-established findings take on considerable new interest now that the locations of many of the modified nucleotides are known. For example, it may be asked whether nucleotide modifications in regions that are involved in long-range interactions in mature rRNA become modified before or after these interactions occur in the primary transcript. More generally, what is the precise temporal relationship between transcription, modification and folding?

A related group of questions concerns the functions of the many modified nucleotides. Again since most of the modifications occur on ribosomal precurosr RNA, it may be asked whether the modifications play an essential role in ribosome biosynthesis. Early experiments in which starvation of HeLa cells for methionine resulted in submethylation of ribosomal precursor RNA and complete inhibition of ribosome maturation (11) suggest that 2'-O-methylation, the predominant type of methylation, does indeed play a role in ribosome maturation.

The relationship between these biological findings and the newer structural data on rRNA methylation now needs to be explored.

OUTLOOK

Before the precise locations of the modified nucleotides in eukaryotic rRNA began to be discovered it was possible to think of rRNA modification in deceptively simple global terms. Methylation occurred rapidly upon ribosomal precursor RNA and was essential for ribosome maturation. Pseudouridine fitted nebulously into the scheme of things. Moreover the phenomenon was largely confined to eukaryotes and was therefore not of universal interest.

This view has now changed. The most generally interesting aspect of the new findings is that the modifications occur predominantly wihin the core regions of rRNA. Thus although the processes that bring about the modifications are eukaryotic-specific, the general structural locations in which they occur are present in all ribosomes. This brings rRNA modification into the central arena of rRNA structure and function.

The challenge of unravelling the molecular recognition processes that give rise to the modifications and the roles of the modifications in the biosynthesis and working life of the ribosome gains general interest from their locations in the core sequences and the possibility of their relationship to the ribosomal machinery.

ACKNOWLEDGEMENTS

This work was supported by grants from the MRC, SERC and Wellcome Trust.

REFERENCES

1. Maden, B.E.H. (1986). Identification of the locations of the methyl groups in 18S ribosomal RNA from Xenopus laevis and man. J. Mol. Biol. 189:681.
2. Maden, B.E.H. and Wakeman, J.A. (1988). Pseudouridine distribution in mammalian 18S ribosomal RNA: a major cluster in the central region of the molecule. Biochem. J. 249:459.
3. Maden, B.E.H. (1988). Locations of methyl groups in 28S rRNA of Xenopus laevis and man: clustering in the conserved core of molecule. J. Mol. Biol. 201:289.
4. Maden, B.E.H. (1990). The numerous modified nucleotides in eukaryotic ribosomal RNA. Progr. Nucl. Acid. Res. Mol. Biol., in the press.
5. Brimacombe, R., Atmadja, J., Steige, W. and Schuller, D. (1988). A detailed model of the three-dimensional structure of Escherichia coli 16S ribosomal RNA in situ in the 30S subunit. J. Mol. Biol. 199:115.
6. Stern, S., Weiser, B. and Noller, H.F. (1988). Model for the three-dimensional folding of 16S ribosomal RNA. J. Mol. Biol. 204:447.
7. Lake, J.A. (1985). Evolving ribosome structure: domains in archaebacteria, eubacteria, oocytes and eukaryotes. Ann. Rev. Biochem. 54:507.
8. Greenberg, H. and Penman, S. (1966). Methylation and processing of ribosomal RNA in HeLa cells. J. Mol. Biol. 21:527.
9. Maden, B.E.H. and Salim, M. (1974). The methylated nucleotide sequences in HeLa cell ribosomal RNA and its precursors. J. Mol. Biol. 88:133.
10. Liau, M.C. and Hurlbert, R.B. (1975). The topographical order of 18S and 28S ribosomal ribonucleic acids within the 45S precursor molecule. J. Mol. Biol. 98:321.
11. Vaughan, M.H., Soeiro, R., Warner, J.R. and Darnell, J.E. (1967). The effects of methionine deprivation on ribosome biosynthesis in HeLa cells. Proc. Nat. Acad. Sci. USA, 58:1527.

STRUCTURE AND FUNCTION OF YEAST mRNA CAPPING ENZYME

Kiyohisa Mizumoto, Yoshio Shibagaki, Naoto Itoh[1],
Hisafumi Yamada[2], Shigekazu Nagata[3] and Yoshito Kaziro

The Institute of Medical Science, The University of Tokyo,
P. O. Takanawa, Minato-ku, Tokyo 108, Japan

ABSTRACT The highly purified yeast capping enzyme is composed of two separate chains of 52 kDa (α) and 80 kDa (β), responsible for the activities of mRNA guanylyltransferase and RNA 5'-triphosphatase, respectively. This is in contrast with the capping enzyme from animal cells such as rat liver or Artemia salina which consists of a single polypeptide chain with an approximate Mr of 70,000 containing two catalytic domains for each of the mRNA guanylyltransferase and RNA 5'-triphosphatase. A yeast genomic expression library in λgt11 was screened with antibodies against yeast capping enzyme. One of the positive clones, λC3, contained a 3,500 bp yeast DNA insert. From experiments based on the affinity selection of the antibodies by antigens produced with λC3 in E. coli, this clone was found to contain the gene for the α subunit. The identity of the gene was further confirmed by expressing the gene in E. coli to give a catalytically active mRNA guanylyltransferase. The gene is present in one copy per haploid genome, and encodes a polypeptide of 459 amino acid residues. Gene disruption experiments indicated that the gene is essential for the growth of yeast. From the primary structure of this gene as well as its mRNA size, we concluded that the α and β subunits of yeast capping enzyme are encoded by two separate genes.

[1]Present address: DNAX Research Institute, Palo Alto, CA
[2]Present address: Yale University, New Haven, CT
[3]Present address: Osaka Bioscience Institute, Japan

INTRODUCTION

Presence of the 5'-terminal cap structure is a ubiquitous feature of eukaryotic mRNA. Previous studies have shown that the cap structure is required for efficient initiation of translation (1-3) and for protection of mRNA against nuclease attack (4). An important role of the cap structure in the mRNA splicing has also been suggested (7). Capping is an early transcriptional event, occurring at the 5'-triphosphate termini of the nascent RNA chains initiated by RNA polymerase II (6) as well as viral RNA polymerases (7,8).

From studies on viral (7,8) and cellular (9-11) capping enzyme systems, the following sequence of reactions are proposed for the cap formation.

1) pppN- $\xrightarrow{\text{RNA 5'-triphosphatase}}$ ppN- + Pi

2) $\overset{\gamma\beta\alpha}{\text{pppG}}$ + $\overset{\beta'\alpha'}{\text{ppN-}}$ $\xrightarrow{\text{mRNA guanylyltransferase}}$ $\overset{\alpha\beta'\alpha'}{\text{G(5')pppN-}}$ + $\overset{\beta\gamma}{\text{PPi}}$

3) G(5')pppN- + AdoMet $\xrightarrow[\text{mRNA(nucleoside-2'-)methyltransferase}]{\text{mRNA(guanine-7-)methyltransferase}}$

$m^7G(5')$pppNm- + AdoHcy

We have characterized the reactions catalyzed by mRNA capping enzyme (mRNA guanylyltransferase), a key enzyme in the cap formation, from eukaryotic cells to understand the molecular mechanism of steps involved in eukaryotic mRNA synthesis. With mRNA capping enzymes from various eukaryotic cells, we have demonstrated that the capping reaction catalyzed by mRNA guanylyltransferase proceeds through a covalent enzyme-GMP intermediate, in which GMP is linked to the ε-amino group of a lysine residue of the enzyme through a phosphoamide bond (12). An RNA 5'-triphosphatase activity specifically hydrolyzing the γ-phosphate from the 5'-triphosphate terminated RNA was found to be tightly associated with mRNA guanylyltransferase purified from rat liver (13), <u>Artemia salina</u> (14), and yeast (15). The vaccinia virus capping enzyme also posesses an

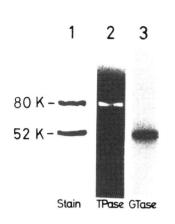

Figure 1. In situ detection of mRNA guanylyltransferase and RNA triphosphatase activities after SDS-polyacrylamide gel electrophoresis of yeast mRNA capping enzyme.
For the triphosphatase activity (lane 2), purified yeast capping enzyme was electrophoresed through a gel containing [γ-^{32}P]pppA(pA)n. Proteins were renatured in the gel and the gel was incubated under the conditions for the RNA 5'-triphosphatase reaction.
For the guanylyltransferase (lane 3), the capping enzyme was electrophoresed through a gel containing unlabeled ppA(pA)n. After renaturation of proteins, the gel was incubated with [α-^{32}P]GTP to allow cap formation. In lane 2, the gel was finally washed with 5% TCA to remove unreacted GTP and to hydrolyze the enzyme-[^{32}P]GMP intermediate. Lane 1 shows silver-staining pattern of the gel. From (15).

RNA 5'-triphosphatase activity (16). Thus a tight association of these two activities may be a general feature of mRNA capping systems. Capping enzyme from rat liver and A. salina consists of a single polypeptide chain with Mrs of 69,000 and 73,000, respectively, which contains catalytic domains for both mRNA guanylyltransferase and RNA 5'-triphosphatase (9,14). In contrast, a highly purified yeast capping enzyme is composed of two separate chains α (52 kDa) and β (80 kDa), responsible for the guanylyltransferase and triphosphatase avtivities, respectively (Fig. 1, 15).
 To see whether the α and β subunits are derived from a single polypeptide or are independent polypeptides encoded by two separate genes, we attempted to isolate the gene(s) coding for the yeast capping enzyme.

RESULTS

Immunoscreening of λgt11 yeast genomic library and identification of mRNA guanylyltransferase clones by epitope selection

To isolate the yeast gene for capping enzyme, a yeast genomic DNA expression library in λgt11 (kindly supplied by R. W. Davis, Stanford University) was screened with the polyclonal antibody raised against yeast capping enzyme. The antibody obtained from guinea pig recognized both the 52 kDa (α) and 80 kDa (β) subunits when analyzed by Western blotting of the 180 kDa yeast capping enzyme (15). Out of 2×10^6 recombinant phages screened, 4 plaques gave positive signals.

To identify which subunit these positive clones code for, immunological cross-reactivity between clone-specified polypeptides and capping enzyme subunits was tested by experiments based on the affinity selection of the antibodies by clone encoded polypeptides. Immobilized proteins from individual clones were used to select epitope-specific antibody which was subsequently eluted and identified by binding to protein blots of purified yeast capping enzyme. As shown in Fig. 2, when a protein blot of yeast capping enzyme was probed with unselected anti-capping enzyme antibodies, both the α and β chains reacted almost equally. However, if the antibodies were affinity selected with the antigens produced in E. coli infected with one of the positive clones, λC3, only the immunoreactivity against the α subunit was detected (Fig. 2, lane 2). These results clearly indicate that the phage clone λC3 contains the sequences specifying an antigenic determinant of the α polypeptide. Other three positive clones, λC11, 12 and 16, gave the same results, indicating that all these positives contain the α subunit gene. From restriction mapping, all

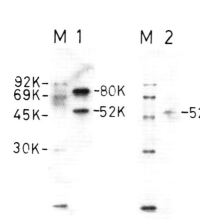

Figure 2. Identification of a recombinant clone by epitope selection.

Yeast capping enzyme was electrophoresed in 10% polyacrylamide gels. Proteins were transferred to nitrocellulose filters and probed with unselected (lane 1) or epitope-specific antibody that had been affinity selected against antigens produced by λC3 (lane 2). M indicates ^{14}C-labeled protein markers. Affinity selection of antibody was carried out as described (17).

four positive clones seemed to contain the same yeast DNA insert which is about 3.5 kb-long.

The structure of λC3 is shown in Fig. 3. λC3 contains a 3,500-bp yeast DNA insert, in which two open reading frames, ORF-1 (459 amino acids) and ORF-2 (450 amino acids), lie in close proximity, only 200-bp apart in a tail to tail fashion. From sequence analysis of λC3 we could not find a termination codon in the 35-bp upstream region of the ORF-1. We tentatively assigned the first ATG which appeared in this region as the initiator codon for the ORF-1 as depicted in Fig. 3. The entire sequence of the yeast DNA insert in λC3 will be published elsewhere (18).

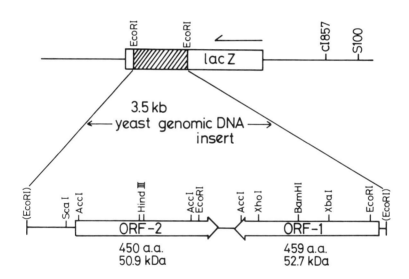

Figure 3. Restriction sites and open reading frames of λC3.

Expression of active mRNA guanylyltransferase in E. coli

Both polypeptides encoded by the ORF-1 and -2 are similar in size which is fairly close to that of the α chain. Furthermore, it is possible that an inserted gene which is oppositely oriented with respect to the the β-galactosidase transcription unit is expressed through a λ phage promoter in the vector. In the first approach to determine which ORF codes for the α subunit, inducibility of the antigen by IPTG in an E. coli strain lysogenized with λC3 was tested. A single 52 kDa protein band was clearly detected on a Western blot of the E. coli extract upon IPTG-induction, while a less intense (about 1/20) band was seen without induction (data not shown). This indicated that the properly oriented open reading frame, ORF-1, codes for the α subunit, and that the reading frame is translated through its own initiation codon but not as a fusion protein with β-galactosidase.

The second approach for determining the ORF for the α subunit was to synthesize catalytically active guanylyltransferase using deletion mutants. It is known that the α subunit catalyzes an enzyme-GMP covalent reaction

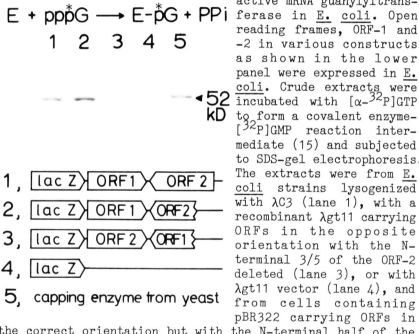

Figure 4. Expression of active mRNA guanylyltransferase in E. coli. Open reading frames, ORF-1 and -2 in various constructs as shown in the lower panel were expressed in E. coli. Crude extracts were incubated with [α-^{32}P]GTP to form a covalent enzyme-[^{32}P]GMP reaction intermediate (15) and subjected to SDS-gel electrophoresis. The extracts were from E. coli strains lysogenized with λC3 (lane 1), with a recombinant λgt11 carrying ORFs in the opposite orientation with the N-terminal 3/5 of the ORF-2 deleted (lane 3), or with λgt11 vector (lane 4), and from cells containing pBR322 carrying ORFs in the correct orientation but with the N-terminal half of the ORF-2 deleted (lane 2). Lane 5 shows an experiment with the purified yeast capping enzyme.

intermediate formation as well as the cap formation in the absence of the β subunit (15, also see Fig.1). Recombinant phages or plasmids carrying truncated ORF-1 or -2 were constructed and tested for the ability to synthesize active guanylyltransferase in E. coli. Crude extracts from strains with these constructs were incubated with [α-^{32}P]GTP to form the enzyme-[^{32}P]GMP complex and analyzed on a SDS-polyacrylamide gel (Fig. 4). A 52 kDa enzyme-GMP complex could be detected only when the ORF-1 was intact. The ORF-1 encoded protein also catalyzed the formation of [^{32}P]GpppG-RNA from ppG-RNA and [α-^{32}P]GTP (data not shown). These results confirm that the α subunit is encoded by ORF-1 and that the protein synthesized in E. coli is catalytically active in transguanylylation.

Yeast capping enzyme subunits are encoded by separate genes

Although the above results indicate that the ORF-1 specifies the α chain sequence, it does not necessarily follow that the α chain is encoded by an independent gene in yeast, since no termination codon is found in the junction region between ORF-1 and the β-galactosidase gene. Therefore, we have attempted to isolate the guanylyltransferase gene that contains the upstream region of ORF-1 by screening a yeast genomic DNA bank in YEp24 that has 10-20 kb yeast DNA inserts (19), using a yeast DNA fragment from λC3 as a probe. Six positive clones were obtained and one of them, pYGT6, contained an approximately 10 kb yeast DNA insert, which includes the entire 3.5 kb insert found in λC3, and additional lengths of 5.4 kb and 0.7 kb at the 5' upstream regions of ORF-1 and -2, respectively. Inspection of sequences 5' to the ORF-1 revealed the existence of termination codons at positions -111, -123 and -156, confirming that ORF-1 is indeed an independent reading frame and could not be part of a larger polypeptide. From Northern blot analysis of yeast poly(A)$^+$ RNA using a DNA fragment containing ORF-1 as a probe, the size of the α subunit mRNA was estimated to be about 1,600 bases, which fairly corresponds to a 52 kDa polypeptide (Fig. 5). These results together with the primary structure described above strongly support that the α and β chains of yeast capping enzyme are encoded by two separate genes.

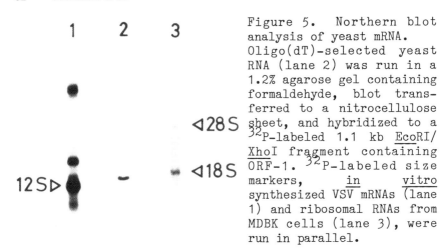

Figure 5. Northern blot analysis of yeast mRNA. Oligo(dT)-selected yeast RNA (lane 2) was run in a 1.2% agarose gel containing formaldehyde, blot transferred to a nitrocellulose sheet, and hybridized to a ^{32}P-labeled 1.1 kb EcoRI/XhoI fragment containing ORF-1. ^{32}P-labeled size markers, in vitro synthesized VSV mRNAs (lane 1) and ribosomal RNAs from MDBK cells (lane 3), were run in parallel.

The α subunit gene is a single copy, essential gene

To test whether the guanylyltransferase gene is essential for growth of S. cerevisiae, we have carried out a gene disruption experiment using fragment mediated transformation (20). Leu2⁻ diploid yeast cells were transformed with a HindIII/PstI restriction fragment carrying the α subunit gene which was separated by the 2.5 kb yeast LEU2 gene as a genetic marker (Fig. 6A). Twenty individual Leu⁺ transformants were selected and one of them was subjected to tetrad analysis. Fig. 6B shows the spores from eight asci. Only two of the four spores were viable and all the growing cells were Leu⁻. That is, LEU2 segregates with lethality and thus suggesting that the α subunit gene is essential for growth in haploid cells.

To ascertain that these genetic events are actually the results of disruption of α subunit gene, the genomic DNA from diploid Leu⁺ transformant and viable spores was prepared and subjected to Southern blot analysis using the 1.1 kb EcoRI/XhoI fragment of ORF-1 (Fig. 6C). When DNA from untransformed diploid cells (d,-) was digested with HindIII, a single 6 kb HindIII fragment was obtained as expected for a single-copy gene. DNA from a Leu⁺ diploid transformant (d,+) gave a new 8.5 kb band in addition to the 6 kb fragment, confirming that the LEU2 fragment was integrated into the α subunit gene of one of the two haploid genomes. On the other hand, when DNA from viable haploids (h1-h4) was analyzed, only a single 6 kb fragment was seen.

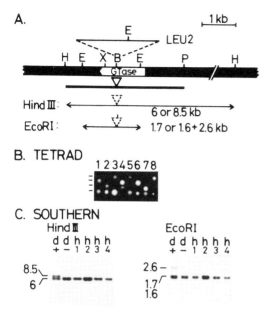

Figure 6. Disruption of mRNA guanylyltransferase gene.
A. Restriction sites of the guanylyltransferase (GTase) gene and LEU2 locus. The hollow arrow indicates the ORF-1. A 2.5 kb BamHI fragment containing the LEU2 gene was inserted in the BamHI site of ORF-1. The bold line is the fragment used for transformation of Leu2⁻ diploid cells. The lengths of restriction fragments relevant to the Southern analysis in C are indicated.
B. Tetrad analysis. Spores from eight individual tetrads of a Leu$^+$ transformant were allowed to germinate on a YPD plate. All of the viable spores were found to be leucine auxotrophs.
C. Genomic Southern blot. DNA from the parental diploid (d,-), diploid after transformation (d,+), or pairs of viable spores from two asci (h1 and h2, h3 and h4) was digested with HindIII or EcoRI. A 1.1 kb EcoRI/XhoI fragment of GTase gene was used as the probe.

These patterns confirm that the viable spores contain the intact α subunit gene. The same conclusion was obtained with EcoRI digested DNA. These observations show that the α subunit gene is essential for yeast cell growth and that one copy of the gene exists per haploid genome.

DISCUSSION

The gene for the α-subunit of yeast mRNA capping enzyme has been isolated by immunological screening of a genomic library in λgt11. Although our anti-capping enzyme antibody recognizes both the α- and β-subunits, all the positive clones we have isolated were identified to contain the α-subunit (guanylyltransferase) sequences by the epitope selection procedure. The reason why we could not obtain the clones for the β-subunit, might be, due firstly to the probability that appropriate regions of β chain can be inserted into the vector to be expressed in E. coli, or secondly, to the instability of antigens in E. coli cells.

One of the positive phage clones, λC3, contained two open reading frames ORF-1 and -2. It was unequivocally shown that the ORF-1 encodes the α subunit by expressing catalytically active guanylyltransferase in E. coli. Although mRNA for ORF-2 is expressed in yeast cells, no function has yet been assigned to the ORF-2-encoded polypeptide. From a gene disruption experiment the ORF-2 was found to be dispensable for yeast cell growth.

From the primary structure of the α subunit gene and from its mRNA size, we concluded that the α and β chains of yeast capping enzyme are encoded by two separate genes. The capping enzyme from mammalian and Artemia salina consists of a single polypeptide chain with an approximate Mr of 70,000 containing two catalytic domains for each of the mRNA guanylyltransferase and the RNA 5'-triphosphatase, both of which can be isolated in active state after limited trypsinization (9). Comparison of the gene structures of mRNA capping enzymes of yeast and animal cells will be of interest in connection with the process of gene fusion that may have occurred during evolution.

The cap structure is synthesized at an initial stage of transcription by RNA polymerase II (21,22) and is known to serve as important signals for the initiation of protein synthesis (1-3) and for RNA splicing (5). To substantiate the essential function of the cap structure, gene disruption experiment was carried out. Disruption of one copy of the α subunit gene in a diploid yeast created a recessive lethal mutation, indicating that the single guanylyltransferase gene is essential for growth. Nuclear localization of the capping enzyme, as expected from observations that capping is an early transcriptional event, was directly demonstrated by immunofluorescence microscopy of yeast cells (15).

Finally, isolation of the structural gene of yeast mRNA capping enzyme has made it possible to analyze the regions or amino acid residues required for the enzyme activity and to have more insight into the functions of cap structure in the eukaryotic gene expression.

REFERENCES

1. Shatkin AJ (1976). Capping of eucaryotic mRNAs. Cell 9:645.
2. Banerjee AK (1980). 5'-Terminal cap structure in eucaryotic messenger Ribonucleic acids. Microbiol Rev 44:175.
3. Filipowicz W (1978). Functions of the 5'-terminal m^7 cap in eukaryotic mRNA. FEBS Lett 96:1.
4. Furuichi Y, LaFiandra A, Shatkin AJ (1977). 5'-Terminal structure and mRNA stability. Nature 166:235.
5. Konarska MM, Padgett RA, Sharp PA (1984). Recognition of cap structure in splicing in vitro of mRNA precursors. Cell 38:731.
6. Gidoni D, Kahana C, Cananni D, Groner Y (1981). Specific in vitro initiation of transcription of simian virus 40 early and late genes occurs at the various cap nucleotides including cytidine. Proc Natl Acad Sci USA 78:2174.
7. Furuichi Y, Muthukrishnam S, Tomasz J, Shatkin AJ (1976). Caps in eukaruotic mRNAs: Mechanism of formation of reovirus mRNA 5'-terminal m^7GpppGmC. Prog Nucl Acid Res Mol Biol 19:3.
8. Moss B, Martin SA, Boone RF, Wei CM (1976). Modification of the 5'-terminals of mRNAs by viral and cellular enzymes. Prog Nucl Acid Res Mol Biol 19:63.
9. Mizumoto K, Kaziro Y (1987). Messenger RNA capping enzymes from eukaryotic cells. Prog Nucl Acid Res Mol Biol 34:1.
10. Venkatesan S, Gershowitz A, Moss B (1980). Purification and characterization of mRNA guanylyltransferase from HeLa cell nuclei. J Biol Chem 255:2829.
11. Shuman S (1982). RNA capping by HeLa cell RNA guanylyltransferase. J Biol Chem 257:7237.
12. Toyama R, Mizumoto K, Nakahara Y, Tatsuno T, Kaziro Y (1983). Mechanism of mRNA guanylyltransferase reaction: Isolation of N^ε-phospholysine and GMP(5'-N^ε)lysine from the guanylylenzyme intermediate. EMBO J 2:2195.
13. Yagi Y, Mizumoto K, Kaziro Y (1983). Association of an RNA 5'-triphosphatase activity with RNA guanylyltransferase partially purified from rat liver nuclei. EMBO J 2:611.
14. Yagi Y, Mizumoto K, Kaziro Y (1984). Limited tryptic digestion of mRNA capping enzyme from Artemia salina: Isolation of domains for guanylyltransferase and RNA 5'-triphosphatase. J Biol Chem 259:4695.

15. Itoh N, Yamada H, Kaziro Y, Mizumoto K (1987). Messenger RNA guanylyltransferase from Saccharomyces cerevisiae: Large scale purification, subunit functions, and subcellular localization. J Biol Chem 262:1989.
16. Venkatesan S, Gershowitz A, Moss B (1980). Modification of the 5'-end of mRNA: Association of RNA triphosphatase with the RNA guanylyltransferase, RNA(guanine-7-)methyltransferase complex from vaccinia virus. J Biol Chem 255:903.
17. Weinberger C, Hollenberg SM, Ong ES, Marmon JM, Brower ST, Cidlowski J, Thompson EB, Rosenfeld MG, Evans RM (1985). Identification of human glucocorticoid receptor cDNA clones by epitope selection. Science 228:740.
18. Shibagaki Y, Itoh N, Yamada H, Nagata S, Kaziro Y, Mizumoto K. manuscript in preparation.
19. Carlson M, Botstein D (1982). Two differentially regurated mRNAs with different 5' ends encode secreted and intracellular formes of yeast invertase. Cell 28:145.
20. Rothstein RJ (1983). One-step gene disruption in yeast. Methods Enzymol 101:202.
21. Salditt-Georgieff M, Harpold M, Chen-Kiang S, Darnell JE (1980). The addition of 5' cap structure occurs early in hnRNA synthesis and prematurely terminated molecules are capped. Cell 19:69.
22. Manley JL, Sharp PA, Gefter MK (1979). RNA synthesis in isolated nuclei: In vitro initiation of adenovirus major late mRNA precursor. Proc Natl Acad Sci USA 76:160.

A NUCLEAR (GUANINE-7-)METHYLTRANSFERASE ISOLATED FROM EHRLICH ASCITES CELLS

Guojun Bu and Thomas O. Sitz

Department of Biochemistry, Virginia Tech
Blacksburg, VA 24061

ABSTRACT A RNA (guanine-7-)methyltransferase has been partially purified from the nuclei of Ehrlich ascites cells. The substrate for this methyltransferase was methyl-deficient RNA extracted from the post-polysomal supernate from mouse liver isolated from ethionine treated mice. The methylated RNA product isolated from an enzyme incubation was hydrolyzed with T2 RNase generating a fragment with a -4.5 charge. Digestion of the methylated fragment to nucleosides generated 7-methylguanosine, suggesting the methylated compound was the "cap" structure and that the substrate for this methyltransferase is capped but unmethylated RNA (GpppN----). Commercial cap structures such as GpppN could act as inefficient substrates. The best substrates for the enzyme were either in vitro transcripts of capped but not methylated RNA or yeast 5S rRNA that was capped using guanyltransferase. Post-polysomal RNA from untreated mouse liver was not a substrate but RNA isolated from Ehrlich ascites cells had about half the methyl-accepting activity of RNA from ethionine treated mouse liver. Therefore in hypomethylated tissue (ethionine treatment or neoplastic tissue) capped but unmethylated RNAs are found in the post-polysomal RNA.

INTRODUCTION

When characterizing methyltransferases in nuclear extracts we found an enzyme that methylated hypomethylated "cap" structures found in post-polysomal RNA isolated from liver of ethionine treated mice. This (guanine-7-)methyltransferase has been isolated from a variety of organism (see Table 1) and has been partially purified and characterized (1-7).

TABLE 1
ISOLATED (GUANINE-7-)METHYLTRANSFERASES

Enzyme source	Molecular Weight	References
Vaccinia Virus	127,000	2
Yeast	49,000	3
N. Crassa	not determined	4
Wheat Germ	not determined	5
Hela Cells	56,000	6
Rat Liver	130,000	7
Ehrlich ascites cells	240,000	Current Study

The "cap" structure (^{m}GpppN----) is synthesized during transcription of most eucaryotic mRNAs in the nucleus and is found on the processed mRNA in the cytoplasm (1). The methylation of the guanine-7 position of the "cap" structure is necessary for the binding of the mRNA to the ribosome(8). Thus this methylation is very important for the expression of most mRNAs in eucaryotic cells.

The 5'-terminal cap structure of mRNA's is normally methylated in a number of positions such as the guanine-7-position (^{m}GpppN----, the N-6-position of adenine (Gppp^{m}A---), and at 2'-O-ribose positions generating a "cap 1" (GpppNmpN---) or "cap 2" (GpppNmpNmpN---) structure. It is not clear if there is a specific order by which these modifications occur. For example, we find hypomethylated "cap 1" structures in Ehrlich ascites cells which are not methylated in the guanine-7-position (GpppNmpN---). Therefore, there is no absolute requirement for the guanine-7-methylated to occur first. While the guanine-7-methylation of the cap structure is required for binding to ribosomes(8) the function of the other methylations is not clear but may confer protection from nuclease digestion.

RESULTS AND DISCUSSION

Methyltransferases in Nuclear Extracts.

When a crude nuclear extract (100,000xg) supernate passed over a DEAE-Sephadex column to remove endogenous RNA) was incubated with methyl-deficient post-polysomal RNA isolated from ethionine treated mice (three days of injection with ethionine and adenine) and ^{3}H-S-adenosylmethionine, radioactive RNA could be isolated. When this RNA was digested with NaOH, applied to a DEAE-Sephadex column and eluted with a NaCl gradient, two radioactive peaks were observed (Figure 1). The first peak was base

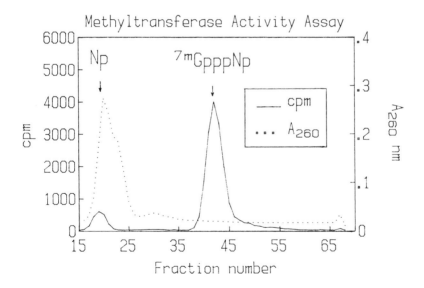

FIGURE 1. Methyltransferase activity assay. Nuclear methyltransferases from Ehrlich ascites cells were assayed with post-polysomal methyl-deficient RNA as substrate. The standard enzyme incubation was carried out at 37°C for

40 minutes. RNA was isolated and digested with 0.3 N NaOH. The digested products were applied to a DEAE-Sephadex column in 7 m urea buffer. The column was eluted with a linear gradient of 0-0.4 M NaCl and 3 ml fractions were collected. For each fraction, radioactivity and UV absorbance was measured. For the radioactive curve, the mononucleotide peak on the left represents base methyltransferase activity while the -5 charge peak on the right represents (guanine-7-)methyltransferase activity, i.e. methylation of "cap" structure.

---

methylated nucleotides (-2 charge) while the second peak was found to be a methylated "cap 0" structure (^{m}GpppNp; -5 charge). Very little cap methylating activity was found in the cytosol of Ehrlich ascites cells demonstrating that this is nuclear methyltransferase. The composition analysis by either TLC (9) or high voltage paper electrophoresis at pH 3.5 of the resulting digestion products of a combined venom phosphodiesterase and alkaline phosphatase of the -5 charged fragment demonstrated that the 7-methylguanosine was the modified nucleoside. If the methylation of capped RNA was inhibited and the guanine-7-methylation of cap structure was necessary for binding to ribosomes then we would expect that the hypomethylated RNA that acts as substrate for the cap methylating enzyme from ethionine treated mice would be found in highest concentration in the post-polysomal fraction. This was what was seen when RNA from different subcellular fractions was isolated and used as a RNA substrate for the methyltransferase in nuclear extract (figure 2). It is clear from these data that while the hypomethylated capped RNA was in highest concentration in the post-polysomal fraction that the base methylating activity seen in the mononucleotide peak was about the same in all three fractions. The RNA that was acting as a substrate for the cap methyltransferase ranged in size from about 50 nucleotides to almost 18S in size. Probably when the hypomethylated capped mRNA appears in the cytosol and does not bind to ribosomes it starts to degrade and generates a heterogeneous size population. This mRNA would also lose its 3'-poly A ends because of this degradation and then would be found in the poly A minus fraction.

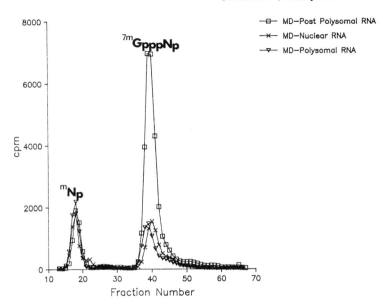

FIGURE 2. Comparison of methyl-deficient RNA isolated from different subcellular fractions. Mice were treated with ethionine and adenine for three days to generate methyl-deficient RNA. The livers of mice were collected and subcellular fractions prepared. Nuclear RNA, polysomal RNA and post-polysomal RNA were isolated and compared as the substrate for the (guanine-7-)methyltransferase by DEAE-Sephadex column technique.

Purification of the (Guanine-7-)Methyltransferase.

This nuclear enzyme was purified about 100 fold as shown in the purification table (Table 2). The DEAE-Sephadex column was mainly used to remove endogenous nucleic acids and the CM-Sepharose column was very efficient in removing most of the ribonuclease activity. The DEAE-Sepharose column step was found to be a good procedure to separate the (guanine-7-)methyltransferase from other base methyltransferases. The enzyme was very stable with good recoveries in the early purification steps but when the total protein concentration was low and the enzyme diluting the recoveries became poor (this can be seen in

TABLE 2
PURIFICATION OF RNA (GUANINE-7-)METHYLTRANSFERASE

Fraction	Total Protein	Total Activity	Specific Activity	Purification	Recovery
	(mg)	(units)	(units/mg)	(fold)	(%)
1. Nuclear Suspension	314	673	2.14	1	100
2. High Speed Centrifuge	61.5	653	10.6	4.95	97
3. DEAE-Sephadex	31.7	567	17.6	8.36	84
4. CM-Sepharose	7.13	511	71.7	33.5	76
5. DEAE-Sepharose	0.82	191	233	109	28

table 2, fraction 5). The enzyme was very stable under the storage conditions (50% glycerol at -20°). One preparation showed almost no loss in activity when stored for over a year. The partially purified enzyme had almost no RNase activity and no other detectable methyltransferase activity. We plan to use these preparations for determining kinetic parameters and hope to overcome the problems in loss of enzyme activity and purify the enzyme to homogeneity.

Substrate Specificity.

While post-polysomal RNA isolated from the livers of ethionine treated mice was the best natural RNA substrate for the guanine-7-methyltransferase, post-polysomal RNA isolated from Ehrlich ascites cells had about 52% of the methyl-accepting activity. RNA isolated from livers not treated with ethionine had no methyl accepting activity for the guanine-7-position. The hypomethylated cap structures isolated from ethionine treated mice and Ehrlich ascites cells were found to be "cap 0" (GpppNp) and "cap 1" (GpppNmpNp) structures respectively. These data with the Ehrlich ascites cells support the observations of a paradox with cancer cells of hypomethylated RNA with the methyltransferase that modifies that position elevated.

We have found that the enzyme activity is elevated in Ehrlich ascites cells about ten fold over the activity found in normal mouse liver.

Some commercial "cap" analogs were substrates for the purified guanine-7-methyltransferase. The following "cap" analogs were tested GpppG, GpppAm, ^{m}GpppG, ^{m}GpppAm and found to have the following relative activities 100%, 84%, 12%, and 0% respectively. It was interesting to find that the ^{m}GpppG cap analog could act as poor substrate to produce the dimethylated product ^{m}GpppG^m. While the symmetrical cap structure GpppG was the best of these analogs it was still an inefficient substrate suggesting that additional nucleotides are needed for optimal activity (GpppGpN---N).

While the post-polysomal RNA isolated from ethionine treated mouse liver was a good substrate it is almost pure tRNA with a small amount of hypomethylated capped RNA. To develop a better substrate we isolated yeast 5S rRNA (a primary transcript with mostly ppN- or pppN- at the 5'-terminal) and capped it with GTP and guanyltransferase. This capped RNA had excellent methyl-accepting activity and will be a useful substrate. An even better substrate was the in vitro transcript using the plasmid pGEM-3Z and the strong T7 promoter. By the selection of the appropriate restriction enzyme cleavage site in the multiple cloning site different size transcripts can be made. When the cap analog GpppG is added in ten fold excess over GTP levels then a capped but not methylated RNA of defined length can be made(10). Figure 3 shows the results of a cleavage with Hind III cleavage which generates a capped run-off transcript 56 nucleotides long. The analysis of the methylated product demonstrates that the only methylation detected was the guanine-7-methylation of the cap structure. This transcription system will be an ideal way to generate capped substrate of various chain lengths. We hope that by the use of these defined RNA substrates that we will be able to characterize the substrate requirements and kinetic properties of this enzyme that is important in mRNA expression.

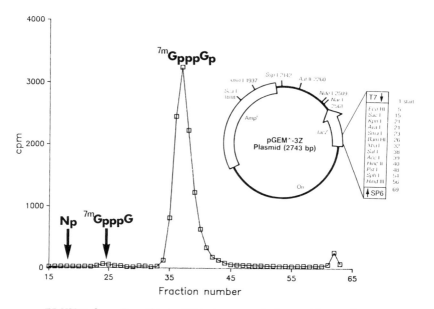

FIGURE 3. In vitro RNA transcript as the substrate for the (guanine-7-)-methyltransferase. Plasmid pGEM-3Z which contains a T7 promoter and a multiple cloning site was used as the DNA template. Hind III restriction endonuclease was used to linearize the plasmid for a run-off transcription. The transcriptional system contained ten times more GpppG than GTP in order to enhance the incorporation of GpppG as the first nucleotide. A 56 nucleotide long RNA transcript with a "capped", unmethylated 5'-terminus was synthesized and used as a substrate for the (guanine-7)-methyltransferase. When RNA was isolated from the enzyme incubation and digested with T2 RNase, the radioactive product m^{7}GpppGp was detected by a DEAE-Sephadex column, which ensured the incorporation of GpppG into the RNA transcript.

REFERENCES

1. Mizumoto K, Kaziro Y (1987). Messenger RNA Capping Enzymes from Eukaryotic Cells. Prog Nucleic Acid Res Mol Biol 34:1.

2. Martin SA, Paoletti E, Moss B (1975). Purification of mRNA Guanylyltransferase and mRNA (guanine-7-)methyltransferase from Vaccinia Virions. J Biol Chem 250:9322.
3. Locht C, Beaudart J-L, Delcour J (1983). Partial Purification and Characterization of mRNA (Guanine-7-)Methyltransferase from the Yeast Saccharomyces cerevisiae. E J Biochem 134:117.
4. Germershausen J, Goodman D, Somberg EW (1978). 5' Cap Methylation of Homologous Poly A(+) RNA by a RNA(guanine-7) Methyltransferase from Neurospora Crassa. BBRC 82:871.
5. Locht C, Delcour J (1985). In Vitro Methylation of Undermethylated Yeast Poly(A)-rich RNA using mRNA (Guanine-7-)methyltransferase Purified from Wheat Germ or Yeast. E J Biochem 152:247.
6. Ensinger MJ, Moss B (1976). Modification of the 5' Terminus of mRNA by an RNA (Guanine-7-)methyltransferase from HeLa Cells. J Biol Chem 251:5283.
7. Mizumoto K, Lipmann F (1979). Transmethylation and Transguanylylation in 5'-RNA Capping System Isolated from Rat Liver Nuclei. PNAS 76:4961.
8. Shatkin AJ, Furuichi Y, Sonenberg N (1979). 5-Terminal Modification and Translation of Eukcaryotic mRNAs. In Usdin E, Borchardt RT, Creveling CR (eds): "Transmethylation", New York: Elsevier North Holland, p 341.
9. Ro-Choi TS, Choi YC, Henning D, McCloskey J, Busch H (1975). Nucleotide Sequence of U-2 Ribonucleic Acid. J Biol Chem 250:3921.
10. Nielsen DA, Shapiro DJ (1986). Preparation of Capped RNA Transcripts using T7 RNA Polymerase. NAS 14:5936.

EFFECTS OF HYPERMETHYLATED AND ABERRANT 5'-CAPS ON TRANSLATION OF β-GLOBIN AND SINDBIS VIRUS 26S mRNAs[1]

Stanley M. Tahara[a], Edward Darzynkiewicz[b], Janusz Stepinski[c], Irena Ekiel[d], Youxin Jin[a,2], Chang Hahn[e], James H. Strauss[e], Teju Sijuwade[a], and Dorota Haber[b]

[a]Dept. of Microbiol., USC School of Medicine, Los Angeles, CA 90033-1054; [b]Dept. of Biophys., Inst. of Exptl. Physics, Univ. of Warsaw, Warsaw, Poland; [c]Dept. of Organic Chem., Inst., of Chem., Univ. of Warsaw, Warsaw, Poland; [d]Biotech. Res. Inst., NRCC, Montreal, Canada; [e]Div. of Biol., Calif. Inst. of Technology, Pasadena CA 91125

ABSTRACT Hypermethylated guanine cap structures occur on cellular snRNAs and togavirus [e.g. Sindbis virus (SV) and Semliki Forest Virus (SFV)] 26S mRNAs. The intramolecular effect of different hypermethylated 5'-cap structures on translation activity of rabbit β-globin and Sindbis virus (SV) 26S mRNAs was examined. It was found that a 2,7-dimethyl guanosine ($Me_2^{2,7}G$) 5'-cap increased β-globin mRNA rate of translation by 50%. In contrast a 2,2,7-trimethyl guanosine ($Me_3^{2,2,7}G$) cap decreased translation by 75% compared to native 7-methyl guanosine (Me^7G) capped mRNA. No translation rate enhancement of SV 26S mRNA by $Me_2^{2,7}G$ cap was observed. It was found that β-Globin mRNA capped with 7-benzyl guanosine (Bn^7G) was stimulated 1.8 fold in translation.

---

[1]Supported by the American Cancer Society (MV-257), the Life and Health Insurance Medical Research Fund, the Margaret E. Early Research Trust, and the NIH (GM38512). E. D. was supported in part by the Polish Academy of Sciences (CPBR-3.13).
[2]Permanent address: Shanghai Institute of Biochemistry, Academia Sinica, Shanghai, People's Republic of China

INTRODUCTION

The role of the 5'-Me^7G cap structure of eukaryotic mRNAs ($Me^7G^{5'}ppp^{5'}N\cdots$) has been extensively studied over the past 15 years (1,2). Although it was first shown to be important for initiation of translation, it participates in other cellular and viral processes as well. For example, the cap structure is known to be important in: mRNA stability (3,4), transcription initiation (5), RNA splicing (6), termination of transcription (7), and 3'-end mRNA processing (8).

Two other physiologically occurring capping nucleotides are known. These are $Me_2^{2,7}G$ and $Me_3^{2,2,7}G$. The latter is found on the 5'-ends of snRNAs (9) whereas both dimethyl and trimethyl guanosines are found to cap togavirus 26S mRNAs (10,11).

Recently, modified cap structures have been shown to be a part of complex mRNA splicing processes in trypanosomes (12) and *Caenorhabditis elegans* (13,14) where $Me_2^{2,7}G$ or $Me_3^{2,2,7}G$ capped leader segments are spliced onto protein coding regions. A related mechanism for cap dependent mRNA synthesis is observed in influenza virus, where Me^7G-capped RNA oligomers derived from cellular mRNAs are used to prime viral mRNA transcription (15).

Several key interactions between the 5'-cap of eukaryotic mRNAs and cap binding protein (eIF-4F) are required for optimum binding to the factor in translation (for review of initiation see ref. 16). As we and others have shown, the chemical features of the cap which are recognized include: 1) the N7 substituent of the capping guanine nucleotide (17-19); 2) the purine C2 amino and C6 keto substituents (17); and 3) the α-phosphate of the capping nucleotide (20-22).

In this study we examined the translational properties of hypermethylated cap structures on cellular (β-globin) and viral (SV) mRNAs.

METHODS

Plasmid Constructions.

pT7rG4. Construction of this plasmid was described earlier (23). It contains a 489 bp *Hind*III/*Bgl*II fragment containing an intact rabbit β-globin coding region subcloned in pT7/T3-19 (Bethesda Research Laboratories).

pTS22Δ34. The 2.0 kb SalI/SalI fragment of pT7SVSF (C. Hahn and J. Strauss, in preparation) was subcloned into the SalI site of pGEM9Zf to give pTS22. Deletion of vector sequences (34 nts) between the authentic 5'-cap site of the 26S mRNA and the first transcribed nucleotide of the T7 promoter gave pTS22Δ34. The details of this construction will be described elsewhere. Transcription of this plasmid resulted in an mRNA molecule of 742 nts excluding the cap, with the normal cap site extended in the 5'-direction by a single guanosine. The insert in pTS22Δ34 represents the first 442 nts of native Sindbis 26S mRNA from pT7SVSF. The remaining 388 nts to the StuI site, result from inversion of an 828 nt NcoI/NcoI fragment, used to construct pT7SVSF from pT7SVSP (C. Hahn and J. Strauss, in preparation). (The latter is a cDNA clone of 26S mRNA (~4 kb)). This was done to create a new translation termination site for the capsid coding region. Translation of this mRNA gave a truncated capsid protein of M_r 19,000 (data not shown).

Synthesis of Mono and Dinucleotide Cap Analogs.

Chemical syntheses of nucleotide cap analogs used in this study have been described (19,23). A full description of synthesis and physical characterization of the extended cap dinucleotides will be reported elsewhere (Darzynkiewicz et al., in preparation; ref. 23).

In Vitro Transcription.

5'-Capped RNA transcripts were synthesized from AvaI-linearized pT7rG4 and StuI-linearized pTS22/pTS22Δ34, using T7 RNA polymerase as described earlier (23). All templates used in this study initiated transcription with G (see above), thus the 5'-terminal base was capped by priming with the capped dinucleotide **NpppG**, where **N** is a G derivative. 5'-End analyses of T7 transcribed RNAs were performed as described earlier (23) using the method of Konarska et al. (6).

In Vitro Translation.

Preparation of rabbit reticulocyte lysate and in vitro translation assays were performed as described earlier (22,24). Translation assays measured the incorporation of [^{35}S]Met (1347 Ci/mmol, Amersham) into hot trichloroacetic acid (TCA) precipitable material.

RESULTS

We examined the properties of $Me_3^{2,2,7}GMP$ and the related compound $Me_2^{2,7}GMP$ as competitive inhibitors of translation. Cap analogs inhibit translation if they successfully compete with mRNA 5'-caps for eIF-4F. We and others have used this approach to study the molecular interactions between the cap structure and eIF-4F (14,19-21). A titration of the hypermethylated cap analogs into a reticulocyte translation assay indicated that $Me_2^{2,7}GMP$ was an effective inhibitor of translation

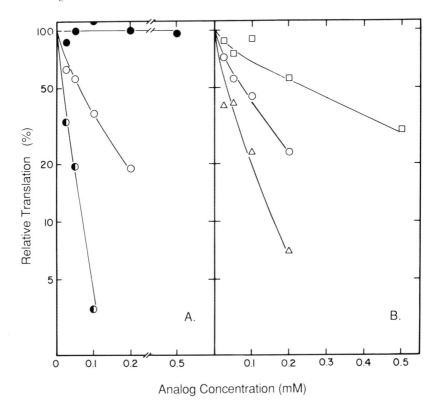

FIGURE 1. Inhibition of β-globin synthesis in reticulocyte lysate by cap analogs. Translation was performed in micrococcal-treated rabbit reticulocyte lysate with different cap analog concentrations, as described earlier (22). A--Me^7GMP (○); $Me_2^{2,7}GMP$ (◐); $Me_3^{2,2,7}GMP$ (●). B--Me^7GMP (○); Et^7GMP (□); Bn^7GMP (△).

(50% inhibition at 0.02 mM), whereas $Me_3^{2,2,7}$GMP was inactive as an inhibitor (Fig. 1A). In comparison to Me^7GMP (50% inhibition at 0.06 mM), $Me_2^{2,7}$GMP was the better inhibitor. This result suggested that it possessed a higher affinity for eIF-4F than did Me^7GMP. For further comparison two other N7 substituted cap analogs [7-Ethyl GMP (Et^7GMP) and 7-benzyl GMP(Bn^7GMP)] were assayed as translation inhibitors. Both cap analogs, as nucleoside diphosphates, were previously reported to be as inhibitory as Me^7GDP (17,18). By contrast, we observed that Et^7GMP and Bn^7GMP differed greatly as inhibitors compared to Me^7GMP (Fig. 1B). Et^7GMP was relatively poor as an inhibitor of translation (50% inhibition at 0.25 mM), whereas Bn^7GMP was a potent inhibitor of translation (50% inhibition at 0.03 mM), comparable to $Me_2^{2,7}$GMP. We recently showed that Me^7GMP analogs with substituents larger than ethyl are essentially inactive as inhibitors, however other 7-phenyl cap analog derivatives may be active inhibitors (19).

The results of the cap analog inhibition studies were somewhat surprising since $Me_2^{2,7}$GMP and Bn^7GMP were more inhibitory than Me^7GMP. We were unsure if the effect was due to an increased affinity for eIF-4F or because of inhibition at another step of initiation. In order to help address this, we examined the effect on translation of non-standard cap structures in cis with mRNA. We expected that $Me_2^{2,7}$G and Bn^7G capped mRNAs would be more active due to an increased affinity of the cap moiety as a

FIGURE 2. 5'-End analyses of artificial mRNAs. β-Globin transcripts were subjected to RNase T2 digestion and analyzed by two-dimensional TLC as described (6). Positions of NpppG and pppGp markers on each chromatogram are shown as dotted circles. The arrows refer to capped 5'-end dinucleotide (NpppGp*) released after RNase T2 treatment. The position of pppGp is denoted by an asterisk in each panel. Autoradiograms are shown. A. Uncapped mRNA; B. GpppG capped mRNA; C. Me^7GpppG capped mRNA; D. Et^7GpppG capped mRNA; E. Bn^7GpppG capped mRNA; F. $Me_2^{2,7}$GpppG capped mRNA; G. $Me_3^{2,2,7}$GpppG capped mRNA; and H. schematic of relative migration of 3'-mononucleotides and expected 5'-end dinucleotides. The oligonucleotides are labeled as follows: pppGp (1); GpppG (2); GpppGp (2a); Me^7GpppG (3); Me^7GpppGp (3a); Et^7GpppG (4); Et^7GpppGp (4a); Bn^7GpppG (5); Bn^7GpppGp (5a); $Me_2^{2,7}$GpppG (6); $Me_2^{2,7}$GpppGp (6a); $Me_3^{2,2,7}$GpppG (7); $Me_3^{2,2,7}$GpppGp (7a).

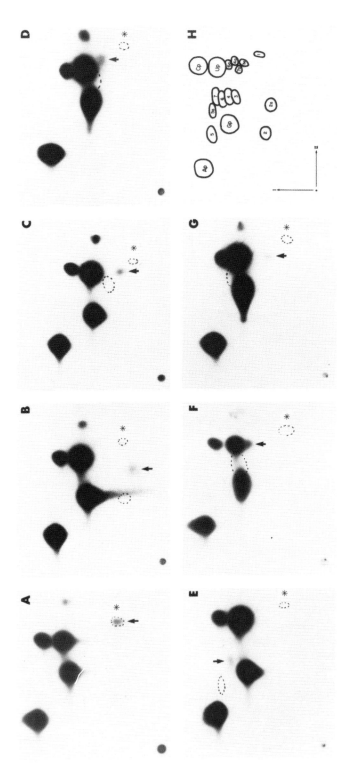

FIGURE 2.

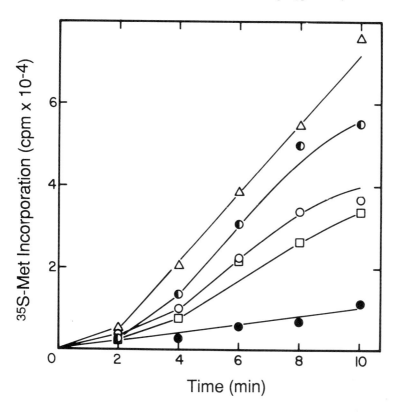

FIGURE 3. Translation activity of globin mRNAs with nonstandard caps. Translation assays were performed as described earlier (23). Synthetic mRNAs terminated with Me^7G or nonstandard 5'-caps were assayed for incorporation of radioactivity into hot-TCA precipitable material. Me^7GpppG-mRNA (O); $Me_3^{2,2,7}GpppG$-mRNA (●); $Me_2^{2,7}GpppG$-mRNA (◐); Et^7GpppG-mRNA (□); Bn^7GpppG-mRNA (△). Relative rates of translation for the various capped mRNAs were: Me^7GpppG (1.0); $Me_2^{2,7}GpppG$ (1.5); $Me_3^{2,2,7}GpppG$ (0.24); Et^7GpppG (0.75); Bn^7GpppG (1.8).

ligand for eIF-4F, if the effect were limited to eIF-4F dependent events. In order to perform these translation studies, extended caps of the form **NpppG** (N=Me^7G, Et^7G, Bn^7G, $Me_2^{2,7}G$, and $Me_3^{2,2,7}G$) were chemically synthesized. The intent was to incorporate these dinucleotides into the 5'-ends of synthetic transcripts using T7 RNA polymerase.

However we were unsure if modified capped dinucleotides, other than Me^7GpppG, could initiate transcription by phage RNA polymerases (6,25). In an effort to determine this, synthetic mRNAs were synthesized from pT7rG4 with different extended cap dinucleotides as primers, and internally labeled with $[\alpha-^{32}P]GTP$. Presumptively capped transcripts were subjected to RNase T2 digestion and analyzed for transfer of radiolabel into the 3'-position of the extended cap (NpppGp*). As shown in Fig. 2, mRNA transcripts were observed to have quantitatively incorporated a 5'-cap corresponding to the synthetic extended cap which was used for transcription initiation. An undetectable fraction of mRNA 5'-ends corresponded to uncapped 5'-ends, as indicated by the absence of pppGp*. Furthermore, transcripts initiated with different extended caps were homogeneous and equal in expected size to globin transcripts initiated with Me^7GpppG (not shown). Thus transcriptional capping by T7 RNA polymerase could be used with the extended cap dinucleotides synthesized for this study, to make full length and appropriately 5'-terminated mRNAs.

β-Globin mRNAs with different capping nucleotides were assayed in rabbit reticulocyte lysate for relative rates of translation. Messenger RNAs capped with $Me_2^{2,7}G$ and Bn^7G were respectively 1.5 and 1.8-fold more active than Me^7G capped mRNA. Replacement of Me^7G by Et^7G diminished activity about 25% whereas $Me_3^{2,2,7}G$ caps were essentially inactive (Fig. 3). GpppG and pppG ended mRNAs tested as translation templates under these conditions were found to have the same activity as $Me_3^{2,2,7}G$ capped mRNA. Polypeptides synthesized from the different capped templates corresponded to full length β-globin; no other products were observed. In a control experiment the differently capped globin transcripts were incubated in reticulocyte lysate under mock translation conditions. All transcripts were observed to undergo slight degradation (≤10%) during the standard incubation period, however there was no significant difference in degradation rates of the variously capped mRNAs. Thus the observed differences in translation rates were not due to differences in mRNA stability.

Given these results we decided to examine the effect of hypermethylated cap structures on the activity of SV mRNA. SV and Semliki Forest virus (SFV) 26S subgenomic mRNAs are capped with $Me_2^{2,7}G$ and $Me_3^{2,2,7}G$ structures in

late infection (10,11), however the effects of such
modifications on translation are unknown. In an effort to
examine the translational properties of SV mRNA terminated
with hypermethylated caps, we synthesized such derivatives
of 26S mRNA from pTS22Δ34. The transcripts made from this
plasmid DNA template represent the native 5'-end of SV 26S
RNA and the 5'-proximal coding region of the capsid gene.
We expected that the effect of the $Me_2^{2,7}G$ cap structure on
SV mRNA would be identical to that observed for β-globin
mRNA.

When the parental cDNA clone of SV 26S mRNA (pTS22),
containing 5'-vector sequences, was used for transcription
and translation, we observed that $Me_2^{2,7}G$ capped mRNA was
2.3-fold more active than Me^7G capped TS22 mRNA (Fig. 4).
This enhancement in mRNA activity was comparable to the
effect of the cap structure on translation as observed for
β-globin mRNA (cf. Fig. 3). In comparison, TS22Δ34 mRNA
was ~5.2-fold more active overall, presumably due to
removal of the unrelated vector sequences; however no rate
enhancement, by replacement of the Me^7G cap with $Me_2^{2,7}G$,
was observed (see Fig. 4). Assay of $Me_2^{2,7}G$ vs. Me^7G capped
TS22Δ34 mRNA translation under conditions of increasing
ionic strength or increasing mRNA concentrations revealed
no significant differences in relative activity. Control
experiments also showed no significant difference in mRNA
stability or product synthesis by either differently
capped TS22 and TS22Δ34 mRNAs (data not shown). The
absence of rate enhancement by $Me_2^{2,7}G$ cap was thus a
general property of TS22Δ34 mRNA.

In a separate experiment, the effect of other 5'-end
structures on TS22 and TS22Δ34 mRNAs was assayed for
relative activity. As shown in Fig. 5, a comparison of
the Me^7G and $Me_2^{2,7}G$ capped forms of TS22 SV mRNA showed
that a $Me_2^{2,7}G$ cap structure afforded a 1.7 fold rate
enhancement over Me^7G capped mRNA, this was slightly less
than that observed for TS22 mRNA in Fig. 4. As observed
earlier, the mRNA construction with an authentic 5'-end
sequence (TS22Δ34 mRNA) showed very little rate
enhancement when capped with $Me_2^{2,7}G$ instead of Me^7G.
Other than this difference in activity of the two related
mRNAs, transcripts made from either pTS22 or pTS22Δ34
behaved similarly when capped with $Me_3^{2,2,7}G$, GpppG or if
left uncapped (pppG). The latter 5'-end structures
decreased mRNA activity to less than 20% of the control.
Thus the absence of rate enhancement of TS22Δ34 mRNA by

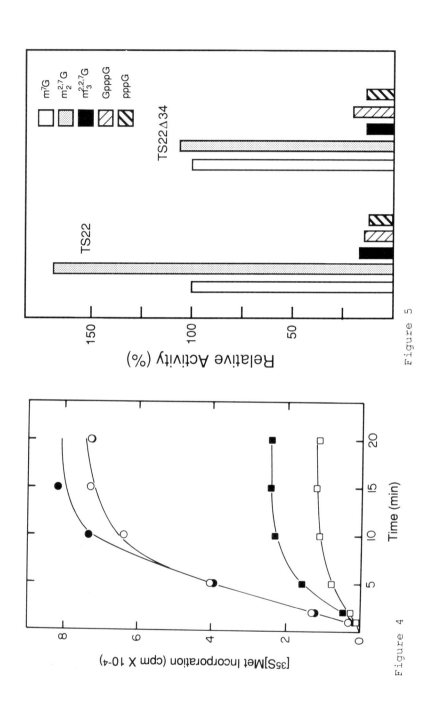

Figure 4

Figure 5

$Me_2^{2,7}G$ must be related to removal of the 34 nt non-viral sequences.

DISCUSSION

Our observation that $Me_3^{2,2,7}G$ capped mRNAs were inefficiently translated is consistent with their cellular localization *in vivo*. Trimethylated caps ($Me_3^{2,2,7}G$) are found on snRNAs which participate in nuclear processes (26). The N7 methylation occurs in the nucleus during transcription initiation; however further cytoplasmic processing is required for N2 exocyclic dimethylation (27). During this phase of snRNA maturation and snRNP assembly, accumulation of Me^7G capped snRNAs in the cytoplasm could be detrimental to normal translation. These presumably could be recognized as potential "mRNAs." Thus a hypermethylated cap might serve to functionally compartmentalize snRNAs from mRNA dependent processes and vice versa. The hypermethylated cap may also be required for retrograde transport of snRNPs from the cytoplasm to the nucleus (27).

Our results indicated that SV 26S mRNA translation is not enhanced by a $Me_2^{2,7}G$ cap structure. Computer aided analysis of potential RNA secondary structures (28) in TS22 and TS22Δ34 mRNAs indicated that the former had a very stable, complex stem loop structure (ΔG= -28.4 kcal/mol), beginning with the penultimate G residue through the next 115 nts. The corresponding 5'-untranslated region (UR) of TS22Δ34 was considerably less stable, with two hairpin loops of approximately equal stability, but with a combined stability of only ΔG= -11 kcal/mol. The latter structures are probably denatured

---

Figure 4 --Effect of hypermethylated caps on SV 26S mRNA translation. Synthetic SV mRNAs (200 ng) terminated with Me^7GpppG or $Me_2^{2,7}GpppG$ were assayed as described in the legend to Fig. 3. TS22 mRNA (□); TS22Δ34 mRNA (○). Open symbols indicate Me^7G capped mRNAs, closed symbols denote $Me_2^{2,7}G$ capped mRNAs.

Figure 5 --Activity of TS22 and TS22Δ34 mRNAs with different 5'-cap structures. Translation rates of differently capped mRNAs were measured as in Fig. 4 and shown here as a ratio, relative to Me^7G capped mRNA, which was given a value of 100.

under physiological conditions. Thus the 34 nts in pTS22 contributed negatively to translation of the resultant T7 transcript.

In the current model of translation initiation, eIF-4F binds to the 5'-cap structure and melts out local secondary structure in an ATP-dependent process (1,16). Further mRNA melting is effected by factors eIF-4A and eIF-4B (29,30). Messenger RNAs can differ in their requirements for either eIF-4A or eIF-4F (31). This difference has been attributed to the degree of structure present in the 5'-UR (29-33).

We explain the rate enhancement observed with $Me_2^{2',7}G$ caps to be due to stabilization of eIF-4F binding to the mRNA via the cap. This would promote subsequent eIF-4A/4B binding (16,29,30). It should be noted that although there was no enhancement of TS22Δ34 translation by $Me_2^{2',7}G$ cap, translation of this mRNA was still cap dependent, as evident from the data in Fig. 5. An absence of 5'-proximal secondary structure in TS22Δ34 may preclude requirement for factors eIF-4A and -4B. Our results are generally consistent with this interpretation, in that the rate determining step is probably 5'-distal RNA melting. This is currently being analyzed.

Our translation results with TS22Δ34 mRNA generally correlated with the reported subcellular distribution of SFV 26S mRNA. van Duijn et al. reported that ~20% of $Me_3^{2,2',7}G$ capped 26S mRNA was polysomal, whereas up to half of the $Me_2^{2',7}G$ capped mRNA was polysome associated (11). In contrast we demonstrated $Me_2^{2',7}G$ capped TS22Δ34 mRNA was as active as Me^7G capped mRNA, but $Me_3^{2,2',7}G$ capped transcript was only ~20% as active.

Hypermethylated caps on normal cellular mRNAs have not yet been observed. This type of modification could be a basis for translational regulation. In this case mRNA activity would be increased by one N2 methyl moiety and decreased by a second. The increased activity of Bn^7G capped mRNA is consistent with the higher affinity of this cap analog for eIF-4F (23). Although at present we have not fully identified the type of interaction between this cap structure and eIF-4F.

Previous studies which examined the effect of exchanging the native Me^7G nucleoside of the cap with another base were limited to replacement by either G (34), 7-methyl inosine (35) or Et^7G (18). By the approach we have described it should be possible to substitute the capping nucleotide with virtually any base. The approaches we have described may also be useful in studying other 5'-cap related phenomena, such as parasite or nematode RNA trans-splicing.

ACKNOWLEDGMENTS

Oligonucleotides were supplied by the Microchemical Core Facility of the Norris Cancer Center which is funded in part by NIH grant 5 P30 CA14089. Some data in this paper have been previously published (19,23).

REFERENCES

1. Shatkin AJ (1985). mRNA cap binding proteins: essential factors for initiating translation. Cell 40:223-224.
2. Banerjee AK (1980). 5'-Terminal cap structure in eucaryotic messenger ribonucleic acids. Microbiol Rev 44:175-205.
3. Furuichi Y, La Fiandra A, Shatkin AJ (1977). 5'-Terminal structure and mRNA stability. Nature 266:235-239.
4. Green MR, Maniatis T, Melton DA (1983). Human β-globin pre-mRNA synthesized *in vitro* is accurately spliced in *Xenopus* oocyte nuclei. Cell 32:681-694.
5. Furuichi Y (1981). Allosteric stimulatory effect of S-adenosylmethionine on the RNA polymerase in cytoplasmic polyhedrosis virus. A model for the positive control of eukaryotic transcription. J Biol Chem 256:483-493.
6. Konarska MM, Padgett RA, Sharp PA (1984). Recognition of cap structure in splicing *in vitro* of mRNA precursors. Cell 38:731-736.
7. Beaton AR, Krug RM (1986). Transcription antitermination during influenza viral template RNA synthesis requires the nucleocapsid protein and the absence of a 5'-capped end. Proc Natl Acad Sci USA 83:6282-6286.
8. Georgiev O, Mous J, Birnstiel MS (1984). Processing and nucleo-cytoplasmic transport of histone gene transcripts. Nucleic Acids Res 12:8539-8551.
9. Reddy R, Ro-Choi TS, Henning D, Busch H (1974). Primary sequence of U-1 nuclear ribonucleic acid of Novikoff hepatoma ascites cells. J Biol Chem 249:6486-6494.

10. HsuChen C-C Dubin DT (1976). Di- and trimethylated congeners of 7-methylguanine in Sindbis virus mRNA. Nature 264:190-191.
11. van Duijn LP, Kasperaitis M, Ameling C, Voorma HO (1986). Additional methylation at the N(2)-position of the cap of 26S Semliki Forest Virus late mRNA and initiation of translation. Virus Res 5:61-66.
12. Freistadt MS, Cross GAM, Branch AD, Robertson HD (1987). Direct analysis of the mini-exon donor RNA of Trypanosoma brucei; detection of a novel cap structure also present in messenger RNA. Nucleic Acids Res 15:9861-9879.
13. Thomas JD, Conrad RC, Blumenthal T (1988). The *C. elegans* trans-spliced leader RNA is bound to Sm and has a trimethylated guanosine cap. Cell 54:533-549.
14. Krause M, Hirsh D (1987). A trans-spliced leader sequence on actin mRNA in *C. elegans*. Cell 49:753-761.
15. Ulmanen I, Broni B, Krug RM (1983). Influenza virus temperature-sensitive cap (m^7GpppNm)-dependent endonuclease. J. Virol. 45:27-35.
16. Pain V (1986). Initiation of protein synthesis in mammalian cells. Biochem J 235:625-637.
17. Adams B, Morgan M, Muthukrishnan S, Hecht SM, Shatkin AJ (1978). The effect of "cap" analogs on reovirus mRNA binding to wheat germ ribosomes. J Biol Chem 253:2589-2595.
18. Furuichi Y, Morgan MA, Shatkin AJ (1979). Synthesis and translation of mRNA containing 5'-terminal 7-ethylguanosine cap. J Biol Chem 254:6732-6738.
19. Darzynkiewicz E, Stepinski J, Ekiel I, Goyer C, Sonenberg N, Temeriusz A, Jin Y, Sijuwade T, Haber D, Tahara SM (1989). Inhibition of eukaryotic translation by nucleoside 5'-monophosphate analogues of mRNA 5'-cap: changes in N7 substituent affect analogue activity. Biochemistry 28:(in press).
20. Darzynkiewicz E, Antosiewicz J, Ekiel I, Morgan M A, Tahara SM, Shatkin AJ (1981). Methyl esterification of $m^7G^{5'}p$ reversibly blocks its activity as an analog of eukaryotic mRNA 5'-caps. J Mol Biol 153:451-458.
21. Darzynkiewicz E, Ekiel I, Tahara SM, Seliger L S, Shatkin AJ (1985). Chemical synthesis and characterization of 7-methylguanosine cap analogues. Biochemistry 24:1701-1707.
22. Darzynkiewicz E, Ekiel I, Lassota P, Tahara SM (1987). Inhibition of eukaryotic translation by analogues of messenger RNA 5'-cap: chemical and biological consequences of 5'-phosphate modifications

of 7-methylguanosine 5'-monophosphate. Biochemistry 26:4372-4380.
23. Darzynkiewicz E, Stepinski J, Ekiel I, Jin Y, Haber D, Sijuwade T, Tahara SM (1988) Nucleic Acids Res 18:8953-8962. β-Globin mRNAs capped with m^7G, $Me_2^{2,7}G$, or $Me_3^{2,2,7}G$ differ in intrinsic translation efficiency.
24. Tahara SM, Morgan MA, Shatkin AJ (1981). Purified m^7G-cap binding protein with different effects on capped mRNA translation in poliovirus-infected HeLa cell extracts. J Biol Chem 256:7691-7694.
25. Edery I, Sonenberg N (1985). Cap-dependent RNA splicing in a HeLa nuclear extract. Proc Natl Acad Sci USA 82:7590-7594.
26. Maniatis T, Reed R (1987). The role of small nuclear ribonucleoprotein particles in pre-mRNA splicing. Nature 325:673-678.
27. Mattaj IW (1986). Cap trimethylation of U snRNA is cytoplasmic and dependent on U snRNP protein binding. Cell 46:905-911.
28. Zuker M, Stiegler P (1981). Nucleic Acids Res 9:133-148. Optimal computer folding of large RNA sequences using thermodynamics and auxiliary information.
29. Lawson TG, Ray BK, Dodds JT, Grifo JA, Abramson RD, Merrick WC, Betsch DF, Weith HL, Thach RE (1986). Influence of 5'-proximal secondary structure on the translational efficiency of eukaryotic mRNAs and on their interaction with initiation factors. J Biol Chem 261:13979-13989.
30. Abramson RD, Dever TE, Lawson TG, Ray BK, Thach RE, Merrick WD (1987). The ATP-dependent interaction of eukaryotic initiation factors with mRNA. J Biol Chem 262:3826-3832.
31. Lawson TG, Cladaras MH, Ray BK, Lee KW, Abramson RD, Merrick WC, Thach RE (1988). Discriminatory interactions of purified eukaryotic initiation factors 4F plus 4A with the 5' ends of reovirus messenger RNAs. J Biol Chem 263:7266-7276.
32. Kozak M (1986). Influences of mRNA secondary structure on initiation by eukaryotic ribosomes. Proc. Natl. Acad. USA 83:2850-2854.
33. Pelletier J, Sonenberg N (1985). Insertion mutagenesis to increase secondary structure within the 5' noncoding region of a eukaryotic mRNA reduces translational efficiency. Cell 40:515-526.
34. Both GW, Banerjee AK, Shatkin AJ (1975). Methylation-dependent translation of viral messenger RNAs *in vitro*. Proc Natl Acad Sci USA 72:1189-1193.

35. Morgan MA, Shatkin AJ (1980). Initiation of reovirus transcription by inosine 5'-triphosphate and properties of 7-methyl inosine-capped, inosine-substituted ribonucleic acids. Biochemistry 19:5960-5966.

INTERNAL N[6]-METHYLADENOSINE SITES IN ROUS SARCOMA VIRUS RNA[1]

Tunde Csepany,[2] Susan E. Kane,[3] Alison Lin,[4] Carl J. Baldick, Jr. and Karen Beemon

Department of Biology, Johns Hopkins University
Baltimore, Maryland 21218

ABSTRACT Twelve N[6]-methyladenosine (m[6]A) residues have been precisely localized on the virion RNA of Rous sarcoma virus. The m[6]A residues were predominantly located in the *src* and *env* genes, which are in the 3' half of the 9312-base RNA. The stoichiometry of methylation varied for the different sites. The methylation was sequence-specific with a preference for RG**A**CU as the substrate; however, this sequence was not methylated in some parts of the genome. Site-specific mutagenesis of this consensus sequence at specific m[6]A sites abolished methylation at those sites.

---

[1]This work was supported by NIH research grant CA 33199 from the National Cancer Institute.
[2]Present address: Department of Neurology and Psychiatry, Medical University of Debrecen, Debrecen, Hungary
[3]Present address: Laboratory of Molecular Biology, National Cancer Institute, Bethesda, Maryland 20892
[4]Present address: Baylor College of Medicine, Houston, Texas 77030

INTRODUCTION

Most mRNAs of higher eukaryotes have methylated nucleotides both in the 5'-terminal cap structures and at internal sites (1). The cap contains a terminal m^7G and usually one or two adjacent 2'-O-methylated nucleotides (1,21). The presence of a methylated cap is important for mRNA translation, splicing, and stability (1,13,21,22).

N^6-methyladenosine (m^6A) is the major internal modification of higher eukaryotic mRNAs (1,8,18,26). Pre-mRNAs are internally methylated in the nucleus prior to their polyadenylation, and the bulk of the methylated bases appear to be conserved during processing and to be present in mature cytoplasmic mRNA (7). If m^6A residues were distributed randomly between mRNAs, an average cellular mRNA would contain approximately three m^6A residues (8,14,18,26). The distribution of m^6A residues in specific mRNAs has been studied in only a few cases; however, it is definitely nonrandom. Both the number and positions of m^6A residues vary with different messages. A total of about 12 m^6A residues (4,11,25) are predominantly located in the 3' half of Rous sarcoma virus (RSV) virion RNA (2,4), whereas bovine prolactin mRNA is methylated at a single site in 3' noncoding sequences (12,17). In contrast, simian virus 40 late mRNAs are methylated near their 5' ends (6), and most m^6A is within the 5' two-thirds of adenovirus messages (23). The number of m^6A residues in different influenza mRNAs ranges from one to eight and is not proportional to the length of the RNA (16). Finally, some mRNAs are not internally methylated at all; globin and histone mRNAs are in this class (15,19). Internal methylation has also not been detected in the mRNAs of lower eukaryotes, such as yeast and slime molds (1).

A specific sequence, RAC, has been observed as the universal substrate for internal methylation (9,20,27). RAC would occur in a random sequence every 32 nts, but only one m^6A per approximately 1000 nts is seen in total cell mRNA (8,14,18,26). Therefore, it appears that either methylation occurs at many sites to a low extent or there is additional enzyme specificity for methylation, perhaps at the level of secondary structure or of extended sequence specificity.

We are studying internal RNA methylation in the oncogenic avian retrovirus, RSV (2,3,4). During an RSV infection, the RNA genome is reverse transcribed and integrated into cellular DNA, where it is transcribed and processed by normal cellular machinery (28). The 9.3 kb primary transcript of RSV has three different fates within the cell. It can be packaged into virions as genomic RNA; it can be translated into *gag* and *pol* gene products, or it can be spliced to form *env* and *src* mRNAs (28). No structural differences are known between RNAs having these different functions, but the possibility of differential methylation remains to be examined.

RSV has about 12 m^{6}A residues per genomic RNA subunit (4,11,25). We have precisely localized a number of methylated bases in RSV virion RNA. Comparison of sequences containing these sites has revealed a more extended consensus sequence for methylation: RG**A**CU. We have observed heterogeneity of methylation at different RG**A**CU sites, ranging from 0 to about 90%. Site-specific mutagenesis of the G, C, or U in this consensus sequence inhibited methylation.

RESULTS

Heterogeneity of Methylation.

The m^{6}Ap content of in vivo ^{32}P-labeled 35S RSV virion RNA was 0.13%, suggesting an average of 12 m^{6}Ap residues per 9312-base RNA. Determination of the m^{6}Ap content in each RNase T$_1$-digested oligonucleotide from a fingerprint of the same RSV RNA revealed more than 12 methylated oligonucleotides, which varied in their extent of methylation. The oligonucleotide containing the highest fraction of m6Ap was (G)ACUG. No m^{6}Ap was observed in the sequence (G)ACG.

Localization of m^6A in RSV RNA.

By analysis of small RNA fragments, we were able to precisely localize 12 m^6A sites on RSV virion RNA by the method described previously (2). We analyzed the region from nt 4715 to 9312 (the 3' terminus) since this had previously been found to contain the bulk of the methylated bases (2,4). Chicken embryo fibroblasts infected with RSV were radiolabeled with ^{32}P-orthophosphate, and virion 70S RNA was hybridized to RSV DNA fragments bound to nitrocellulose filters. After elution and digestion with RNase T_1, the RNA was fingerprinted on cellulose acetate strips and DEAE-cellulose paper. Oligonucleotides were digested with RNases A and T_2, and the resulting mononucleotides were separated by electrophoresis on 3MM paper at pH 3.5. The Ap/m^6Ap spot was eluted, and the two components were separated by chromatography on cellulose thin layer plates in isopropanol-5% ammonium acetate (pH 3.5) (40:25). The sites identified as containing m^6Ap are listed in Table 1. We again observed variation in the extent of methylation between the different sites, as summarized in the table.

In addition to the sites localized precisely (Table 1), we also observed methylation at nt 8366 and/or 8369 in the sequence AGG**A**CA**A**CG. We have not determined if one of these A residues is the preferred methylation site or if both can serve as substrates. We also detected another methylated base at either nt 4976 (GGG**A**CACUG) or nt 5044 (GUG**A**CUAAG) but have not yet distinguished between the two sites.

Sequence Specificity of N^6-methyladenosine Sites.

By comparing the 12 different methylated sequences found in RSV RNA (Table 1), we noticed that the GAC consensus sequence was invariant. Further, a U was present at position +3 (the methylated A is at +1) at 10 out of 12 sites. The exceptions at nt 8413 and 8633 had an A or a C at +3, respectively, and both were weakly methylated.

TABLE 1
METHYLATION AT m6A SITES

Site(nt)	Sequence	Extent[a]
6394	G**A**GACUAG	low
6447	GG**GACU**UAUUG	low
6507	UU**GACU**UCUUG	low
6718	A**GGACU**G	high
7414	A**GGACU**G	high
7424	C**GGACU**UG	high
7890	GG**GACU**CG	low
8014	A**AGACU**CUG	high
8339	CU**GACU**UCG	low
8413	U**GGACA**G	low
8485	CU**GACU**G	high
8633	A**GGACC**CUG	low
CONSENSUS SEQUENCE	**RGACU**	

[a]High methylation denotes 50-90% methylation at a specific site; low methylation means approximately 20 to 30% of all RSV genomic RNAs are methylated at this nucleotide.

At position -2, we observed G seven times, A two times and U three times. At three out of five of the highly methylated sites, we observed GGACU. The other two strong sites were AGACU and UGACU. At positions -3 and +4, the sequence appeared to be random. In conclusion, there seems to be a strong preferred sequence of methylation: RG**A**CU.

Mutagenesis of the m6A Consensus Sequence.

Site-specific mutagenesis studies (3,29) have been carried out with RSV RNA to study further the sequence specificity of methylation and to begin to address the functional significance of m6A methylation. We mutated the four strong m6A sites localized in the *src* oncogene at nts 7414, 7424, 8014, and 8485. Mutagenesis of bases at sites -1

(G to A or U), +2 (C to U or A), or +3 (U to A or G) inhibited methylation. We have not analyzed effects of mutations at positions +1 or -2.

We have begun to assess the affect of these mutations on RSV RNA metabolism. Our earliest mutagenesis studies involved mutations at two sites: 7414 and 7424. This double mutation did not have any observable effect on RSV infectivity, focus formation, intracellular RNA levels or their subcellular location (3). From this original double mutant, additional mutations were created at sites 8014 and 8485, generating viruses with 3 or 4 mutations in the *src* gene. Preliminary results suggested that the viruses lacking all four of the major m^{6}A sites in the src gene had decreased levels of *src* mRNA and *src* protein. The steady-state levels of unspliced and *env* mRNA were not appreciably affected by these mutations.

DISCUSSION

Localization of m^{6}A residues in the RSV genome led to the derivation of a new consensus sequence for m^{6}A: RG**A**CU. The m^{6}A site localized in bovine prolactin mRNA was at an AG**A**CU sequence (17), which also fits this consensus sequence. This extended sequence is better than the limited R**A**C sequence in accounting for the total distribution of m^{6}A observed in many mRNAs. Further, the extent of methylation was found to be heterogeneous at different RSV RNA sites. Some specific RGACU sites were not detectably methylated in vivo, while others were methylated in up to 90% of all RSV molecules.

We have begun to assess the relevance of this consensus sequence for methylase specificity by performing oligonucleotide-directed mutagenesis. These studies have revealed the importance of conservation of the G (-1), C (+2), and U (+3) residues, since methylation was inhibited when these bases were mutated. We assume the A is also obligatory but have not mutated it. As predicted by our consensus sequence, alteration of the C at +2 or changing the G at -1 to a pyrimidine prevented methylation. However, it was somewhat surprising that the G to A mutation at nt 8485, which generated the sequence UAACU, also inhibited methylation. Although AAC, as well as GAC, sequences are reportedly methylated in a variety of mRNAs (9.20.27), methylated GAC sites are much more abundant than AAC sites in RSV RNA. It is possible that methylation of an AAC requires a different context than that for GAC (6).

Previous studies using methylation inhibitors have suggested a possible role for m^6A in RNA processing or transport of mRNA from the nucleus to the cytoplasm (5,10,24). We have generated mutants of RSV that are undermethylated at specific sites to assess the functional significance of the methylation. While undermethylation at two sites had no noticeable affect on RNA phenotype (3), preliminary results suggest that four mutations in the *src* gene may lead to a decrease in the steady-state level of *src* mRNA. The decrease in *src* mRNA could be due to inhibition of mRNA splicing, transport, or stability.

REFERENCES

1. Banerjee AK (1980). 5'-Terminal cap structure in eucaryotic messenger ribonucleic acids. Microbiol Rev 44:175.
2. Beemon K, Kane SE (1985). Precise localization of m^6A in Rous sarcoma virus RNA reveals clustering of methylation sites: implications for RNA processing. Mol Cell Biol 5:2298.
3. Beemon K, Kane SE (1987). Inhibition of methylation at two internal N^6-methyladenosine sites caused by GAC to GAU mutations. J Biol Chem 262:3422.
4. Beemon K, Keith J (1977). Localization of N^6-methyladenosine in the Rous sarcoma virus genome. J Mol Biol 113:165.
5. Camper SA, Albers EJ, Coward JK, Rottman FM (1984). Effect of undermethylation on mRNA cytoplasmic appearance and half-life. Mol Cell Biol 4:538.

6. Canaani D, Kahana C, Lavi S, Groner Y (1979). Identification and mapping of N^6-methyladenosine containing sequences in simian virus 40 RNA. Nucleic Acids Res 6:2879.
7. Chen-Kiang S, Nevins JR, Darnell JE (1979). N^6-methyladenosine in adenovirus type 2 nuclear RNA is conserved in the formation of messenger RNA. J Mol Biol 135:733.
8. Desrosiers RC, Friderici KH, Rottman FM (1975). Characterization of Novikoff hepatoma mRNA methylation and heterogeneity in the methylated 5' terminus. Biochemistry 14:4367.
9. Dimock K, Stoltzfus CM (1977). Sequence specificity of internal methylation in B77 avian sarcoma virus RNA subunits. Biochemistry 16:471.
10. Finkel D, Groner Y (1983). Methylations of adenosine residues (m^6A) in pre-mRNA are important for formation of late simian virus 40 mRNAs. Virology 131:409.
11. Furuichi Y, Shatkin AJ, Stavnezer E, Bishop JM (1975). Blocked, methylated 5'-terminal sequence in avian sarcoma virus RNA. Nature 257:618.
12. Horowitz S, Horowitz A, Nilsen TW, Munns TW, Rottman FM (1984). Mapping of N6-methyl-adenosine residues in bovine prolactin mRNA. Proc Natl Acad Sci USA 81:5667.
13. Konarska MM, Padgett RA, Sharp PA (1984). Recognition of cap structure in splicing in vitro of mRNA precursors. Cell 38:731.
14. Lavi U, Fernandez-Munoz R, Darnell JE (1977). Content of N-6 methyl adenylic acid in heterogeneous nuclear and messenger RNA of HeLa cells. Nucleic Acids Res 4:63.
15. Moss B, Gershowitz A, Weber LA, Baglioni C (1977). Histone mRNAs contain blocked and methylated 5' terminal sequences but lack methylated nucleosides at internal positions. Cell 10:113.
16. Narayan P, Ayers DF, Rottman FM, Maroney PA, Nilsen TW (1987). Unequal distribution of N^6-methyladenosine in influenza virus mRNAs. Mol Cell Biol 7:1572.

17. Narayan P, Rottman FM (1988). An in vitro system for accurate methylation of internal adenosine residues in messenger RNA. Science 242:1159.
18. Perry RP, Kelley DE, Friderici K, Rottman F. (1975). The methylated constituents of L cell messenger RNA: evidence for an unusual cluster at the 5' terminus. Cell 4:387.
19. Perry RP, Scherrer K (1975). Methylated constituents of globin mRNA. FEBS Lett 57:73.
20. Schibler U, Kelley DE, Perry RP (1977). Comparison of methylated sequences in mouse L cells. J Mol Biol 115:695.
21. Shatkin AJ (1976). Capping of eucaryotic mRNAs. Cell 9:645.
22. Shatkin AJ (1985). mRNA cap binding proteins: essential factors for initiating translation. Cell 40:223.
23. Sommer S, Salditt-Georgieff M, Bachenheimer S, Darnell JE, Furuichi Y, Morgan M, Shatkin AJ (1976). The methylation of adenovirus-specific nuclear and cytoplasmic RNA. Nucleic Acids Res 3:749.
24. Stoltzfus CM, Dane RW (1982). Accumulation of spliced avian retrovirus mRNA is inhibited in *S*-adenosylmethionine-depleted chicken embryo fibroblasts. J Virol 42:918.
25. Stoltzfus CM, Dimock D (1976). Evidence for methylation of B77 avian sarcoma virus genome RNA subunits. J Virol 18:586.
26. Wei CM, Gershowitz A, Moss B (1976). 5'-terminal and internal methylated nucleotide sequences in HeLa cell mRNA. Biochemistry 15:397.
27. Wei CM, Moss B (1977). Nucleotide sequence at the N^6-methyladenosine sites of HeLa cell messenger ribonucleic acid. Biochemistry 16:1672.
28. Weiss R, Teich N, Varmus H, Coffin J (1984). Molecular biology of tumor viruses: RNA tumor viruses, 2nd ed., vol. 2. Cold Spring Harbor Laboratory.
29. Zoller MJ, Smith M (1984). Oligonucleotide-directed mutagenesis: a simple method using two oligonucleotide primers and a single-stranded DNA template. DNA 3:479.

THE RELATIONSHIP BETWEEN PROTEIN SYNTHESIS AND RIBOSOMAL RNA METHYLATION [1]

Gary A. Clawson, Julie Sesno, and Kathy Milam

Department of Pathology, George Washington University, Washington, DC 20037

ABSTRACT   Many hepatotoxins produce early "mysterious" defects in protein synthesis. We have found that (in a wide variety of settings) ribosomal RNA methylation is related to protein synthetic capacity with a product moment correlation coefficient of r = .95, with high significance ($p < .003$). The decreases in ribosomal RNA methylation may involve microsomal demethylations (cytochrome P_{450} mediated?) and cytosolic remethylations. We propose that the hepatic protein synthetic defects are due to a net hypomethylation of 2'O-ribose moieties in ribosomal RNA.

INTRODUCTION

Many diverse hepatotoxins, which produce various patterns of liver damage, produce early, unexplained defects in protein synthesis. These early defects have generally been first observed as a "degranulation" of the rough endoplasmic reticulum and later correlated with loss of protein synthesis. The degranulation represents dissociation of membrane-bound polysomes into free ribosomes (and thus seems to represent a global effect) and is associated with a decreased capacity of the ribosomes to perform protein synthesis in vitro. In initial investigations, we examined the inhibition of protein synthesis by CCl_4 (1), which is one of the earliest functional deficits produced by this agent. We have found that the protein synthetic deficit is related to

[1] This work was supported by grants CA21141 and CA01003 from the NIH, and by grant 2115 from the Council for Tobacco Research.

hypomethylation of purine 2'-O-ribose moieties in rRNA, and that function could be restored to ribosomal subunits by methylating them in vitro with cell sap (S-100) fractions and S-adenosyl-L-methionine (1). For a variety of reasons, the hypomethylation appeared to be a cytoplasmic process: a) the specific activity of cytoplasmic rRNA was considerably higher than that of nucleolar RNA (in spite of the fact that the cytoplasmic pool is much larger than the nucleolar), b) the absence of a chase period and the linearity of the labeling over the time examined, c) the fact that the nucleolar rRNA methylation pattern was unaltered, and d) the fact that S100 fractions could indeed methylate rRNA in purified ribosomal subunits. We have examined the generality of the role of rRNA methylation as a control for protein synthesis using two other hepatotoxins, galactosamine (GAL) and ethionine (ETN).

In contrast to the centrilobular injury produced by CCl_4, the pattern of injury produced by GAL is diffuse in nature (2-4), producing only scattered liver cell death at later time points. GAL is taken up by hepatocytes, phosphorylated, and it subsequently depletes the uridine nucleotide pools (2,3,5); its mechanism of toxicity is therefore thought to relate to inhibition of RNA synthesis, and in fact administration of uridine significantly protects against the extent of injury (5). However, the early inhibition of protein synthesis is not related to RNA synthesis (6), either temporally or mechanistically; as was the case with CCl_4, it proceeds independently of any transcriptional effects. More recently, it was found that GAL induces a substantial decrease in hepatocyte SAM levels (7) via inhibition of ATP: L-methioinine-S-adenosyltransferase activity, suggestings its utility in exploring the role of cytoplasmic rRNA methylation in protein synthesis.

The other hepatotoxin investigated, ETN, does not result in liver cell death at the doseage employed. Ethionine is taken up by hepatocytes and acts as an analog of methionine (8,9), and is rapidly converted to S-adenosylethionine. Hepatocyte cellular ATP levels are dramatically reduced by this toxin (10,11), and ATP levels are inversely related to S-adenosylethionine levels (12). Additionally, there is a clear sexual dimorphism with this agent (11,13): Whereas both male and female rats show dramatic declines in ATP levels, males are resistant to changes in protein synthetic ability of ribosomes whereas females are not. Thus, ETN provides a valuable model whose effects on protein synthesis are independent of ATP levels. On the premise that the early protein synthesis defects produced by GAL and ETN (and the sexual dimorphism with ETN) might be mediated through hypomethylation of rRNA, we examined the effects of these agents on rRNA methylation.

MATERIALS AND METHODS

Male or female Sprague-Dawley rats were used with body weights of about 250 g. GAL was administered by gavage at 20 mg/100 g body weight. ETN was given by gavage at 50 mg (613 umol)/100 g body weight in two divided doses (14); half of the total dose was given, and two hours later the second half was administered. For in vivo labeling experiments, rats were given 1.7 mCi ^{3}H-methyl-L-methionine (New England Nuclear, specific activity 70-85 Ci/mol) i.p. in saline; in standard experiments, label was given 2 h before sacrifice. Livers were removed and homogenized in 15 ml of STKM2 buffer (250 mM sucrose, 50 mM Tris-HCl (pH 7.6), 25 mM KCl, 5 mM $MgCl_2$, 5 mM 2-mercaptoethanol). The homogenate was centrifuged at 10,000 x g for 6 min at 4 C. Nuclei were prepared from the pellet (1) and nucleolar RNA was subsequently isolated (15). The supernate was used for preparation of cytoplasmic RNA, microsomal and polysomal fractions, and S-100 fractions (1). Cytoplasmic RNA was separated into poly(A)+ and poly(A)-fractions using oligo(dT) cellulose chromatography. RNA preparations routinely showed $A_{260/280}$ ratios of 2.0. DNA was not detectable (16) in the preparations. For specific activities, RNA was quantitated spectrophotometrically and radioactivity was determined.

Poly(A)- fractions were digested to ribonucleosides using P_1 and T_2 RNases followed by treatment with bacterial alkaline phosphatase (1,17). The mixtures were then extracted with phenol/chloroform, frozen and lyophilized. Nucleosides were analyzed by HPLC, using a Waters 600 HPLC, UV detector and fraction collector, a radial compressed C_{18}-ubondapak reverse-phase column, a flow rate of 1 ml/min, and a linear gradient of 0-30% acetonitrile from 0-40 min. 0.3 ml fractions were collected and counted. In later experiments, we employed a flow rate of 0.5 ml/min, a linear 0-10% gradient over 40 min, and collected 0.25 ml fractions.

Protein synthesis was determined using a 30 min in vitro assay (1), except that reactions contained 2 uCi of ^{3}H-amino acids (NEN). Following incubations, mixtures were precipitated with 10% trichloroacetic acid for 30 min on ice, and precipitable material was collected by centrifugation. Pellets were then boiled in 5% TCA for 1 h, and the mixture was again centrifuged. The supernate was discarded and the residual pellets were rinsed once. The pellet was then hydrolyzed with 1 N NaOH by heating at 70 C for 15 min and the mixture was centrifuged. The supernate was acidified with HCl and counted in liquid scintillant. In some cases, aliquots were precipitated as above, and the synthesized products were examined by 12% polyacrylamide gel electrophoresis and autoradiography.

In vitro methylations using ^{3}H-SAM were as described (1), using ribosomal subunits (100 ug protein, determined using the Lowry assay [18]) and S-100 fractions (200 ug protein). For studies on the effects of in vitro methylation on the protein synthetic capacity of ribosomes, 10 mg aliquots of purified ribosomes were methylated with 20 mg of S-100 protein and 0.85 mM SAM for 30 min at 37 C. Ribosomes were subsequently reisolated as described (1) and tested for in vitro protein synthesis in the assay described above.

For studies on demethylation, 1 mg of in vivo methyl-labeled rRNA was incubated with 2 mg S-100 or microsomal protein (18) from control or 4 h GAL-treated animals, for 1 h at 37 C, in the presence of 10 mM vanadyl ribonucleoside complex. Following incubation, the mixtures were extracted with phenol/chloroform, and precipitated with ethanol overnight at 4 C. Aliquots were used for quantitation, and additional aliquots were examined on 1.2% agarose gels (19) to verify that degradation of rRNA did not occur.

RESULTS

At various intervals following galactosamine treatment, ribosomes were isolated and tested for protein synthesis (20). There was a substantial decrease in their capacity to perform protein synthesis by 1 h, and the defect (a 60% decrease) remained essentially constant for 6 h. We selected the 4 h period for further study, employing a 2 h in vivo labeling period (cytoplasmic rRNA labeling is linear for 4 h). When in vivo methylation of rRNA was assessed, we observed an 85% decrease in methylation (from 7.7 $\pm$ 1.3 to 1.5 $\pm$ 0.8 cpm/ug; see 20). In contrast to previous results with CCl_4, we also observed substantially decreased methylation of the poly(A)+ and nucleolar RNA fractions; the decreases were from 15.6 $\pm$ 2.7 to 6.2 $\pm$ 1.1 cpm/ug for the poly(A)+ fractions and 1.8 $\pm$ 0.7 to 0.5 $\pm$ 0.1 cpm/ug for the nucleolar fractions.

Cytoplasmic rRNA was digested to nucleosides and examined via HPLC. There was no strict site specificity for decreased methylation observed with GAL treatment, with both base and 2'0-ribose sites affected roughly proportionally (see 20), although there was a slight increase in the proportion of pyrimidine 2'-O-ribose methylation. These results seem consistent with methylation effects mediated through decreased SAM levels. When isolated ribosomal subunits were incubated with S-100 cell sap fractions and ^{3}H-SAM, we observed a substantially greater methylation of subunits from GAL-treated preparations, presumably reflecting their relative hypomethylation (20). Further, when treated subunits were incubated in vitro with unlabeled SAM and S-100 fractions, reisolated, and tested for protein synthesis, the in vitro methylation appeared to restore their functional capacity in many experiments (20). These results are analogous to those obtained with cytoplasmic rRNA preparations after CCl_4 treatment (1).

In extending our investigations of hepatotoxins with early effects on protein synthesis, we examined RNA methylation at 4 h after ETN treatment (Table 1). We noted a marked differential response to the agent by male or female

TABLE 1.
SEXUAL DIMORPHISM IN EFFECTS OF ETHIONINE ON HEPATOCYTE RIBOSOMAL RNA METHYLATION: RELATIONSHIP TO PROTEIN SYNTHESIS[a]

Sample	Specific Activity Ribosomal RNA Methylation (cpm/ug)	Cellular ATP level (% Control)	Protein Synthesis in vitro (% Control)
Control	7.5± 1.5	100± 05	100± 06
Male-ETN	6.1± 1.5	21± 02	90
Female-ETN	1.0± 0.2	15± 02	12± 02

[a] Male or female rats were treated with ETN and methylation of rRNA was quantitated as described. ATP levels and protein synthesis values are from reference 11 and our results for in vitro protein synthesis.

rats. In female rats, we observed a decrease of 87% in methylation of rRNA, coincident with the 88% decrease in protein synthesis associated with ETN treatment (Table 1). In male rats, the decrease in rRNA methylation was only 21%, paralleling the relative resistance of protein synthesis to this agent in males (Table 1).

While these results clearly establish a relationship between protein synthetic ability and methylation state of rRNA, the observed hypomethylation (if cytoplasmic) could conceivably result from increased demethylation or decreased methylation. In initial experiments, we documented methylation activities in S-100 fractions, and this methylation activity did not change following CCl_4 treatment (1). Similarly, we have not observed differences in S-100 methylation activities following GAL treatment (It should be noted, however, that the methylation activity is quite labile, and seems to be destroyed even by freeze-thaw cycles).

TABLE 2.
DEMETHYLATION OF RIBOSOMAL RNA BY MICROSOMAL FRACTIONS[a]

Fraction Added for incubation	^{3}H-Methylated rRNA recovered (% Control)
None	100
Control S100	94
4 h GAL S100	100
Control Mic	85± 07
4 h GAL Mic	78± 10

[a] 1 mg of in vivo ^{3}H-methyl-labeled rRNA was incubated with S100 or microsomal fractions (2 mg protein) from treated or control animals for 1 h at 37 C in the presence of 10 mM vanadyl ribonucleoside complex. Following incubation, the mixture was extracted with phenol, precipitated with ethanol, and aliquots were assessed for radioactivity. Additional aliquots were analyzed by agarose gel electrophoresis to insure that degradation did not occur.

We also performed experiments to assess "demethylase" activity, using in vivo methylated rRNA and microsomal or S-100 fractions (from control or GAL-treated rats), in the presence of RNase inhibitors (Table 2). Under the conditions employed, there was no demethylation of rRNA with S-100 fractions (Table 2), suggesting that preliminary results with S-100 fractions (1) might have reflected RNase activity. With microsomal preparations we observed a significant demethylation of in vivo methyl-labeled rRNA (Table 2); the extent of demethylation did not differ significantly between preparations from control and GAL-treated animals.

DISCUSSION

Our results with 3 unrelated hepatotoxins (CCl_4, GAL, and ETN) clearly implicate hypomethylation of rRNA in the early defects produced in protein synthesis by these agents. When our accumulated data are analyzed statistically (Figure 1), we obtain a product-moment correlation coefficient of $r = .95$ with high statistical significance ($p < .003$). It is also important to stress that there is no correlation with nucleolar rRNA labeling.

It is therefore difficult to reconcile our data with the existing dogma that all rRNA methylation occurs in pre-rRNA in the nucleolus, and it seems logical to suggest that the hypomethylation we observe (at least to a considerable degree) appears to represent a cytoplasmic event. In this regard, essentially all early nucleolar rRNA methylation studies utilized rapidly growing cultured cell populations: Not only are such populations distinctly different from that in mature liver, the unique metabolic status of hepatocytes may play a major role in determining cytoplasmic methylation. We have now obtained data suggesting that demethylase activity is associated with microsomal fractions, although we have thus far not found any change in demethylase activity following GAL treatment. We speculate that cytochrome P_{450} may be involved in the demethylation phenomenon, and suggest that demethylation of fully-methylated rRNA is a necessary prelude to subsequent cytoplasmic labeling. Indeed, there is a general parallelism between the responses of cytochromes P_{450} and rRNA methylation to the various agents, most notably with ETN. In this regard, the ability of estrogens/progesterone to alter the susceptibility of males and females to ETN changes (11) might be mediated through alterations in P_{450} populations.

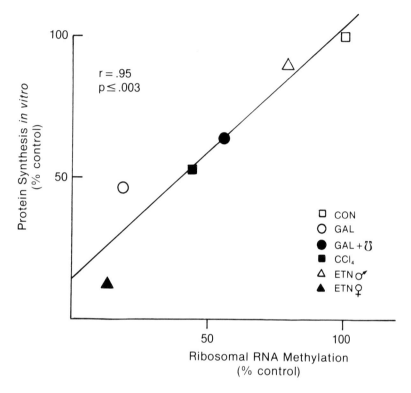

Figure 1. Correlation between in vitro protein synthetic capacity and rRNA methylation. Pearson's correlation coefficient was calculated using the True Epistat statistical program.

While these characterizations are preliminary, they at least suggest that net hypomethylation may relate to microsomal 2'-O-ribose demethylations which may occur with intact ribosomal subunits, coupled with subsequent cytoplasmic remethylations by cytosolic factors. Such a scheme is certainly consistent with the locations of the ribosomes (active and inactive) relative to their methylation state. This suggestion is consistent with known nucleolar methylation (see contribution by Maden, this volume) if the site-specificity of the methylations roughly mirrors that found in the nucleolus. The "degranulation" of ER observed with these hepatotoxins suggests a global process rather than a defect in "finishing-up" methylation of newly synthesized ribosomal subunits. In this regard, on the basis of changes in absorbance profiles, we estimate that the process involves roughly 5-10 methyl groups per ribosome. Further studies are underway to more clearly define the potential role of cytoplasmic demethylation in controlling the methylation state of rRNA. It seems clear, however, that rRNA methylation is related to protein synthetic capacity.

REFERENCES

1. Clawson G, MacDonald J, and Woo C (1987) Early hypomethylation of 2'O-ribose moieties in hepatocyte cytoplasmic ribosomal RNA underlies the protein synthetic defect produced by CCl_4. J. Cell Biol. 105: 705.
2. Decker K, and Keppler D (1974) Galactosamine hepatitis Key role of the nucleotide deficiency period in the pathogenesis of cell injury and cell death. Rev. Physiol. Biochem. Pharmacol. 71: 77.
3. Decker K, Keppler D, Rudigier J, and Domschke W (1971) Cell damage by trapping of biosynthetic intermediates. The role of uracil nucleotides in experimental hepatitis. Hoppe-Seyler. Z. Physiol. Chem. 352: 412.
4. Keppler D, and Decker K (1969) Studies on the mechanism of galactosamine hepatitis: Accumulation of galactosamine-1-phosphate and its inhibition of UDP-glucose pyrophosphorylase. Eur. J. Biochem. 10: 219.
5. Keppler D, Pausch J, and Decker K (1974) Selective uridine triphosphate deficiency induced by D-galactosamine in liver and reversed by pyrimidine nucleotide precursors. Effect on ribonucleic acid synthesis. J. Biol. Chem. 249: 211.
6. Anukarahanota T, Shinozuka H, and Farber E (1973) Inhibition of protein synthesis in rat liver by D-galactosamine. Res. Commun. Chem. Path. Pharm. 5: 481.
7. Ozturk M, Lemonnier F, Cresteil D, and Lemonnier A (1986) Changes in methionine metabolism induced by D-galactosamine in isolated rat hepatocytes. Biochem. Pharm. 35: 4223.
8. Farber E (1973) ATP and cell integrity. Fed. Proc. 32: 1534.
9. Shinozuka H, Reid I, Shull K, Liang H, and Farber E (1970) Dynamics of liver cell injury and repair. Lab. Invest. 23: 253.
10. Okazaki K, Shull K, and Farber E (1968) Effects of

ethionine on adenosine triphosphate levels and ionic composition of liver cell nuclei. J. Biol. Chem. 243: 4661.
11. Oler A, Farber E, and Shull K (1969) Resistance of liver polysomes to ATP deficiency in the male rat. Biochim. Biophys. Acta 190: 161.
12. Brada Z, Hrstka I, and Bulba S (1988) Formation of S-Adenosylethionine in liver of rats chronically fed with DL-ethionine. Cancer Res. 48: 4464.
13. Chen W, and Smuckler E (1978) Ethionine-induced alterations in hepatic microsomal cytochromes in vivo and in vitro: Sexual dimorphism. Tox. Appl. Pharmacol. 44: 269.
14. Smuckler E, Koplitz M, and Sell S (1976) Alphafetoprotein in toxic liver injury. Cancer Res. 36:4558.
15. Muramatsu M, and Busch H (1964) Studies on nucleolar RNA of Walker 256 carcinosarcoma and the liver of the rat. Cancer Res. 24: 1028.
16. Burton K (1956) A study of the conditions and mechanism of the diphenylamine reaction for the colorimetric estimation of deoxyribonucleic acid. Biochem. J. 62: 315.
17. Campers S, Alberts R, Coward J, and Rottman F (1984) Effect of undermethylation on mRNA cytoplasmic appearance and half-life. Mol. Cell. Biol. 4: 538.
18. Lowry O, Rosebrough N, Farr A, and Randall R (1951) Protein measurement with the folin phenol reagent. J. Biol. Chem. 193: 165.
19. Maniatis T, Fritsch E, and Sambrook J (1982) "Molecular Cloning A Laboratory Manual" New York, Cold Spring Harbor Laboratories, p150.
20. Clawson G, Sesno J, Milam K, Wang Y, and Gabriel C (1989) The hepatocyte protein synthesis defect induced by galactosamine involves hypomethylation of ribosomal RNA (Manuscript in press).

THE ROLE OF RNA METHYLATION IN THE DIFFERENTIATION OF L5 MYOBLAST CELL LINE

S. Scarpa, L. Di Renzo and R. Strom

Department of Human Biopathology and 1st Institute of Clinical Surgery, Università "La Sapienza", Rome, Italy.

ABSTRACT L5 myoblast cell line can be maintained in the undifferentiated state for many passages in "growth medium" (F14 medium supplemented with 10% foetal calf serum) or be induced to differentiate into multinucleated fibers in "fusion medium" (with only 1% foetal calf serum). We decided to investigate the possible changes in RNA methylation levels, in fusing and non-fusing conditions, upon addition of either 3-deazaadenosine, a potent methylation inhibitor which stimulates L5 myoblast differentiation, or Concanavalin-A, which has been reported to increase RNA methylation in T-lymphocytes and which had been found to inhibit L5 myoblast differentiation. The cells treated with deazaadenosine or with Concanavalin in the first day after plating and kept in the presence of the drugs throughout the duration of the culture, were labeled with [^{3}H-CH$_3$]-methionine and ^{14}C-uridine after inhibition of de novo purine synthesis and of methyl transfer reactions occurring through the tetrahydrofolate pathway, total RNA being then extracted and purified. The results obtained in growth medium demonstrate that, under those conditions, neither an increase or a decrease in the extent of total RNA methylation induce per se the differentiation of L5 myoblasts. On the other hand, in fusion medium the RNA methylation levels appear to be inversely related to the induction of the phenomenon.

INTRODUCTION

The differentiation of myoblasts in culture is widely used as a model to study the pattern and the mechanisms involved in the regulation of gene expression in eukariotes. L5 cell line has proven to be particularly suitable for such studies since clones could be selected (1) which, when cultured in F14 medium supplemented with 10% foetal calf serum (growth medium), continue to grow until they reach confluency without undergoing differentiation. These clones, when transferred to F14 medium supplemented with only 1% foetal calf serum (fusion medium), will differentiate after a few cell divisions, generating multinucleated fibers containing muscle-specific proteins (1). The main appreciable differencies among the various myoblast clones were extent of fusion ability and its decline after a number of passages, varying between 15 and 30.

In previous work (2) we have shown that an adenosine analog, 3-deazaadenosine (DZA), in combination with homocysteine thiolactone (HCY), strongly stimulates the differentiation of clones derived from the L5 myoblast cell line. Due to its chemical structure, DZA is both a potent inhibitor of, and a good substrate for, S-adenosylhomocysteine hydrolase; treatment of cells with DZA results in a variety of biological effects that have been attributed to the inhibition of methyl transfer reactions caused by the accumulation of S-adenosylhomocysteine, the formation of 3-deaza-S-adenosylhomocysteine, or both (3-6). No correlation has so far been shown between a given biological response and the inhibition of a specific methylation reaction by DZA.

Concanavalin-A (Con-A), a plant lectin that inhibits myoblasts differentiation without showing any mitogenic effect (S.Scarpa: unpublished data), has been reported to increase RNA methylation in

T-lymphocytes (7).

We report here that total RNA methylation levels are changed by addition of DZA or of Con-A and that an inverse relationship between RNA methylation and cell differentiation exists, but only if fusion conditions have been respected.

METHODS

Cell culture and cytological examination. F14 medium (8), a modification of Ham's F12, supplemented with either 10% or 1% foetal calf serum, was used to culture cells in all experiments, as previously described (1). Fusion tests were performed in duplicate using 25 cm^2 plastic flasks. Fusion was evaluated, after fixing and staining the cell layer, as the percent of total nuclei that were found in fibers.

Cells. Clone M6 was isolated from L5 myoblast cell line, established by Yaffe (9) from primary cultures of newborn rat thigh muscle.

Chemicals. 3-Deazaadenosine was a kind gift of Bioresearch Co., Milan, Italy. All the other chemicals were obtained from Sigma.

Buffers. TNES lysis buffer: 30 mM Tris-HCl pH 7.4, 100 mM NaCl, 5 mM EDTA, 3% SDS; TM buffer: 30 mM Tris-HCl pH 7.5, 5 mM MgCl$_2$.

RNA methylation. The cells were preincubated with 0.02 mM adenosine for at least 12 hours to suppress purine de novo synthesis. After medium removal, the cells were washed twice with PBS and incubated for one hour with F14 methio-nine-free medium, containing either 10 µCi/ml [^{3}H-CH$_3$]-methionine (80 Ci/mmole) or 0.2 µCi/ml ^{14}C-uridine (50 mCi/mmole) and 0.02 mM each of adenosine, guanosine and sodium formate to inhibit the methyl transfer reaction from methionine via tetrahydrofolic acid.

RNA isolation. Radioactive medium was removed from the flasks and the cell sheet adhering to the flasks was rinsed twice with cold Ca-, Mg-free PBS, the cells were collected with a rubber policeman in TNES buffer + 0.5% Nonidet P-40 and K proteinase 50 µg/ml and the suspension was incubated for 1 h at 37·C. An equal volume of

TNES-saturated phenol was then added to the samples, which were thereafter gently shaken for 30 minutes at room temperature, then centrifuged at 5000 rpm for 15 min. The upper phase was removed and the interphase reextracted at 60·C for 10 min with 1 ml of TNES buffer. Aqueous phases were pooled together and extracted once again with chloroform/isoamyl alcohol (24:1) for 30 min to remove all denaturated proteins. The aqueous phases, which contained the nucleic acids, were additioned with 0.3M sodium acetate pH 5.5 and with absolute ethanol (2.5 volumes) and left overnight at -20·C. After centrifugation at 7000 rpm, the pellets were dried under a nitrogen flow and resuspended in TM buffer with DNAase (RNAase-free) 50µg/ml, for 30 min at 37·C, K-proteinase 5µg/ml being added to each sample. A further chloroform/isoamyl alcohol extraction was performed, followed by a cold ethanol precipitation. RNA was collected by centrifugation, resuspended in H_2O and stored at -20·C. The total amount of extracted RNA was determined by the orcinol method and the radioactivity was measured by liquid scintillation.

RESULTS

Addition of DZA to myoblasts in the fusion medium brings about a stimulation of cell fusion: as shown in the experiment of figure 1, in which an appropriate clone at a high passage fused spontaneously to approximately 20%, the best combination giving an average value as high as 75% fusion. The addition of HCY by itself, either 50 µM or 100 µM, did not produce any effect on myoblast fusion; DZA, when added alone, had to be ten times more concentrated (50 µM) to achieve its maximum effect, which reached however only 43% fusion (not shown in figure 1).

The dose-response curve obtained upon addition of Con-A, demonstrated on the other hand, as shown in figure 2, the inhibitory effect of this drug on myoblast fusion. It should be noticed that the total number of nuclei in the flasks treated with the lectin was not significantly different from the control, unless Con-A concentration exceeded 4 µg/ml, with some toxic effect showing up at higher concentrations. It could thus be excluded that the Con-A inhibitory effect on differentiation, be ascribed to the mitogenic property of the lectin.

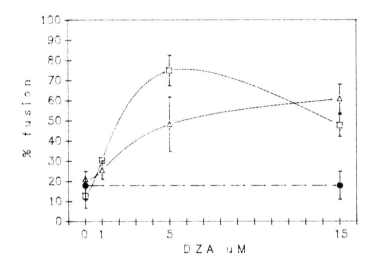

FIGURE 1. Stimulating effect of DZA when added together with HCY on the fusion of L5 myoblasts (clone M6 passage 25). The drugs were added to fusion medium at day 1, the effect being recorded at day 6. Filled circles represent the cultures in absence of HCY. Open triangles the results obtained with 50 µM HCY, open squares those obtained with 100 µM HCY.

The common feature of these two ex-periments was the need, for both drugs, to be present from the first day after plating, i. e. when the cells were being committed to differentiation, in order to produce their effect: this led us to postulate some similarities in their mode of action. DNA methylation has been reported, by many authors, to play a role in the regulation of gene expression. The methylation of specific genes has been inversely related to their transcriptional activity. DZA is however known to be a rather nonspecific metabolic inhibitor of enzymatic methylation processes, while, according to Grunert and Schäfer (7), although with a phenotypic outcome which was opposite to the one occurring in our cell system, Con-A caused a marked increase of RNA methylation.

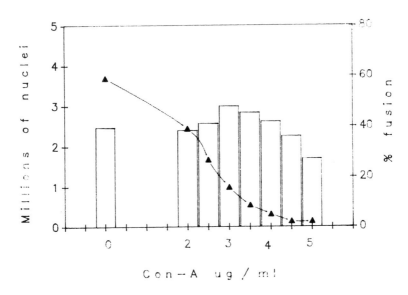

FIGURE 2. Effect of Con-A on L5 myoblast fusion, clone A10 passage 5. Triangles represent the extent of fusion, as reached at day 6 of culture, in the dose-response curve. Columns evidence the total number of nuclei per flask.

Fig. 3 illustrates how, in our system, DZA and Con-A affected total RNA methylation during the differentiation process. In growth medium, incorporation of methyl groups into total RNA from labeled methionone was at fairly high levels and the presence of DZA resulted in a significant, but time-unvarying hypomethylation with respect to controls, while Con-A produced a slight increase of the RNA methylation levels. In fusion medium, as differentiation progressed, there was, already in controls, a gradual decrease of the incorporation into total RNA of methyl groups from labeled methionine, which was sharper until day 4, then leveled off. Addition of DZA at day 1 produced an approximately 50% decrease of methyl groups incorporation, while, in the presence of Con-A, there was a small, but fairly significant, increase; the time-course of the variations of methylation levels showed, however an almost parallel decline from day 2 to day 5. It can be noticed that, as shown in fig. 3, the lowest values reached, at day 5, in the presence of Con-A, were still higher than those found in cells cultured in growth medium in the presence of DZA, an obvious limitation being that these results do not distinguish among RNA species.

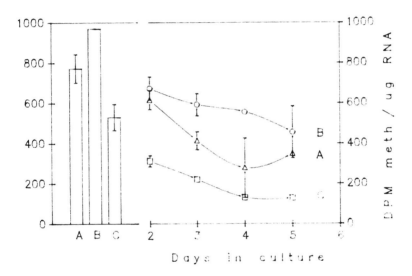

FIGURE 3. Variation of total RNA methylation as methyl groups incorporation per µg RNA. Columns represent the RNA methylation in growth medium; A corresponds to the controls, B to the Con-A-treated cultures, C to the DZA ones. Symbols show the values in fusion medium: triangles represent the untreated controls, circles the Con-A-treated samples, squares the DZA-treated ones. Experimental points correspond to the days of culture in fusion medium.

Since, among all RNA species, the messenger RNA should be the one with the fastest turnover, although it represents only about 4% of the total RNA, we compared, in the course of differentiation, the rate of RNA methylation, as measured by labeling the cells with ^{3}H-methyl-methionine, to the rate of RNA synthesis, as measured by ^{14}C-uridine pulse incorporation. Being aware that both rates depend on several variables, such as the size of methionine S-adenosyl-methionine and nucleotides intracellular pools (which influence the specific radioactivity of these compounds after administration of labeled methionine and of labeled uridine, the activity of RNA methylases and of RNA polymerases, as well as the stability of methylated and unmethylated RNA), we can interpret the results represented in figure 4 only in general terms.

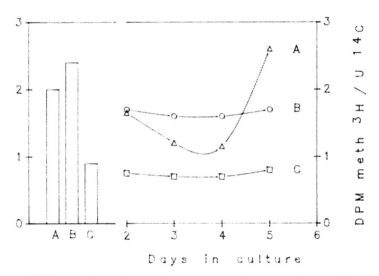

FIGURE 4. Variation of the methylation vs synthesis ratio of total RNA in L5 myoblasts during in vitro differentiation. Symbols as in fig. 3.

It can however be noticed that, while in control myoblasts (undergoing "spontaneous" differentiation and fusion) the values of the methylation-to-biosynthesis ratio exhibit a clearly biphasic pattern (undergoing at first a progressive decrease and then, as fusion takes place, a sudden increase); in the presence of either DZA or Con-A they remain almost constant, at a lower level when DZA has been added, and, respectively, at rather high level after addition of Con-A. As a plausible interpretation of the data reported in fig. 4, it appears that, although the rate of RNA methylation is modified during differentiation, it can hardly be considered to be the determinant factor inducing the process.

DISCUSSION

There is undoubtedly, within a given clone of L5 myoblast, an inverse relationship between total RNA methylation and onset of the differentiation process. During "spontaneous" differentiation (induced only by a lowering of the foetal calf serum concentration in the culture medium), there is a progressive decline of RNA methylation. The day 5 increase, in fig. 4, being essentially due to a drop in the rate of RNA biosynthesis occurring upon completion of fusion. DZA exerts simoltaneously an inhibition of RNA methylation and a stimulation of the differentiation process; the former event being however insufficient, by itself. From the same L5 cell line, clones can indeed be selected, which have lost the ability to undergo fusion, but will fuse up to 50% in the presence of DZA (data not shown). Inhibition of RNA methylation by DZA is however unable, by itself, to achieve myoblasts differentiation, since this inhibition is present upon addition of the drug to the cells in growth medium, but no fusion takes place under these conditions. As for Con-A, the inhibition it exerts on the differentiation process appears indeed to be associated to RNA hypermethylation; although no hypothesis can sofar be advanced on the mechanism(s) of those two effects. Whether either one, or both of them, are mediated by activation or repression of specific regulatory genes is still an open question.

REFERENCES

1. Scarpa S, Uhlendorf W, Cantoni GL (1985). The differentiation of L5/A10 cell line (a subclone of L5 line) is controlled by changes of cultural conditions. Cell Differentiation 17:105.
2. Scarpa S, Strom R, Bozzi A, Aksamit RR, Backlund PS, Chen J, Cantoni GL (1984). Differentiation of myoblast cell line and biological methylation: 3-Deazaadenosine stimulates formation of multinucleated myofibers. Proc Natl Acad Sci USA 81:3064
3. Bader JP, Brown NR, Chiang PK, Cantoni GL (1978). 3-Deazaadenosine an inhibitor of adenosylhomocysteine hydrolase inhibits reproduction of Rous sarcoma virus and transformation of chick embryo cells. Virology 89:494.
4. Leonard EJ, Skeel A, Chiang PK, Cantoni GL (1978). The action of the adenosylhomocysteine hydrolase inhibitor, 3-DZA, on phagocytic function of mouse macrophages and human monocytes. Biochem Biophys Res Commun 84:102.
5. Zimmerman TP, Wolberg G, Duncan GS (1978). Inhibition of lymphocyte-mediated cytolysis by 3-deazaadenosine: evidence for a methylation reaction essential to cytolysis. Proc Natl Acad Sci USA 75:6220.
6. Aksamit RR, Falk W, Cantoni GL (1982). Inhibition of Chemotaxis by S-3-Deazaadenosylhomocysteine in a Mouse Macrophage Cell Line. J Biol Chem 257:621.
7. Grunert B, Schäfer KP (1982). RNA methylation in resting and concanavalin A-stimulated lymphocytes. Exp Cell Res 140:137.
8. Vogel Z, Sytkowski AJ, Niremberg MW (1972). Acetylcholine receptors of muscle grown in vitro. Proc Natl Acad Sci USA 69:3180.
9. Yaffe D (1968). Retention of differentiation potentialities during prolonged cultivation of myogenic cells. Proc Natl Acad Sci USA 61:477.

DNA METHYLATION AND RESTRICTION PROCESSES IN ESCHERICHIA COLI: INSIGHTS BY USE OF BACTERIAL VIRUSES T3 AND T7

Detlev H. Krüger[1], Thomas A. Bickle[2], Monika Reuter[1], Claus-Dietmar Pein[3] and Cornelia Schroeder[1]

Institute of Medical Virology[1] and Dept. of Chemistry[3], Humboldt University, DDR - 1040 Berlin, German Democratic Republic; and [2]Biocenter, University of Basel, CH-4056 Basel, Switzerland

ABSTRACT DNA modification-restriction enzymes of *Escherichia coli* cells can be grouped into 3 families, called type I, II, and III. - The ocr⁺ gene of T7 encodes the first known inhibitor protein specifically blocking a group of methylases and endonucleases which, in particular, belong to the type I (*Eco*B, *Eco*K) enzymes. - The appearance of recognition sites for type II enzymes (e.g., *Eco*RII, methylases *Eco*Dam and *Eco*Dcm) is strongly counterselected in the T7 genome, furthermore, there are additional mechanisms preventing the enzymes from acting on the remaining sites. For instance, *Eco*RII is a restriction enzyme which requires the coordinated presence of at least 2 recognition sites in the substrate DNA for its activity. Sites which do not fulfill this requirement are refractory to *Eco*RII but can be cleaved by this enzyme in the presence of a second, susceptible DNA species or oligonucleotides. - The resistance of T7 DNA towards the type III enzyme *Eco*P15 is explained by the absolute "strand bias" of the non-symmetric sites (in which only 1 strand can be methylated) in the DNA molecule. These data allow some understanding of how the endogenous DNA in EcoP15-encoding cells could be protected against "self-restriction" during replication. - The frequency and polarity of recognition sites in the DNA molecule may play a critical role in the function of certain restriction endonucleases and methylases.

INTRODUCTION

The postsynthetic methylation of DNA in prokaryote cells produces 6-methyladenine (6mA), 5-methylcytosine (5mC) or N4-methylcytosine (N4mC). The responsible methyltransferases (MTases) are strictly sequence-specific. This DNA modification plays a decisive role in the protection of DNA against the corresponding restriction endonuclease. This fact can also be exploited in vitro to modulate the cleavage pattern of DNA molecules (1).

In *E.coli* cells MTases and restriction endonucleases of all three types may occur (2). Besides type I, II and III MTases which are functionally (and in the cases of types I and III also structurally) associated with restriction endonucleases - Hsd enzymes for host specificity of DNA - there are MTases for which a corresponding (i.e. recognizing the same sequence and blocked by its specific methylation) restriction endonuclease does not exist. These are two type II enzymes: the Dam MTase (synthesizing 6mA in the recognition sequence GATC) and the Dcm MTase (synthesizing 5mC from the internal C in 5'-CC[A/T]GG). Particlularly for Dam methylation a number of biological functions are known, among others influences on gene expression, initiation of DNA replication and mismatch repair (3,4,5).

In our lab the closely related phages T3 and T7 which belong to the family podoviridae have been investigated for a long time (6). Our experiments on these phage have revealed some insights into mechanisms of DNA methylation and restriction in *E.coli*.

RESULTS

Ocr as a Specific Inhibitor Protein for Type II Enzymes

Virus-coded effector poteins have a great influence on the methylation processes in the infected cell. The Ado-Met hydrolase (sam$^+$ gene function) which is part of the Ocr protein of phage T3 destroys the intracellular methyldonor Ado-Met and, therefore, prevents all transmethylation processes, including those of DNA (7,8).

The Ocr protein of the phages T7 and T3 is the first recognized inhibitor protein for type I (and possibly also type III) restriction endonucleases (9). It exerts its effect by direct protein-protein interaction with its target enzymes. Investigation of the susceptibility of MTases to blocking by Ocr led to the demonstration that Ocr also represents the first specific inhibitor protein for type I, but not type II, MTases (8,10,11,12).

Specific effector molecules which enhance or lower the activity of a certain group of MTases in the cell could influence global methylation patterns of the genome (13).

Counterselection against Recognition Sequences for Type II Enzymes

Although type II MTases and endonucleases are not susceptible to inhibition by the phage-coded Ocr protein, the cellular Dam and Dcm MTases hardly methylate T7 DNA (8). One reason for this is the strict counterselection against Dam and Dcm recognition sites in the virus genome, both occurring about 20-times less than expected from the frequencies of their respective partial sequences (14).

The 6 Dam sites remaining in T7 DNA are, on the average, methylated in only one site, while both Dcm sites are nonmethylated (8). The underlying causes are probably different for Dam and Dcm MTases (15). In the case of Dcm MTase which has the same sequence and methylation specificity as the *Eco*RII *hsd* enzyme specific substrate requirements could be involved (see below).

Viruses apparently try in various ways to avoid DNA methylation which is detrimental for their reproduction. We confirmed this in transfection studies on T7 DNA which had been methylated in vitro with Dam MTase. Such methylated DNA reproducibly transfected *E.coli* cells less efficiently than normal, unmethylated DNA (our unpubl. data).

Substrate Requirements of *Eco*RII and Artificial Stimulation of Restriction Endonucleases

*Eco*RII requires a specific constellation of target sites. The restriction endonuclease *Eco*RII interacts with the DNA sequence 5'-CC[A/T]GG. T7 DNA contains two of these sites, T3 DNA three. Despite their being nonmethylated they are, however, resistant to *Eco*RII digestion (8).
On the other hand DNAs with a higher density of *Eco*RII sites, like pBR322 and lambda DNA are cleaved normally by *Eco*RII (Table 1). We showed, using *Rsa*I-generated fragments of pBR322, that the susceptibility to *Eco*RII cleavage is proportional to the number of sites in the individual fragment (16).

TABLE 1
COMPARISON OF ECORII-RESISTANT AND ECORII-SENSITIVE DNA MOLECULES

DNA	total length [kbp]	number of *Eco*RII sites	average density of *Eco*RII sites	cleavage by *Eco*RII
T3	38.7	3	1/12.9 kbp	–
T7	39.9	2	1/20 kbp	–
pBR322	4.36	6	1/ 0.7 kbp	+
lambda	48.5	71	1/ 0.7 kbp	+

EcoRII could be the prototype of restriction endonucleases which require the coordinated presence of at least two recognition sites in the target DNA molecule for activity. The distance between two sites as well as the orientation of the central A/T pair could play a critical role.

Stimulation of endonucleolytic action. Co-incubation of primarily resistant T3 or T7 DNA with EcoRII-sensitive plasmid or phage DNA or with site-containing oligonucleotide duplexes makes the former susceptible to cleavage. Following its activation on susceptible DNA, EcoRII can obviously "turn over" to attack resistant DNA species. A direct correlation between the concentration of "activator" DNA or oligonucleotides (or the number of susceptible sites, resp.) and the degree of EcoRII digestion of the primarily resistant DNA is observed (16,17).

In this way originally uncleavable DNAs can be made susceptible to site-specific cleavage by certain restriction endonucleases.

Comparison with other enzyme systems. Enzymes which catalyze intramolecular recombination processes also require the coordinated presence of 2 DNA sites. By interacting with the protein, the 2 DNA sites can "sense" each others presence and, in some cases, even their mutual orientation in the DNA molecule (18,19). Apart from the processes of site-specific recombination their are several examples of functional interactions between distant sites in the same DNA molecule. Various models explaining this have been put forward, e.g. DNA scanning or looping (20,21).

The distance between two DNA sites can play a role in the efficiency of the interaction too, as has been postulated for the influence of Dam sites on the effficiency of mismatch repair in a different part of the DNA molecule (22).

The thymine in the central nonsymmetric A/T pair of the EcoRII recognition sequence 5'-CC[A/T]GG is especially important for the interaction with the restriction endonuclease (23); conceivably this also is true of the relative orientation of A/T pairs in two sites. Thymine residues are critical in quite a number of

DNA-protein interactions, e.g. those of *lac* repressor *E.coli* RNA polymerase, Cro repressor (24).

Protection of Non-Symmetrical Recognition Sites against Restriction

Site Specificity of Type III Restriction/ Modification Enzyme. Type III enzymes recognize short, nonsymmetrical DNA sequences. These are 5'-AGACC for *Eco*P1 and 5'-CAGCAG for *Eco*P15. The specific modification of the recognition sites consists in adenosine methylation. Fully modified recognition sites, therefore, can only be methylated in one strand, since the opposite strand does not contain adenosine (2,25).

Usually recognition sequences for restriction endonucleases are methylated in both strands so that, during DNA replication, at least the parental strands are modified and nascent DNA is protected against the cognate endonucleases. In the case of type III enzymes, however, totally unmethylated recognition sites arise (FIG. 1A). How newly synthesized DNA in cells with such *hsd* systems escapes degradation was not understood.

Orientation and Sensitivity of *Eco*P15 Sites in the Genomes of T7 and T3. In *Eco*P15-coding cells T7 is not restricted whereas sam^- derivatives of T3 are (14). When checking the total T7 DNA sequence (26) we got a surprising result: The sequence CAGCAG occurs with the expected frequency - 36 times (as calculated from the frequency of its subsequences) but all the adenine-containing sequences are confined to the same DNA strand. We called this uniform site orientation "strand bias" (14). In contrast to the situation with T7, an analysis of sequenced parts of T3 DNA (27,28) showed *Eco*P15 sites in both orientations (FIG. 1B).

We have examined the *in vitro* sensitivity of the different genomes to *Eco*P15 digestion. Unlike T7 DNA, T3 DNA is cleaved into several fragments. Recombinants of T7 containing small pieces of T3 with an inverted *Eco*P15 site become susceptible to *Eco*P15 digestion (our unpubl. data).

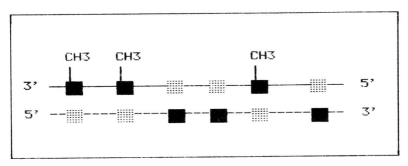

FIGURE 1. A) Configuration of type III recognition sites during replication of a modified genome

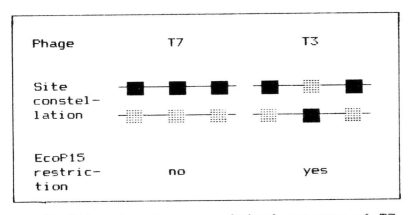

B) Situation in non-modified genomes of T7 and T3

■ recognition sequence, methylated strand, e.g. CAGCAG
▦ recognition sequence, complementary strand, e.g. CTGCTG
— parental DNA strand
-- daughter DNA strand

Model for the Protection of Hemimethylated DNA Molecules. Our observations support the conclusion that one recognition site is not sufficient for the activity of *Eco*P15 endonuclease but that functional interaction with two sites within the same DNA molecule is required. If these are orientated identically (i.e. both A's in one, both T's in the other strand) restriction is precluded even if both are nonmethylated. Nonmethylated sites in reverse orientation are restricted by *Eco*P15.

In other words, the type III restriction enzyme is unable to attack unmethylated recognition sites when they are in "strand bias". This is exactly the case during DNA replication where the unmethylated adenines are always confined to one, namely the newly synthesized DNA strand, while all the thymines complementary to the unmethylated A's are in the parental strain (cp. FIG. 1).

Our model also explains why artificial heteroduplexes of lambda DNA which were constructed from an *Eco*P1-modified and a nonmodified strand are resistant to *Eco*P1 digestion (29): The constellation of sites in the heteroduplex is equivalent to that arising during cellular DNA replication.

It will be interesting to check whether this model is also valid for the protection of recognition sites of certain typeII-S enzymes during DNA replication. For example, the recognition sequence of *Mbo*II is 5'-GAAGA and the cognate methylation is that of the right-most adenine (1). Since there is no adenine in the complementary strand, hemimethylation should protect against restriction. During DNA replication of *Moraxella bovis* cells, therefore, totally unmethylated sites arise.

General Conclusions

We suggest that *Eco*RII and *Eco*P15 - similar to other enzymes such as site-specific recombinases - require the functional interaction with more than one recognition site for their activity. The frequency and polarity of recognition sites in the DNA molecule may play a critical role for the function of restriction endonucleases (and MTases) with nonpalindromic recognition sites.

ACKNOWLEDGEMENTS

We thank Sigrid Hansen (Berlin) for help in the experiments, H.O. Smith and G.J. Barcak (Baltimore, MD) for stimulating support in the *Eco*RII studies and R. Kahmann (Berlin-West) for critical suggestions.

REFERENCES

1. McClelland M, Nelson M (1988). The effect of site-specific DNA methylation on restriction endonucleases and DNA modification methyltransferases - a review. Gene 74:291.
2. Bickle TA (1987). DNA restriction and modification systems. In Ingraham J et al (eds): "*Escherichia coli* and *Salmonella typhimurium*: Cellular and Molecular Biology", Washington DC: ASM Publ, p 692.
3. Sternberg N (1985). Evidence that adenine methylation influences DNA-protein interactions in *Escherichia coli*. J Bacteriol 164:490.
4. Marinus MG (1987). DNA methylation in *Escherichia coli*. Annu Rev Genet 21:113.
5. Messer W, Noyer-Weidner M (1988). Timing and targeting: the biological functions of Dam methylation in *E. coli*. Cell 54:735.
6. Krüger DH, Schroeder C (1981). Bacteriophage T3 and bacteriophage T7 virus-host cell interactions. Microbiol Rev 45:9.
7. Krüger DH, Presber W, Hansen S, Rosenthal HA (1975). Biological functions of the bacteriophage T3 SAMase gene. J Virol 16:453.

8. Krüger DH, Schroeder C, Reuter M, Bogdarina IG, Buryanov YI, Bickle TA (1985). DNA methylation of bacterial viruses T3 and T7 by different DNA methylases in *Escherichia coli* K12 cells. Eur J Biochem 150:323.
9. Krüger DH, Bickle TA (1983). Bacteriophage survival: multiple mechanisms for avoiding DNA restriction systems of their hosts. Microbiol Revs 47:345.
10. Bogdarina IG, Reuter M, Krüger DH, Buryanov YI, Bayev AA (1983). Methylation of DNA of phages T3 and T7 by various types of DNA-adenine methylases and inhibition of the EcoK methylase by the Ocr protein. Proc Acad Sci USSR 273:234.
11. Schroeder C, Reuter M, Krüger DH (1984). DNA methylation of T3 virus ocr^+ and ocr^- strains in *Escherichia coli* cells harbouring the EcoK DNA host specificity system. Biomed Biochim Acta 43:K1.
12. Bandyopadhyay PK, Studier FW, Hamilton DL, Yuan R (1985). Inhibition of the type I restriction - modification enzymes EcoB and EcoK by the gene 0.3 protein of bacteriophage T7. J Mol Biol 182:567.
13. Krüger DH (1988). Methylierung der DNA als molekularbiologisches Regulationssignal. Biol Zentralbl 107:257.
14. Schroeder C, Jurkschat H, Meisel A, Reich JG, Krüger DH (1986). Unusual occurrence of EcoP1 and EcoP15 recognition sites and counterselection of type II methylation and restriction sequences in bacteriophage T7 DNA. Gene 45:77.
15. Krüger DH, Schroeder C, Santibanez-Koref M, Reuter M (1989). Avoidance of DNA methylation: a virus-encoded methylase inhibitor and evidence for counterselection of methylase recognition sites in viral genomes. Cell Biophys 15:in press.
16. Krüger DH, Barcak GJ, Reuter M, Smith HO (1988). EcoRII can be activated to cleave refractory DNA recognition sites. Nucl Acids Res 16:3997.

17. Pein CD, Reuter M, Cech D, Krüger DH (1989). Oligonucleotide duplexes containing CC(A/T)GG stimulate cleavage of refractory DNA by restriction endonuclease EcoRII. FEBS Letters: in press.
18. Sadowski P (1986). Site-specific recombinases: changing partners and doing the twist. J Bacteriol 165:341.
19. Gellert M, Nash H (1987). Communication between segments of DNA during site-specific recombination. Nature 325:401.
20. Müller MM, Gerster T, Schaffner W (1988). Enhancer sequences and the regulation of gene transcription. Eur J Biochem 176:485.
21. Wang JC, Giaever GN (1988). Action at a distance along a DNA. Science 240:300.
22. Bruni R, Martin D, Jiricny J (1988). d(GATC) sequences influence Escherichia coli mismatch repair in a distance-dependent manner from positions both upstream and downstream of the mismatch. Nucl Acids Res 16:4875.
23. Yolov AA, Vinogradova MN, Gromova ES, Rosenthal A, Cech D, Vetko VP, Metelev VG, Kosykh VG, Buryanov YI, Bayev AA, Shabarova ZA (1985). Interaction of EcoRII restriction and modification enzymes with synthetic DNA fragments: the binding and cleavage of substrates containing nucleotide analogs. Nucl Acids Res 13:8983.
24. Caruthers MH, Gottlieb P, Bracco L, Cummins L (1987). The thymine 5-methyl group: a protein - DNA contact site useful for redesigning Cro repressor to recognize a new operator. In Oxender DL (ed): "Protein Structure, Folding, and Design 2" (UCLA Symp Mol Cell Biol, New Series, Vol 69), New York: Liss Inc, p 9.
25. Bickle TA (1982). The ATP-dependent restriction endonucleases. In Linn SM, Roberts RJ (eds): "Nucleases", Cold Spring Harbor: CSH Press, p 85.
26. Dunn JJ, Studier FW (1983). Complete nucleotide sequence of bacteriophage T7 DNA and the locations of T7 genetic elements. J Mol Biol 166:477.

27. Schmitt MP, Beck PJ, Kearney CA, Spence JL, DiGiovanni D, Condreay JP, Molineux IJ (1987). Sequence of a conditionally essential region of bacteriophage T3, including the primary origin of DNA replication. J Mol Biol 193:479.
28. Yamada M, Fujisawa H, Kato H, Hamada K, Minagawa T (1986). Cloning and sequencing of the genetic right end of bacteriophage T3 DNA. Virology 151:350.
29. Hattman S, Brooks JE, Masurekar M (1978). Sequence specificity of the P1 modification methylase (M.*Eco*P1) and the DNA methylase (M.*Eco*dam) controlled by the *Escherichia coli dam* gene. J Mol Biol 126:367.

ERM METHYLASES IN STAPHYLOCOCCI AND THEIR MODE OF SPREAD[1]

Donald T. Dubin,[2] Lucia E. Tillotson,[2] Leslie Huwyler,[3] and Ellen Murphy[3]

[2]Department of Molecular Genetics and Microbiology, UMDNJ-Robert Wood Johnson Medical School, Piscataway, New Jersey 08854 and [3]The Public Health Research Institute, New York, New York 10016

ABSTRACT We present findings on the nature of genetic elements that serve as vectors for two commonly occurring methylase gene classes in staphylococci, ermA and ermC, that encode MLS resistance. ErmC occurs in a small plasmid, pNE131, different from its initially described vector. pNE131 has enjoyed wide interspecific, intergeneric and geographical dissemination.

ErmA occurs in a chromosomal transposon, Tn554, with several unusual features. Analysis of secondary insertions revealed a unique mode of information transfer, whereby 6-7 bp at the 5' end (the "5'-junction") translocate to the 3' junction during transposition. The occurrence of many 3'-junction variants in nature indicates that such "junction translocation," and hence transposition to or from secondary sites, has been common on an evolutionary time scale. A particular secondary insertion was closely associated with methicillin resistance in S. aureus, and may be part of the extra DNA that characterizes methicillin-resistant strains.

---

1. This work was supported in part by grant 20.88 from the Foundation of the University of Medicine and Dentistry of New Jersey to D.T.D. and grant GM27253 from the National Institutes of Health to E.M. L.E.T. received a stipend from National Cancer Institute Research Service Award CA 009069.

INTRODUCTION

We shall describe studies on the DNA that flanks certain methylase genes causing antibiotic resistance— DNA considered important in maintenance and spread of such resistance. The genes belong to the *erm* family, which encodes methylases that convert the A2058-equivalents of bacterial 23S rRNA to m_2^6A, rendering ribosomes resistant to the MLS (macrolide, lincosamide and streptogramin B) groups of antibiotics (1,2). Three evolutionarily distinct *erm* subclasses, *A*, *B*, and *C*, were discovered in *Staphylococcus aureus* isolates a number of years ago (3,4,5). The *erm* genes from these isolates, and flanking regions of DNA that serve as replicons and/or vectors for these genes, have been studied extensively as prototypes, after transfer to standard laboratory *S. aureus* strains (see refs. 6-9). To assess the contemporary prevalence of these classes, we surveyed series of clinical isolates in 1984/5 (10) and 1987 (11), not only in *S. aureus* but also in coagulase negative staphylococci, in view of the current appreciation of the importance of these latter groups as reservoir of resistance determinants and as nosocomial pathogens. The predominant classes in local (NJ) hospitals proved to be *A* and *C*, no *ermB* being found.

RESULTS

ermC Isolates.

In contrast to classical *ermC*, which was found in the 3.7 kb plasmid pE194 (6), the *ermC* determinants of our isolates proved to reside in a 2.4 kb plasmid lacking homology to pE194 aside from the respective *erm* genes. A 2.4 kb MLS resistance-conferring plasmid, pNE131, had recently been described in *S. epidermidis* (12), and Southern hybridization analysis showed close similarity between it and our 2.4 kb plasmids (10). In 1986, sequences appeared for pNE131 (13), and a pNE131-like plasmid from *Bacillus subtilis*, pIM13 (14). As shown in Fig. 1, pIM13 lacks a 121 bp fragment present at the end of a presumed maintenance gene of pNE131. Perusal of the relevant sequences suggests that the deletion resulted from a reciprocal crossover event related to a 10 bp repeat in pNE131 (or in a relatively

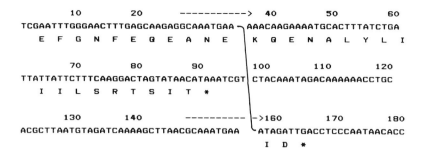

FIGURE 1. Generation of pIM13 from pNE131. We show a portion of the pNE131 sequence, numbered as in (13). The arrows designate 10 bp repeats, reciprocal recombination between which could generate the pIM13 sequence (14), as illustrated by the line connecting pNE131 residue 35 with 157. The two amino acids beyond the postulated crossover terminate the pIM13 protein homologous to the pNE131 one.

close ancestor of this plasmid). We infer that the pNE131 class is indeed native to staphylococci, an idea supported also by homology to a silent *S. aureus* plasmid, pSN2 (15). pNE131-class plasmids have been noted to serve as *ermC* replicons also in recent *S. aureus* isolates from Brazil (16), and from France, Germany and Russia (14).

ermA Isolates and Junction Translocation.

ErmA occurs in a chromosomal transposon, Tn554. Earlier work had shown that, in standard *S. aureus* laboratory strains, Tn554 has extraordinary site specificity, entering only a single chromosomal site, *att554*, in one ["(+)"] orientation (8,17). Insertions into other sites could be engineered in plasmids, however, in which case inserts occurred in either (+) or (-) orientation, just 3' or 5' to a 6-7 bp core, for (+) or (-) inserts, respectively (8,18).

Our *ermA* clinical isolates all contained chromosomal Tn554, as expected. However, restriction patterns of junctional segments showed that many *S. aureus* isolates had inserts in addition to the classical primary insert; and all coagulase negative *ermA* isolates had one or more secondary

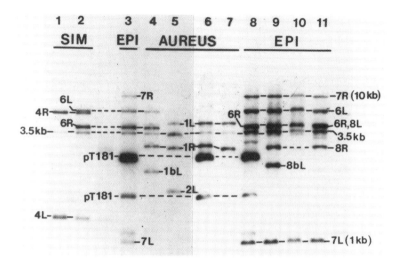

FIGURE 2. Restriction patterns of Tn554 HindIII/PstI junctional fragments of prototype strains. Samples of total cellular DNA were digested with HindIII plus PstI. Digests were subjected to Southern blot analyses using as probe Tn554 Segments A and then C (see Fig. 5). Bands corresponding to junctional fragments are designated as to restriction pattern class and whether right ("R") or left ("L") junction. Bands arising from an internal 3.5 kb fragment (which hybridizes weakly with Tn554 segment C), or from the tet-containing plasmid pT181 (which hybridizes with a vector moiety of this probe), are so designated. Lanes are as follows: 1 and 2, single- and double-insert S. simulans isolates (10); 3 and 8, double-insert S. epidermidis isolates (tc^r); 4, Mc^r S. aureus isolate 155 (see Text); 5, double-insert S. aureus isolate WJ137 (see Text); 6, single-insert S. aureus isolate (tc^r); 7, single-insert S. aureus laboratory strain RN2863; 9-11, double- and triple-insert S. epidermidis isolates. The right fragment of insert class 2 (lane 5) runs with the 3.5 kb internal piece, and is not labeled. The right fragment of insert class 6 runs with the left fragment of class 8. (Modified from Ref. 11, with permission.)

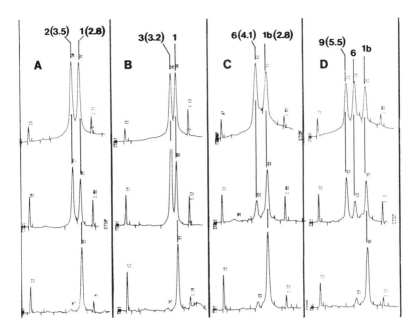

FIGURE 3. Southern blot analysis of right junctions and transposon ends of *S. aureus* inserts: DNA and oligonucleotide hybridizations. We show tracings of autoradiographs generated by probing *Hind*III/*Pst*I digests with right transposon probe D (Fig. 5) (top tracings), or with 16-20 residue oligonucleotide probes corresponding to the right primary junction (bottom) or the right end of Tn554 (middle) (11). Blots were probed successively with oligonucleotides and transposon segment; hybridization removed essentially all label from prior blots. For precise alignment lines were drawn across autoradiograms 2.5 and 7.0 cm from the origin; these appear as spikes on the tracings. Panel A shows strain WJ137, with inserts 1 and 2; B, strain DR3, with inserts 1 and 3. Panels C and D show Mc^r isolates with inserts 1b and 6; one has in addition yet another insert class, 9 (see ref. 11). Numbers in parentheses show sizes in kb.

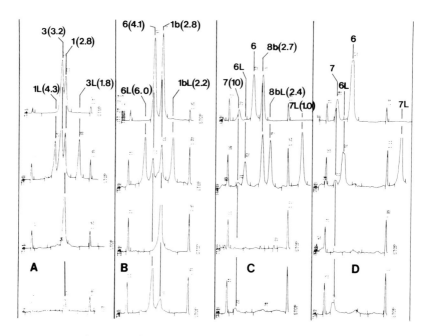

FIGURE 4. Southern blot analysis of junctions and transposon ends of *S. aureus* and *S. epidermidis* inserts. The general scheme was as for Fig. 3. The bottom tracings show results for hybridization with a right *S. aureus* insert 6 junctional oligonucleotide; the next shows results for a right primary insert junctional oligonucleotide; the next, results for combined hybridization with right and left transposon terminal oligonucleotides; and the top tracings, results for transposon segment D. Column A, *S. aureus* WJ137; column B, Mc^r *S. aureus* containing inserts 1b and 6; Columns C and D, *S. epidermidis* with inserts 6, 7 and 8b (C) or 6 and 7 (D). Designations of peaks are as for Fig. 3, "L" denoting left junctions.

inserts while lacking primary inserts (Figs. 2-4; see also ref. 10). These and comparable patterns yielded maps defining 7 secondary insert classes from our 1984/5 survey (Fig. 5), and encouraged further attempts to obtain secondary chromosomal inserts in laboratory strains of defined genetic background. We succeeded, by using as transduction recipients strains deleted for *att554* (Murphy et al, in preparation). We recovered 15 such transductants, scattered among 9 different secondary sites as monitored by

Erm Methylase Gene Vectors 131

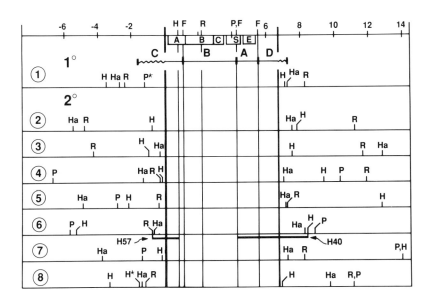

FIGURE 5. Restriction pattern classes of Tn554 insertions. The top line shows Tn554 with transposase genes (A,B,C) and resistance genes *spc* (S) and *ermA* (E), noted (19); distances from the left end are indicated in kb. 19). The second line shows the plasmid, pEM715, used as source of Tn554 DNA segment probes. The wavy lines indicate vector (pT181) moieties; we have arbitrarily linearized at a vector *Fnu*4HI site. The remaining lines show restriction patterns for vicinal chromosomal DNA of the various classes (encircled numbers) of Tn554 insert. Restriction enzymes are abbreviated as follows: H, *Hind*III; Ha, *Hae*III; R, *Eco*RI; P, *Pst*I; F, *Fnu*4HI. P* and H* represent restriction site polymorphisms; the corresponding classes are designated 1b and 8b (ref. 11). (Reprinted from ref. 11 with permission.)

junctional segment restriction patterns. However, 2 patterns, represented by strains RN6028 and RN6029, were most abundant; and they were strikingly similar to those of our initial classes of *S. aureus* secondary inserts (#'s 2 and 3), represented by isolates WJ137 and DR3 (10). Further, the patterns were compatible with the inserts being in the same site (designated *att137*) but in differing orientations. As shown in Fig. 6, sequence analysis of the

```
                    CHROMOSOME OR VECTOR                          5'-JOINT      Tn554
                    ----------------------                        --------      -----
   WJ137     GATCTTGCAT TGTGGAATAA AGTCAATCCA TCCTTCGCGCtaaaat    TTAAC...
   RN6029    *TG******* ********** ********** ****************    ********

   DR3       CACACAGGTT CCATCAATTC ATCATAACCA CCAAACATTGatttta    TTAAC...
   RN6028    ********** ****#***** ********** ****************    ********

   PRIMARY + TGCCAATGCA ATCATCGCAG TTCATAATCA TCCATCCGGTsatsta    TTAAC...
           - TCAACCTCAT TGTTGTTATG ATATCTTCTT GTGAGGGCGTtacatc    ********

   PLASMID + GCATAAAATT TTAAACTTAG ACCGATGCGG AAGATGGTAAtttaatt   TTAAC...
     2°    - TTGCTTCTTT ATCTGTTATT TTATCTTCAT CACTTGTATTaattaaa   ********


             Tn554    3'-JOINT        CHROMOSOME OR VECTOR
             -----   --------         --------------------
   WJ137     ...GATAAsatsta    CAATGTTTGG TGGTTATGAT GAATTAATGG AACCTGTGTG
   RN6029    ********aattaaa   ********** ********** *****G**** **********

   DR3       ...GATAAsatsta    GCGCGAAGGA TGGATTGACT TTATTCCACA ATGCAAGATC
   RN6028    ********aattaaa   ********** ********** ********** *******CA*

   PRIMARY + ...GATAAsatsta    ACGCCCTCAC AAGAAGATAT CATAACAACA ATGACGTTGA
           - **************    ACCGGATGGA TGATTATGAA CTGCGATGAT TGCA

   PLASMID + ...GATAAsatsta    AATACAAGTG ATGAAGATAA AATAACAGAT GAAGAAGCAA
     2°    - **************    TTACCATCTT CGGCATCGGT CTAAGTTTAA AATTTTATGC
```

FIGURE 6. Junctional sequences of Tn554 inserts. We show sequence obtained for inserts 2 and 3 (Fig. 5), as represented by isolates WJ137 and DR3, and laboratory strains RN6028 and RN6029 (see Text); and, for comparison, sequence from primary and secondary plasmid inserts (18,19). Junctions ("JOINTS") are in lower case letters. The 5' junction of the parent insert in pRN4177 ("PLASMID -") and the daughter 3'junctions of strains RN6028/9 are underlined.

respective junctions bore this out. WJ137 and RN6029 contained secondary inserts [arbitrarily designated (+)] just 3' of a TAAAAT core; and DR3 and RN6028 contained (-) inserts just 5' of this core (or, when transposon sequences are presented in the same orientation, 3' of the reverse complement of the core). The flanking left chromosomal sequence of the RN6029 insert (beyond the core) was a perfect reverse complement of the right flanking chromosomal sequence of RN6028, and *vice versa*, over the range sequenced, in keeping with the isogenic origin of these strains and the inverted orientations of their secondary inserts. Two altered nucleotides in the clinical isolates vis a vis the laboratory strains [TG to AT in the 5' arm of WJ137; CA to AT in the 3' arm of DR3] retained the reverse complementarity, indicating, as would be predicted from the temporal and geographical proximity of their origins, that the clinical isolates are more closely related to each other than to the laboratory strains; i.e., that ancestor(s) of DR3 and WJ137 (but not of the laboratory strains) served as recipient for the secondary inserts of the clinical isolates. The G to A change present in the right arm of WJ137 but not reflected in the left arm of DR3 presumably is related to minor evolutionary divergence between these natural isolates.

Our most surprising sequence result was the replacement in the inserts of RN6028/9 of what had been considered the terminal 6 residues of the transposon, GATGTA, with AATTAAA. This latter heptanucleotide corresponds to the core of the att site of the parent transposon of 6028/9, a secondary insertion in plasmid pRN4177 (8). We suggested on this basis a "junction translocation" rule for transposition of Tn554: namely, that the 5' junctional chromosomal 6 or 7 residues of the parent insert (i.e., the core of the parent att site) becomes the 3'-terminal 6 or 7 residues (the 3'-junction) of the daughter (20). The presence of GATGTA as the 3'-junction of the naturally occurring secondary inserts in isolates WJ137 nd DR3 would then indicate that they arose from parental transposons with GATGTA 5'-junctions. This is as expected, as such junctions are by far the most frequently occuring in natural isolates of S. aureus. Subsequent transductions from a variety of secondary sites with 5'-junctions different from GATGTA, to unoccupied att554, has established this rule for transposition of Tn554 (Murphy et al, in preparation).

Analyses of methicillin-resistant ("Mc^r") staphylococcal isolates collected during an outbreak of multi-resistant S. aureus infections (11) extended these observations. The outbreak S. aureus isolates all had, in addition to a variant of the primary Tn554 insert, a secondary insert that yielded junctional restriction fragment patterns resembling a class ("6") previously found only in coagulase negative staphylococci (ref. 10; see also Fig. 5). One of 8 S. simulans isolates and all of 10 S. epidermidis ermA isolates tested contained this insert, making it the most common class among coagulase negative staphylococci. The S. epidermidis version was of special interest in the light of the junction translocation rule

		5'JUNCTION	TN554	3'JUNCTION
#6.	S. aureus	AATCATCCAAGTGGA	GATGTA TT.AA	GATGTA ACACCTTCTCCAGGA
	S. epidermidis	AATCATCCAAGTGGA	GATGTA TT.AA	GACATC ACACCTTCTCCAGGA
#1.	S. aureus	AATCATCCATCCGGT	GATGTA TT.AA	GATGTA ACGCCCTCACAAGAA

FIGURE 7. Junctional sequences of Tn554 insert 6. We show sequences for insert 6 as obtained for *S. aureus* and *S. epidermidis* and, for comparison, the primary insert.

generated from laboratory *S. aureus* strains, since right junctional restriction fragments failed to hybridize with oligonucleotide probes based on the "normal" GATGTA 3' end of Tn554 (compare Fig. 4C, D with Fig. 3 and 4A, B). Junctional sequences obtained for the *S. aureus* and *S. epidermidis* versions of insert 6 are shown in Fig. 7. The corresponding *att* site, designated *att155* after a prototype isolate, was indeed found to be the same for the inserts from the 2 species. *att155* proved to be highly similar to *att554*, and retains the GATGTA core. *S. aureus* insert 6 also retains the GATGTA 3'-junction, suggesting that it arose from an insert in an *att554* (or *att155*) class site. The junctional sequences obtained for the *S. epidermidis* version of insert 6 were the same as for *S. aureus*, except for the 3'-junction. In keeping with the oligonucleotide hybridization results, this was different: namely, GACATC. The reverse complement of this junction, GATGTC, closely resembles the standard GATGTA. We propose that the *S. epidermidis* junction is a natural example of junction translocation, and that *S. epidermidis* insert 6 arose from a parent transposon inserted in (-) orientation next to a GATGTC core. Search for such a parent insert may provide information on staphylococcal evolution, and on the evolution and spread of antibiotic resistance in this genus.

We have recently begun a systematic check of *ermA* isolates by right transposon-terminal oligonucleotide hybridization, as in Figs. 3 and 4. A substantial minority primary inserts found among NJ isolates, and primary and secondary inserts of isolates from Australia (provided by P. Stewart and P. Matthews) and England (provided by W.C. Noble), failed to hybridize. Thus, manifestations of what we believe to be junction translocation occur at reasonably high frequency in staphylococci.

DISCUSSION

The findings we have presented show that, at least on an evolutionary time scale, there have been many instances of transfer among staphylococci, both intra- and interspecific, of pNE131 as vector for *ermC* and Tn554 as vector for *ermA*. Neither pNE131 nor Tn554 is conjugative. Presumably their transfer relies on mobilization, such as by a conjugative plasmid or a transducing phage, but the precise mechanisms involved in the natural spread of these elements remain unclear.

Likewise, the molecular basis for the epidemiologic association in *S. aureus* between Tn554 insert 6 and methicillin resistance requires clarification. The gene encoding methicillin resistance, *mec*, occurs within a stretch of "extra" DNA, the "mec-associated" DNA (21-23). It is possible that mec-associated DNA acts as a resistance determinant-accepting cassette, and that among its features is *attl55*. We have obtained some evidence supporting this conjecture (11).

ACKNOWLEDGMENTS

We thank P. Stewart, P. Matthews, and W.C. Noble for sending strains, and L. Moon-McDermott for expert technical assistance.

REFERENCES

1. Weisblum B (1975). Altered methylation of ribosomal ribonucleic acid in erythromycin-resistant *Staphylococcus aureus*. In Schlessinger D (ed): "Microbiology, 1974" Amer Soc Microbiol, Washington, D.C., p 199.
2. Thakker-Varia S, Ranzini AR, Dubin DT (1985). Mechanism of erythromycin-resistance in *Staphylococcus aureus* and *E. coli*. The "MLS" oligonucleotide of 23S RNA. Plasmid 14:152.
3. Mitsuhashi S, Hashimoto H, Kono M, Morimura M (1965). Drug resistance of staphylococci. II. Joint elimination and joint transduction of the determinants of penicillinase production and resistance to macrolide antibiotics. J Bacteriol 89:988.
4. Weisblum B, Demohn V (1969). Erythromycin-inducible resistance in *Staphylococcus aureus*: survey of antibiotic classes involved. J Bacteriol 98:447.

5. Iordanescu S (1976). Three distinct plasmids originating in the same *Staphylococcus aureus* strain. Arch Roum Pathol Exp Microbiol 35:111.
6. Iordanescu S, Surdeanu M, DellaLatta P, Novick R (1978). Incompatibility and molecular relationships between small staphylococcal plasmids carrying the same resistance marker. Plasmid 1:468.
7. Novick RP, Edelman I, Schwesinger MD, Gross AD, Swanson EC and Pattee PA (1979). Genetic translocation in *Staphylococcus aureus*. Genetics 76:400.
8. Murphy E, Phillips S, Edelman I, Novick RP (1981). Tn554: isolation and characterization of plasmid insertions. Plasmid 5:292.
9. Horinouchi S, Weisblum B (1982). Nucleotide sequence and functional map of pE194, a plasmid that specifies inducible resistance to macrolide, lincosamide, and streptogramin type B antibiotics. J Bacteriol 150:804.
10. Thakker-Varia S, Jenssen W, Moon-McDermott L, Weinstein M, Dubin DT (1987). Molecular epidemiology of macrolides-lincosamides-streptogramin B resistance in *Staphylococcus aureus* and coagulase-negative staphylococci. Antimicrob Agents Chemother 31:735.
11. Tillotson LE, Jenssen WD, Moon-McDermott L, Dubin DT (1989). Characterization of a novel insertion of the macrolides-lincosamides-streptogramin B resistance transposon Tn554 in methicillin-resistant *staphylococcus aureus* and *staphylococcus epidermidis*. Antimicrob Agents Chemother 33:541.
12. Lampson BC, Parisi JT (1986). Naturally occurring *Staphylococcus epidermidis* plasmid expressing constitutive macrolide-lincosamide-streptogramin B resistance contains a deleted attenuator. J Bacteriol 166:479.
13. Lampson BC, Parisi JT (1986). Nucleotide sequence of the constitutive macrolide-lincosamide-streptogramin B resistance plasmid pNE131 from *Staphylococcus epidermidis* and homologies with *Staphylococcus aureus* plasmids pE194 and pSN2. J Bacteriol 167:888.
14. Monod M, Denoya C, Dubnau D (1986). Sequence and properties of pIM13, a macrolide-lincosamide-streptogramin B resistance plasmid from Bacillus subtilis. J Bacteriol 167:138.
15. Kahn SA, Novick RP (1982). Structural analysis of plasmid pSN2 in *Staphylococcus aureus*: No involvement in enterotoxin B production. J Bacteriol 149:642.

16. Bastos MCF, Bonaldo MD, Penido EGC (1980). Constitutive erythromycin resistance plasmid in *Staphylococcus aureus*. J Gen Microbiol 121:513.
17. Krolewski JJ, Murphy E, Novick RP, Rush MG 91981). Site specificity of the chromosomal insertion of *Staphylococcus aureus* transposon Tn554. J Mol Biol. 152:19.
18. Murphy E, Lofdahl S (1984). Transposition of Tn554 does not generate a target duplication. Nature (London) 307:292.
19. Murphy E, Huwyler L, Bastos MC (1985). Transposon Tn554: complete nucleotide sequence and isolation of transposition-defective and antibiotic-sensitive mutants. EMBO J 4:3357.
20. Tillotson LE, Moon-McDermott L, Dubin DT (1988). Sequence of a 2° insertion site of the macrolide-lincosamide-streptogramin (MLS) transposon, Tn554. Abstr Annu Meet Am Soc Microbiol p 179.
21. Beck WD, Berger-Bachi B, Kayser FH (1986). Additional DNA in methicillin-resistant *Staphylococcus aureus* and molecular cloning of *mec*-specific DNA. J Bacteriol 165:373.
22. Matsuhashi M, Song MD, Ishimo F, Wachi M, Dol M, Inoue M, Ubukata K, Yamashita N, Konno M (1986). Molecular cloning of the gene of a penicillin-binding protein supposed to cause high resistance to beta-lactam antibiotics in *Staphylococcus aureus*. J Bacteriol 167:975.
23. Matthews PR, Reed KC, Stewart PR (1987). The cloning of chromosomal DNA associated with methicillin and other resistances in *Staphylococcus aureus*. J Gen Microbiol 133:1919.

TRANSCRIPTION OF A METHYLATED DNA VIRUS[1]

Dawn B. Willis,[2] Karim Essani, Rakesh Goorha, James P. Thompson,[3] and Allan Granoff

Department of Virology and Molecular Biology
St. Jude Children's Research Hospital
Memphis, Tennessee 38101

ABSTRACT The *Iridovirus* frog virus 3 is unique among animal viruses in containing a double-stranded DNA genome in which every cytosine in the dinucleotide sequence dCpdG is methylated by a viral-encoded DNA methyltransferase. The DNA methyltransferase has been purified and consists of two subunits, both of which are required for activity. Infection with frog virus 3 induces a *trans*-acting factor that facilitates the transcription of highly-methylated DNA templates by the host RNA polymerase II. This factor is not required for the transcription of immediate-early viral genes, but may play a role in the transcription of delayed-early or late viral genes.

---

[1]This work was supported by research grant CA 07055 from the National Cancer Institute, grant GM 23639 from the Institute of General Medical Sciences, and Cancer Center Support (CORE) grant CA 21765 from the National Cancer Institute, and by American Lebanese Syrian Associated Charities.

[2]Present address: American Cancer Society, Atlanta, GA 30329

[3]Present address: Department of Medicine, University of Tennessee, Memphis, TN

INTRODUCTION

Since its discovery by Granoff over 23 years ago (1), frog virus 3 (FV3) has proved to be a virus with many unusual properties. Most relevant to the present discussion is the high degree of genomic DNA methylation, first suspected following the observation that many common restriction enzymes, notably those inhibited by cytosine methylation at the dCpdG dinucleotide, failed to cut FV3 DNA (2). Chemical analysis of purified viral DNA revealed that over 22% of the cytosines were methylated at the 5-carbon position (2), and the cytoplasm of infected cells contained a DNA methyltransferase (MeTase) activity with different substrate specificity from that of the host cell nuclear enzyme (3). A mutant virus resistant to 5-azacytidine (azaC^r), an inhibitor of DNA MeTase (4), did not induce any cytoplasmic DNA MeTase, showing that the enzyme was viral-encoded (5).

Originally thought to have a strictly cytoplasmic site of replication, FV3 DNA is now known to first enter the nucleus, where the incoming methylated parental genome is transcribed by the host RNA polymerase II (6) and undergoes an initial round of DNA replication into unit length molecules (7). Despite the presence of cellular DNA MeTase in the nucleus (8), the viral DNA synthesized in the nucleus is unmethylated (3). After the viral DNA is transported to the cytoplasm, it replicates in the form of large concatemers (7) and is methylated de novo in that site (3).

The presence of such a high degree of DNA methylation was difficult to reconcile with the fact that the parental viral genome was known to be transcribed by the host RNA polymerase II (6). By that time, the 1980 theory of Razin and Riggs (9)--that DNA methylation was inhibitory to transcription--was widely accepted. If their theory was correct, the FV3 genome was either (a) not methylated in regions critical for transcription, or (b) the virus had evolved a way to enable the host enzyme to transcribe methylated DNA.

This manuscript will first review the recent progress that has been made in the purification and cloning of the FV3-encoded DNA MeTase, and then discuss the implications that a highly-methylated genome has for the transcriptional regulatory process.

RESULTS

Purification of FV3 DNA Methyltransferase.

Within 4 hours after infection, the cytoplasm of FV3-infected cells contains a high level of DNA MeTase activity (3). Using an assay that measured the transfer of [^{3}H]methionine from S-adenosyl methionine (SAM) to DNA substrates, we observed that, in contrast to the substrates preferred by the cellular enzyme, double-stranded unmethylated DNA was a better substrate than either hemi-methylated DNA or single-stranded DNA (3). The best substrate of all the ones tested was a poly-(dCdG).poly(dCdG) alternating copolymer; C's juxtaposed to A or C were essentially inactive substrates (3). As the azacytidine-resistant mutant FV3 did not induce any DNA MeTase activity in the cytoplasm (3), we concluded that the enzyme was coded for by the virus--the first reported instance of DNA MeTase coded for by a eukaryotic DNA virus.

When DNA from wild-type or mutant virions was subjected to restriction endonuclease analysis with the isoschizomeric enzymes (HpaII or MspI), we observed that wild-type DNA was completely cut by MspI, but totally refractory to cutting with HpaII; mutant DNA was cut equally well with both enzymes (10). This result indicated that every C in the internal sequence of the tetranucleotide sequence CCGG was methylated in the wild-type, but not in the mutant virus, confirming our hypothesis that viral DNA was methylated in the cytoplasm by the viral-encoded enzyme.

An inspection of radio-labeled viral proteins from cells infected by the azaC^r virus in a one dimensional SDS polyacrylamide gel showed a novel protein band not present in the extracts of cells infected with wild-type FV3 (10); this band replaced a normal band of slightly increased molecular weight (~26 kD). In a series of independently produced single-step reversions, there was a perfect correlation between the reappaearance of DNA MeTase activity and the original 26 kD polypeptide (10), strongly suggesting that this band was associated with the enzyme activity.

Using both the enzyme assay and SDS-polyacrylamide gel analysis to track the purification process, the enzyme was purified by DNA affinity chromatography, cibacron blue chromatography, and glycerol gradient centrifugation (11). An endonuclease activity capable of nicking PM 2 DNA was also found to co-purify with the DNA methyltransferase activity (11). At the end of the purification procedure, three bands of approximate molecular weights 18 kD, 26 kD, and 30 kD remained. Any attempt to separate these three bands resulted in a total loss of both enzyme activities (11).

To determine if the enzyme required more than one polypeptide subunit for activity, the three bands were eluted from the gel, renatured, and reconstituted in all possible combinations. No polypeptide had activity on its own; however the combination of the 26 kD and 18 kD polypeptides resulted in a complete restoration of DNA MeTase activity (11). We were unable to reconstitute the endonuclease activity, even with all three subunits; however, the purification conditions (boiling in SDS and mercaptoethanol, electrophoresis on polyacrylamide, elution and renaturating) may have been too harsh on a labile enzyme.

The original $azaC^r$ mutant had been isolated by serial passage in 10 μg/ml 5-azacytidine rather than by mutagenesis and selection of drug resistance in a single step. At the time of isolation, we noticed that a number of the resistant mutants still had methylated DNA, but these mutants were discarded because we were looking for an unmethylated DNA to use for cloning. With the awareness that the enzyme complex might be a multisubunit complex similar to that of the type I bacterial restriction endonucleases (5, 11), but methylating C instead of A, we decided to re-investigate the MeTase activity of single-step azaC resistant mutants (Table 1).

TABLE 1
CLASSIFICATION OF FV3 azaC^r MUTANTS[a]

Wild-type FV3	R+ M+
azaC^r, Class 1	R- M+
azaC^r, Class 2	R- M-
Lethal	R+ M-

[a]Suspensions of virus were treated with 5 μg/ml nitrosoguanidine and plated on fathead minnow cells in the presence of 5 μg/ml 5-azacytidine. Plaques of drug-resistant mutants were picked and used to prepare stock virus. Cytoplasmic extracts of cells infected from the plaque-purified stocks were assayed for methylase activity. R = endonuclease; M = MeTase.

Two classes of azaC-resistant mutants were routinely found, one with and one without DNA MeTase activity. Our previous work has shown that wild-type FV3 is unable to replicate in the presence of azacytidine because viral DNA is cleaved in the presence of the drug (12), whereas mutant DNA is not. Therefore the phenotype of azaC resistance is due to the lack of an endonuclease rather than a lack of MeTase. Also, since a mutation to both R- and M- could occur in a single step, the mutation is probably in the DNA recognition subunit of the molecule. Our working hypothesis is that the 26 kD polypeptide is the DNA recognition subunit, and the 18 kD polypeptide contains DNA MeTase activity. Presumably, the 30 kD polypeptide bears the endonuclease activity. A necessary corollary to this hypothesis is that the 18 kD subunit from the azaC^r Class 2 mutant will display MeTase activity when combined with the 26 kD polypeptide from the wild-type virus. Experiments are presently underway to test this hypothesis.

Also in progress is the cloning and sequencing of the 26 kD polypeptide. About 40 amino acids from cyanogen bromide-cleaved internal sequences of the polypeptide have been identified. Degenerate oligonucleotides corresponding to these amino acid sequences have been prepared, end-labeled with [^{32}P], and hybridized to a Southern blot of restriction endonuclease digests of FV3 DNA. A single signal, located on HindIII fragment C (or XbaI A, and spanning the junction between KpnI fragments K and G_2), has been observed (unpublished observations). This region contains a high degree of homology with the adenovirus 72 kD DNA-binding protein, supporting our hypothesis that this subunit is the DNA recognition portion of the complex.

Interestingly, no signal whatever is seen when FV3 DNA is probed with oligonucleotides complementary to the conserved regions of sequenced bacterial or mammalian DNA MeTases (unpublished observations). Therefore, the FV3 DNA MeTase complex appears different from other bacterial or eukaryotic enzymes with similar activities that have been sequenced to date, not only in its size and multi-subunit structure, but in the sequence of the presumed active site.

Transcription of Methylated FV3 DNA

Site-specific DNA methylation has a strong silencing effect on genes transcribed by eukaryotic RNA polymerase II (9), yet FV3 uses this enzyme to transcribe immediate-early mRNA from its highly methylated genome (6). Since virtually every cytosine in the dinucleotide sequence dCpdG is methylated in the FV3 genome (3), either the promoters do not contain such sequences in critical sites, or the virus has evolved a mechanism for overriding the normal inhibitory effect of cytosine methylation on RNA polymerase II activity. The evidence to be described below supports a combination of these two mechanisms, in which the immediate-early (IE) promoters are not methylated at critical sites, and an FV3 IE gene product allows transcription of genes normally inhibited by methylation.

To test for the existence of an FV3 trans-acting protein that would permit transcription from methylated templates, we made use of a plasmid containing the adenovirus 12 E1A promoter attached to the gene for the bacterial enzyme, chloramphenicol acetyltransferase

(CAT), pAd12E1A-cat. This plasmid had been previously shown to be incapable of promoting transcription when it was methylated in vitro with the bacterial enzymes HpaII and HhaI before being transfected into mammalian cells (13). We found that if HeLa cells were transfected with the methylated pAd12E1A-cat and then infected with FV3, CAT activity equivalent to that seen in cells transfected with the unmethylated plasmid could be induced (14). This effect was only seen when the transfected, infected cells were incubated at 30°C, the optimum temperature for FV3 growth; no trans-activation of the methylated promoter was observed when the cultures were incubated at 37°C, a temperature that is non-permissive for FV3 replication (14).

There were several possible explanations for the revival of transcription from a template silenced via methylation. First, the template could have become "demethylated," perhaps by excision of the methylated cytosine and repair with the unmethylated base, as Razin had shown to take place during the induction of Friend leukemia cells (15). Second, a transcription factor or a new polymerase could be induced by FV3 that recognized a cryptic site within the plasmid or the E1A promoter.

To rule out the first possibility, DNA was extracted from transfected, infected cells and digested with HpaII or MspI. The digestion products were electrophoresed on an agarose gel, blotted to a nylon membrane, and hybridized with a radioactive probe of the E1A region. The results showed that the input methylated template remained resistant to HpaII digestion even after infection with FV3, indicating that no wholesale demethylation followed FV3 infection (14). However, since HpaII cannot cut hemi-methylated DNA (16), we do not know if perhaps one of the two template strands was demethylated while the other remained methylated.

The second possibility (a cryptic promoter) was investigated in two ways. First, the methylated plasmid without the adenovirus promoter sequences could not be activated by FV3, so no sequences within the plasmid were being used to promote transcription (14). Primer extension analysis of mRNA with a complementary oligonucleotide 40 nucleotides from within the cat coding region demonstrated that transcription from the methylated template was initiated at the same sites after FV3 infection as in the unmethylated template (14).

Methylation of the pAd12E1A-cat template with MspI methylase, which methylates the external C in the CCGG sequence, was also inhibitory to transcription; infection with FV3 could overcome this inhibition as well (17). However, methylation with the adeninine-specific methylase EcoRI, while inhibitory to transcription, could not be overridden by FV3 infection (18). This result indicated that the FV3 trans-acting factor was specific for C methylation, although the methylated C could be in the dinucleotide sequence CC as well as CG (17)--recall, however, that the FV3 DNA MeTase did not methylate C in the CC sequence (3). The ability to transcribe methylated DNA was not limited to the E1A promoter, which has a canonical TATA box sequence (19); expression of CAT driven from the early E2A promoter, which lacks a TATA box (19) was also inhibited by methylation, and this inhibition could be overcome by infection with FV3 (14).

Purified FV3 DNA is not infectious; however, it can be non-genetically reactivated by UV-inactivated virus (20). A protein present in the virus particle itself is required for the initiation of transcription of at least one, and presumably all, of the FV3 immediate-early (IE) genes (21). A plasmid containing 78 base pairs of the promoter for the IE gene ICR169, when attached to the cat gene, will not induce the synthesis of CAT mRNA when transfected into susceptible cells; this can only be accomplished if the cells are subsequently treated with FV3 (21). However, UV-inactivated FV3 is equally effective in turning on transcription, and CAT mRNA is synthesized in the presence of inhibitors of protein synthesis (21).

Additional experiments with the FV3 induction of CAT synthesis from methylated AdE1A promoters showed that no CAT enzyme or CAT-specific mRNA was made in FV3-infected cells that had been transfected with the methylated pAd12E1A-cat in the absence of protein synthesis, or in transfected cells treated with UV-inactivated virus (14). Therefore, the FV3 protein responsible for overriding the inhibitory effect of methylation on transcription was not the same as the virion protein required to initiate IE viral mRNA synthesis, but was instead a viral-induced gene, probably an FV3 IE gene product.

Since the synthesis of IE mRNA talkes place in the absence of protein synthesis or after infection with UV-inactivated virus--conditions under which the input genomic DNA remains methylated (3)--the methylated genome must serve as a template for IE FV3 mRNA synthesis. Therefore, the IE genes cannot require the protein that overrides the inhibitory effect of methylation, either because the promoters are not methylated, or because the dCpdG sequences are located at sites that are not critical to transcription. The promoter of ICR169 contains three dCpdG doublets (21), the two proximal of which have been shown by genomic sequencing to be methylated (22). By changing one base surrounding each of the three dCpdG sequences, we were able to convert all of them to either HpaII or HhaI restriction sites. Similar to what we saw with the wild-type ICR169 promoter attached to the cat gene, the mutated ICR169 promoter was transcribed only when the transfected cells were subsequently infected with live or UV-inactivated virus, showing that the mutations did not alter the ability of the virion-associated transcriptional activator to promote transcription from the ICR169 promoter. More importantly, methylation of the ICR169 promoter with the bacterial methylases before transfection did not affect the ability of these genes to be activated by FV3 (22). Therefore, the dCpdG sequences are not in critical positions in the ICR169 promoter, and the response of the promoter to the virion-associated protein, although sequence-specific, is independent of its methylation status.

DISCUSSION

We are thus left with the problem of what role the trans-acting protein that overrides the inhibitory effect of methylation plays in the regulation of FV3 transcription, and what its mechanism of action might be. We know that separate virus-induced proteins are needed to turn of delayed early and late transcription (23). A direct effect on specific promoters is dubious, since the nonmethylated FV3 mutant undergoes a normal temporal transcriptional program (5). However, the growth cycle of the mutant is substantially prolonged, and the final yield of virus is from 20-30% of that obtained with the wild-type virus. The role of the _trans_-acting protein for transcribing methylated DNA may be simply that of a facilitator, acting either directly with the methylated template, or by altering the host RNA polymerase II so that it is no longer inhibited by methylation of the template.

Recent work from the laboratory of Cedar has shown that completely methylated plasmids that have been stably integrated into non-expressing cells enter a transcriptionally inactive state as measured by DNAse I hypersensitivty (24). This state can be overcome in expressing cell lines, even without "demethylation", suggesting that the expressing cells contain a _trans_-acting factor that overcomes repressor molecules coating the gene. This is one possible mechanism for the action of the FV3-induced _trans_-acting factor.

On the other hand, it may be that the FV3 factor alters the host cell polymerase. Evidence supporting this hypothesis comes from early studies of Campidelli-Fiume et al. (25), who showed that RNA polymerase II isolated from FV3-infected cells has lost its ability to bind radioactive α-amanitin. FV3 does code for both protein kinase and protein phosphatase (reviewed in 26), and the role of phosphate groups in altering enzyme activity is well known.

FV3 is clearly a most unusual eukaryotic virus, and its DNA MeTase enzyme complex is very different from that of methyltransferases from all other animal sources. The enzyme complex appears to be a restriction-modification system, but one with only a dinucleotide sequence specificity. The real function of FV3 DNA methylation, in fact, may be in DNA replication and recombination, for double mutants (temperature-sensitive plus $azaC^r$) are extremely defective in recombination (5).

Nevertheless, whatever the function of FV3 DNA methylation may be, the virus had to evolve a mechanism for overcoming the inhibitory effect of such methylation on transcription if it was to use the host RNA polymerase II. FV3 seems to have accomplished this feat by eliminating dCpdG sequences from the critical sequences in at least the IE promoters, and by producing a _trans_-acting factor that facilitates transcription from promoters, DE or late, that might retain important methylated sequences in critical regions. Techniques are now available to give a definitive answer to the mechanistic questions that have been raised by these observations.

ACKNOWLEDGMENTS

The authors express their appreciation to Glenith D. White for preparation of this manuscript.

REFERENCES

1. Granoff A, Came PE, Breeze DC (1966). Viruses and renal carcinoma of Rana pipiens. I. The isolation and properties of virus from normal and tumor tissue. Virology 29:133.
2. Willis DB, Granoff A (1980). Frog virus 3 DNA is heavily methylated at CpG sequences. Virology 107:250.
3. Willis DB, Goorha R, Granoff A (1984). DNA methyltransferase induced by frog virus 3. J Virol 49:86.
4. Jones PA, Taylor, SM (1980). Cellular differentiation, cytidine analogs, and DNA methylation. Cell 20:85.
5. Goorha R, Antol, K (1989). Methylation of frog virus 3 DNA: Its role in restriction-modification and genomic replication. Virology (in press).
6. Goorha R (1981). Frog virus 3 requires RNA polymerase II for its replication. J Virol 37:496.
7. Goorha R (1982). Frog virus 3 DNA replication occurs in two stages. J Virol 43:519.
8. Adams RLP, McKay EL, Craig LM, Burdon, RH (1979). Mouse DNA methylase:methylation of native DNA. Biochim Biophys Acta 561:345.
9. Razin A, Riggs AD (1980). DNA methylation and gene function. Science 210:604.
10. Essani K, Goorha R, Granoff A (1987). Mutation in a DNA-binding protein reveals an association between DNA-methyltransferase activity and a 26,000-Da polypeptide in frog virus 3 infected cells. Virology 161:211.
11. Essani K, Goorha R, Granoff A (1988). An animal virus-induced DNA methyltransferase. Gene 74:71.

12. Goorha R, Granoff A, Willis DB, Murti KG (1984). The role of DNA methylation in virus replication: Inhibition of frog virus 3 replication by 5-azacytidine. Virology 138:94.
13. Kruczek I, Doerfler W (1983). Expression of the chloramphenicol acetyltransferase gene in mammalian cells under the control of adenovirus type 12 promoters: Effect of promoter methylation on gene expression. Proc Natl Acad Sci USA 80:7586.
14. Thompson JP, Granoff A, Willis DB (1986). Transactivation of a methylated adenovirus promoter by a frog virus 3 protein. Proc Natl Acad Sci USA 83:7688.
15. Razin A, Szyf M, Kafri T, Roll M, Giloh H, Scarpa S, Carott D, Cantoni GL (1986). Replacement of 5 methylcytosine by cytosine--a possible mechanism for transient DNA demethylation during differentiation. Proc Natl Acad Sci USA 83:2827.
16. Saluz HP, Jiricny J, Jost JP (1986). Genomic sequencing reveals a positive correlation between the synthesis of a strand-specific demethylation in the regulatory region of the chicken vitellogenin II gene. Proc Natl Acad Sci USA 83:7167.
17. Thompson JP, Granoff A, Willis DB (1987). Infection with frog virus 3 allows transcription of DNA methylated at cytosine but not adenine residues. Virology 160:275.
18. Corden J, Wasylyk B, Buchwalder A, Sassone-Corsi P, Kedinger C, Chambon P (1980). Promoter sequences of eukaryotic protein-coding genes. Science 209:1406.
19. Mathis D, Elkaim R, Kedinger C, Chambon P (1982). Specific *in vitro* initiation of transcription on the adenovirus type 2 early and late EII transcription units. Proc Natl Acad Sci USA 78:7383.
20. Willis DB, Goorha R, Granoff A (1979). Nongenetic reactivation of frog virus 3 DNA. Virology 98:476.
21. Willis D, Foglesong D, Granoff A (1984). Nucleotide sequence of an immediate-early frog virus 3 gene. J Virol 52:905.
22. Thompson JP, Granoff A, Willis DB (1988). Methylation of a promoter for an immediate-early frog virus 3 gene does not inhibit transcription. J Virol 62:4680.

23. Willis DB, Goorha R, Miles M, Granoff A (1977). Macromolecular synthesis in cells infected by frog virus 3. VII. Transcriptional and post-transcriptional regulation of virus gene expression. J Virol 24:326.
24. Keshet I, Lieman-Hurwitz J, Cedar H (1986). DNA methylation affects the formation of active chromatin. Cell 44:535.
25. Campadelli-Fuime G, Costanzo F, Foa-Tanasi L, LaPlaca M (1975). Modification of cellular RNA polymerase II after infection with frog virus 3. J Gen Virol 27:341.
26. Willis DB, Goorha R, Chinchar VG (1985). Macromolecular synthesis in cells infected by frog virus 3. Curr Topics Micro Immunol 116:77.

THE ROLE OF DNA METHYLATION IN HIV LATENCY

D.P. Bednarik, J.A. Cook, and P.M. Pitha[1]

The Johns Hopkins University
Oncology Center
[1]Department of Molecular Biology and Genetics
Baltimore, Maryland 21205

ABSTRACT Infection of cells by HIV can result in a period of quiescence or latency which may be obviated by inducing agents or <u>trans</u>-acting factors. The mechanism of virus reactivation and release from latency are not completely understood, but are clinically relevant to the progression of AIDS disease. Evidence from these experiments have established the existence of two CpG sites mapped to -146 and -218 base pairs in the HIV LTR which can silence transcription of both reporter genes (CAT) and infectious virus products when enzymatically methylated. This transcriptional block was consistently overcome by the presence of <u>tat</u>. Two putative nuclear transcriptional factors (NPI, NPII) have been identified by mobility shift analysis. NPI binds specifically to these sites and is inhibited by CpG methylation. NPII is produced in cells expressing the HIV <u>trans</u>activator <u>tat</u>, and appears to prevent methylation-mediated inhibition of NP binding. Emphasis is stressed on the understanding of the modulatory role played by the methylation of HIV DNA, a process implicated in the establishment of viral latency.

INTRODUCTION

In humans, HIV infection is characterized by a

This work was supported by the American Foundation for AIDS Research (000639 DPB).

period of latency followed by progression to AIDS or AIDS-related complex (ARC) in some cases (1,2). Factors that influence HIV latency are poorly understood. Several models have been proposed which might explain how HIV, when harbored in a latent form, can be induced by physiochemical stimuli and be expressed as infectious virus particles (3-5). These models include transcriptional repression of integrated proviral DNA by either DNA-binding proteins, chromatin conformation or DNA hypermethylation.

HIV-infected T-cells can be maintained *in vitro* for extended periods without release of reverse transcriptase activity (4,5). Only after antigenic stimulation by agents such as phytohemagglutinin or phorbol esters, does virus replication and release of reverse transcriptase occur (4). The expression of *tat* is believed to play a role in this escape from latency to lytic replication.
The release of HIV-infected cells from latency after antigenic activation must result first in the initial induction of viral mRNA synthesis, initiated from the viral LTR, in order for the *tat* gene to be expressed. If the viral LTR was inactivated early after integration into host chromatin, viral mRNA would not be synthesized and the *tat* gene product would not accumulate. In this state, latency would be preserved. An environmental stimulus may release the transcriptional block, thereby permitting induction of mRNA synthesis and escape from latency.

Control of cellular and viral gene expression has been shown to be modulated by DNA methylation (6-8). Extensive evidence has implicated the enzymatic conversion of cytosine to 5-methylcytosine in the dinucleotide sequence CpG as a mechanism for transcriptional inhibition of genes (9,10). Evidence from *in vitro* methylation/transfection experiments demonstrate that transcription of genes is inhibited when CpG islands are methylated (9,11). What determines whether a CpG island is methylated or unmethylated? Little is known about the molecular events leading to *de novo* methylation of DNA, but data imply that most cell types are capable of methylating accessible CpGs. Only methylation of sequences in the upstream promoter region render genes transcriptionally inactive (9).

Viruses are relevant to this concept since they contain sequences which are rich in CpG islands. Retroviral LTRs, adenovirus and Herpes viruses are G + C rich, and moreover, both adenovirus and Moloney retrovirus are inactivated by DNA methylation (9,12-14). Reactivation of latent viruses have been demonstrated by experiments which employ the use of methylation antagonists, such as 5-azacytidine. Many gene sequences as well as virus promoters depend upon induction of transcription by a protein which acts in trans. Promoter sequences which are hypermethylated demonstrate a greatly reduced capacity to bind some of these nuclear proteins (15), however, the same sequences are able to form complexes with protein and show a nuclear protein binding capacity of 100% when the DNA is hypomethylated by 5-azaC treatment (15). Recently, trans-acting factors of the adenovirus system have recently been shown to overcome transcriptional inhibition by methylation without changing the methylation pattern (16).

Although considerable progress has been made in defining the genetic structure of HIV and delineating the complex array of genes encoded by the 9.7 kilobase RNA transcript of this virus, the precise role these genes play during HIV infection is unknown. The methylation-sensitive sites in the HIV promoter are within the boundaries of the enhancer/TAR region and are therefore influenced by trans-acting factors which interact with these regions of DNA. Infection/transformation of cells by Herpes Simplex virus has recently been shown to cause hypomethylation of host cell DNA (17). We have demonstrated that tat, Herpes virus IE110 (18,19), and ultraviolet light (20,21) can reactivate latent HIV with kinetics similar to treatment with 5-azaC (22) and thereby implicate hypomethylation as a possible common mechanism of reactivation.

METHODS

Cells and Viruses. Vero cells (simian), SW480 (human) adenocarcinoma), and A3.01 ($CD4^+$, human CEM T-cell line) were grown in either DMEM (Gibco) supplemented with 5% fetal bovine serum (Vero, SW480) or OPTI-MEM (Gibco) supplemented with 2% fetal bovine serum (A3.01). A3N92.2 cells, containing a single copy of HIVCATκB mutant and pSV_2Neo, were constructed by electroporation of cells with each plasmid DNA (10 μg and 1.0 μg, respectively) followed by selection of colonies in soft agar containing G418 sulfate (1 mg/ml), and maintenance in OPTI-MEM media as described above. Stocks of HIV were generated by electroporation of A3.01 cells with 10 µg of infectious clone DNA (pHCBX2). HIV concentration was determined by p24 antigen capture (Abbott) (23). De novo infection of A3.01 cells was initiated by inoculating 5 x 10^6 cells with 1000 pg p24 antigen.

Plasmid DNA and transfection/electroporation of cells. The plasmids pU3RIIICAT (HIVLTRCAT), pIIIextatIII (HIV LTRtat), pHCBX2, and pHIVκBCAT (courtesy Drs. G. Nable and D. Baltimore) were constructed as previously described (18,19,22). Fibroblast cell lines were transfected by either calcium phosphate coprecipitation (18,19,22) or lipofection (24). Transfection of lymphoid cells was performed by either electroporation (Promega X-cell 450) or lipofection (24).

Chloramphenicol acetyltransferase assay (CAT) and RNA analysis. Cell extracts were assayed for CAT enzyme activity as described (18,19,22). Isolation of total RNA, and analysis was performed by S1 nuclease and Northern hybridization as described previously (18,19,22).

In vitro methylation of plasmid DNA. Plasmid DNA was enzymatically methylated by protocols prescribed by the vendor (New England Biolabs).

Preparation of nuclear protein extracts, nuclei, and analysis of nuclear protein binding (mobility shift assay). Isolation of nuclei and preparation of nuclear extracts was performed by the method of Raj et al. (25). The DNA-binding assay and gel electrophoresis was performed by the methods of Baldwin and Sharp (26).

Polymerase chain reaction (PCR). PCR was performed by employing a GeneAmp DNA amplification reagent kit (Perkin Elmer Cetus Corp., Norwalk, CT). Reactions were accomplished by protocols prescribed by the vendor. Oligonucleotide primers complementary to regions spanning the +46 to +66 nucleotides of the 3' coding strand and -352 to -332 nucleotides of the 5' non-coding strand were synthesized chemically (Genetic Designs, Inc., Houston, TX).

RESULTS

Reactivation of the latent HIV LTR. Previous studies have tested the possibility that methylation of the HIV LTR was involved in suppression of LTR expression

(22). In cell lines permanently transfected with the HIV LTR directing CAT expression (pU3RIIICAT), LTR sequences were demonstrated to be inactivated by hypermethylation, which could be reactivated by treatment with 5-azacytidine (22), or by exposure to ultraviolet light (20,21). Both of these agents are well known DNA methylation antagonists and each site within the HIV LTR sensitive to reactivation by either agent was mapped by Bal31 digestion (21). The inductive effect of 5-azacytidine was lost upon deletion of the 5' region of the LTR to -57/+80 nucleotides (data not shown). Deletion of the 3' untranslated region (TAR) had no effect on induction by UV light or 5-azacytidine, thereby eliminating the possibility that the mechanism of activation by each inducer was similar to that of the HIV-encoded tatIII gene product (21). The physiological relevance of DNA methylation as a virus latency preservation mechanism is evident from data shown in Figure 1. The introduction of provirus into the human

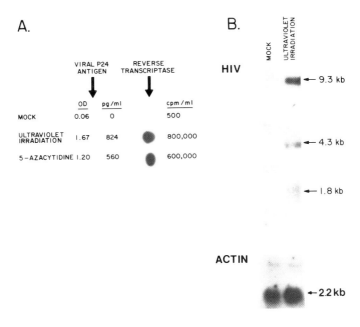

FIGURE 1. Activation of latent HIV in a clonal CEM T-cell line.

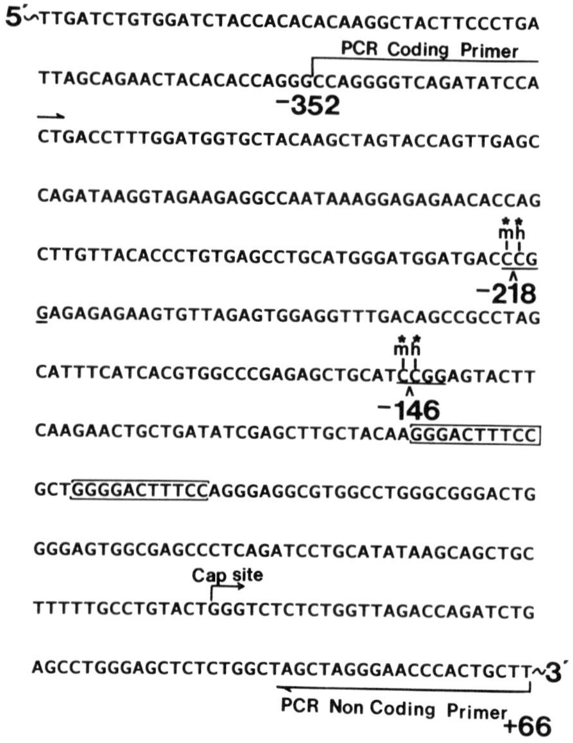

```
5'⌒TTGATCTGTGGATCTACCACACACAAGGCTACTTCCCTGA
                            PCR Coding Primer
TTAGCAGAACTACACACCAGGGCCAGGGGTCAGATATCCA
     →              -352
CTGACCTTTGGATGGTGCTACAAGCTAGTACCAGTTGAGC

CAGATAAGGTAGAAGAGGCCAATAAAGGAGAGAACACCAG
                                      m h
                                      | |
CTTGTTACACCCTGTGAGCCTGCATGGGATGGATGACCCG
                                      ^
                               -218
GAGAGAGAAGTGTTAGAGTGGAGGTTTGACAGCCGCCTAG
                           m h
                           | |
CATTTCATCACGTGGCCCGAGAGCTGCATCCGGAGTACTT
                              ^
                     -146
CAAGAACTGCTGATATCGAGCTTGCTACAAGGGACTTTCC

GCTGGGACTTTCCAGGGAGGCGTGGCCTGGGCGGGACTG

GGGAGTGGCGAGCCCTCAGATCCTGCATATAAGCAGCTGC
                  Cap site
                      →
TTTTTGCCTGTACTGGGTCTCTCTGGTTAGACCAGATCTG

AGCCTGGGAGCTCTCTGGCTAGCTAGGGAACCCACTGCTT⌒3'
                            PCR Non Coding Primer
                                             +66
```

FIGURE 2. PCR-amplified target region within the HIV LTR for in vitro methylation and probe synthesis.

CEM T-cell line A2.01, and subsequent cloning resulted in the establishment of the latent cell population denoted A2.HIV.7 (21). While essentially no virus production was observed constitutively, virus production was readily observed after treatment of cells for 24 hours with 10 μM 5-azacytidine or after irradiation with ultraviolet light ($12J/m^2$) (21). Virus expression was measured at the protein level (reverse transcriptase, p24 antigen; Fig. 1, panel A), and at the transcriptional level (mRNA panel B).

Determination of methylation sensitive sites in the HIV LTR. As a logical extension of these studies, methylation sensitive CpG and non-CpG sites were examined for their ability to inactivate HIV LTR expression subsequent to in vitro enzymatic methylation. Methylated plasmid DNA for transient transfection assay was prepared

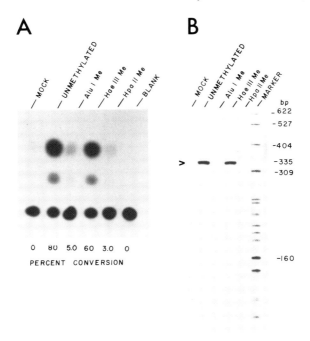

FIGURE 3. Inhibition of HIV LTR transactivation and expression by site-specific DNA methylation.

by methylation of 20 µg pU3RIIICAT (HIV CAT) or pIIIextatIII (HIV tat) DNA with commercially available DNA methylase enzymes (New England Biolabs) as prescribed by the vendor. DNA fragments used as probes (Fig. 2) for transcription factor-binding assays were synthesized by ^{32}P-end labeling the 5' coding and 3' non-coding synthetic oligomers (solid arrows) followed by DNA amplification via the polymerase chain reaction (PCR; Perkin-Elmer Cetus Corp.) to generate a 418 base pair DNA probe. Probe DNA was subsequently methylated enzymatically at MspI (m) or HpaII (h) sites as described above. The boxed region denotes the sequences of NF-κB protein binding sites. The use of PCR to generate probe DNA is convenient since this process guarantees a DNA fragment free of 5-methylcytosine. The sequence depicted in Figure 2 contains eight CpG sites bordered by the PCR primer regions, two of which are MspI/HpaII restriction sites. Plasmid DNA encoding the HIV trans-acting factor tat (pIIIextatIII/HIV LTRtat) was enzymatically methylated at either AluI (AGCT), HaeIII (GGCC), or HpaII (CCGG) sites under conditions according to the vendor (New England Biolabs). The efficiency of the methylation reaction was determined by restriction with the corresponding endonuclease. Methylated plasmid DNA (10 µg) was purified and transfected by electroporation into A3N92.2 cells which is a CEM T-cell line containing a single integrated copy of HIV LTRCAT DNA (pU3RIIICAT). Figure 3, panel A demonstrates the effect of in vitro methylation on specific sites within the LTR. Lanes from left are mock transfection, unmethylated control, AluI methylated, HaeIII methylated, HpaII methylated, and blank. The corresponding percent conversions are indicated below each lane. Panel B shows S1 nuclease transcriptional analysis of CAT mRNA induced in response to methylated or unmethylated HIV LTRtat (pIIIextatIII). Lanes are identical to panel A. The arrow indicates a 335 base pair protected fragment corresponding to a properly initiated LTR CAT transcript. Synthesis of CAT enzyme by this cell line is negligible unless the LTR is activated in trans by the appropriate inducing agent (21).

The function of tat in the presence of methylated, transcriptionally inactive DNA. Previous studies by our laboratory suggested that various trans-acting agents such as tat or herpes virus infection activate the silent HIV LTR with a corresponding change in DNA methylation patterns (18-22). Elegant work by Doerfler et al (16) also indicates that trans-acting factors of the adenovirus system are capable of overcoming a transcriptional block introduced by in vitro methylation. We also studied the function of the HIV transactivator, tat, on LTR sequences which were methylated in vitro. Transient transfection assays were performed on Vero (simian) cell monolayers by transfection of enzymatically methylated HIV LTRCAT (pu3RIIICAT) plasmid DNA (10 µg) either alone or in combination with 2 µg LTRtat (pIIIextatIII). Induction of CAT protein synthesis 24 hours post-transfection in the presence or absence of tat is observed in Figure 4, panel A. Lanes from left are mock transfection, unmethylated control, unmethylated control/tat, HphI methylated, HphI methylated/tat, and blank. The percent conversion of chloramphenicol to acetylchloramphenicol was very similar in those cells transfected with tat regardless of the presence or absence of methyl groups. The overall cis-level

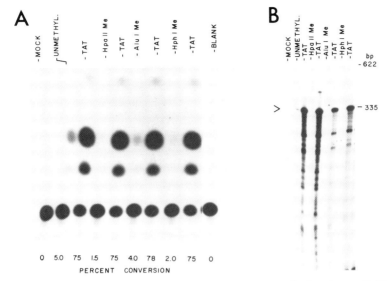

FIGURE 4. The effect of in vitro methylation of HIV LTR sequences on transactivation by tat.

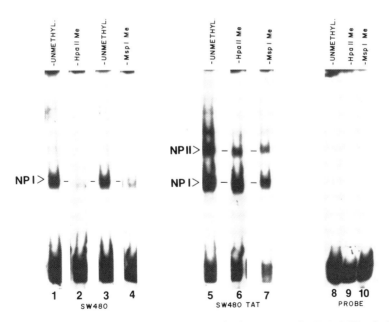

FIGURE 5. The effect of methylation of HIV LTR CpG sites -218/-146 on binding of nuclear proteins.

expression of the LTR was reduced over 3-fold when both HpaII sites (Fig. 2) were methylated. In each case, tat overcame the transcriptional block.

The effect of site-specific LTR DNA methylation on binding of host transcriptional factors. In order to assess the effect of site-specific in vitro methylation on the binding of nuclear factors to the HIV LTR, we employed the mobility shift gel assay (26), in which the mobility of a labeled DNA fragment is retarded on a non-denaturing polyacrylamide gel as a protein-DNA complex. Labeled (^{32}P) DNA fragments were incubated with a partially purified nuclear fraction obtained from SW480 cells or an SW480 cell line expressing the tat protein (SW480 tat). Binding reactions were performed utilizing a 234 base pair fragment, generated by TaqI restriction of the PCR-amplified DNA segment spanning both CpG sites (see Fig. 2). Protein-DNA complexes and uncomplexed probe DNA were resolved on a 16 cm, 4% non-denaturing polyacrylamide gel by the mobility shift assay (Fig. 5). Incubations were completed in a 1360-fold weight excess of poly(dI.dC) relative to probe with 1 µg nuclear extract protein at room temperature. Arrows denote the positions of nuclear binding proteins which retard the migration of labeled DNA probe (NPI, NPII). When LTR DNA fragments were methylated at either -218 or -146 with either HpaII or MspI methylase (Fig. 5; lanes 2, 3) the nuclear protein denoted NPI demonstrated a significantly lower binding affinity when compared to the unmethylated probe DNA (Fig. 5; lanes 1, 3). The binding affinity of the methylated probe DNA was restored when incubated with nuclear extracts isolated from SW480 cells expressing tat protein (Fig. 5; lanes 6, 7). Concomitant with this observation was the appearance of a second nuclear protein (NPII) which retarded the mobility of labeled probe. The identity of either factor is at present, unknown.

To test the specificity of nuclear protein binding to CpG sites at -218/-146, mobility shift experiments were performed as above, however, a 5-fold excess of unlabeled competitor DNA was preincubated with nuclear extracts 15 minutes prior to incubation with unmethylated, labeled DNA probe. When HpaII or MspI methylated competitor DNA was employed, essentially no competition occurred (Fig. 6; lanes 3,4). When unmethylated competitor was co-incubated with labeled probe, no binding of the probe DNA was observed (Fig. 6;

lane 2), thus confirming the specificity of protein binding.

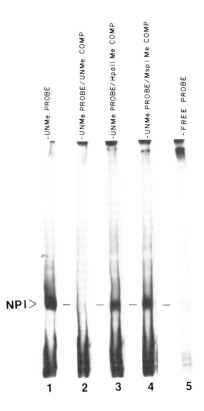

FIGURE 6. Specificity of nuclear protein binding to HIV LTR probe DNA.

DISCUSSION

Infection of individuals by HIV-1 results in an incubation time averaging 2-5 years prior to the onset of severe clinical manifestations (1,2). Early after integration of proviral DNA into the host chromatin, virus expression can become quiescent, or latent (18-22,27). Virus expression can be reactivated by exposure of infected cells to a spectrum of agents or pathogens (18-22,27), and the regions of the HIV LTR sensitive to induction by each agent have been mapped (21). The physiological mechanism of latency is not known, but is believed to involve a complex interplay between both viral and cellular host factors (27).

In the present study, we present evidence that suggests the role of a host cellular function, DNA methylation, to preserve the latent state of HIV. Prior studies have shown that methylation of cytosine residues in the controlling elements of murine retroviruses and adenovirus can create an unexpressed state of either virus (9,12-14). Our earlier work also implicated HIV as being sensitive to methylation antagonistic agents (21, 22). The regions of HIV LTR sequences sensitive to inactivation by methylation were mapped by deletion analysis to include upstream sequences containing both core enhancer/NFκB binding sites and three SpI binding sites. Within this region we have identified two HpaII/MspI CpG sites located at -218 and -146 nucleotides upstream from the cap site which, when enzymatically methylated in vitro, renders the LTR transcriptionally inactive (Fig. 3). Transfection of A3N92.2 cells with tat (pIIIextatIII) resulted in activation, in trans, of the normally silent HIV LTR (Fig. 3, panel A). However, when the HIV LTR directing tat expression (pIIIextatIII) is methylated by HpaII or MspI methylase at either position (-218/-146), production of tat does not occur and the integrated LTRCAT gene remains latent. Non-CpG specific methylases such as AluI were also observed to be effective inhibitors of transcription. In vivo, the presence of 5-methylcytosine in the integrated HIV LTR at these positions, or others, may effectively preserve latency. Such a transcriptional block would prevent cis-level transcriptional initiation, thereby preventing the accumulation of downstream-encoded viral proteins such as tat. This scenario would eliminate the involvement of viral products in regulation of the latent provirus assuming, of course, the cell does not become infected by a subsequent exposure to HIV, or secondary DNA viruses (18,19).

If, indeed, secondary infection or exposure to other trans-acting factors were to occur, would the methylated LTR inhibit the action of the putative transactivator? We have addressed this question by transient transfection of Vero cells with plasmid DNA (pU3RIIICAT) methylated with CpG-specific (HpaII) or non-CpG-specific (HphI, AluI) methylase enzymes. Co-transfection of pIIIextatIII with methylated HIVLTRCAT (pU3RIIICAT) DNA resulted in transactivation of CAT expression (Fig. 4), while cis-

level expression was nearly eliminated in the absence of
tat. This observation is in accord with the results
described by Doerfler et al. in which trans-acting
factors of the adenovirus system overcome a similar
transcriptional block introduced by in vitro methylation
of adenovirus regulatory elements (16). In order to
further examine the mechanism by which tat obviates the
methylation-induced transcriptional block, we prepared
nuclear extracts from SW480 adenocarcinoma cells or SW480
cells expressing tat protein. Incubation of nuclear
extracts with labeled probe DNA containing HpaII/MspI
sites at -218/-146 (Fig. 2) resulted in retardation of
the probe by a nuclear protein (NPI; Fig. 5) when
analyzed by electrophoretic mobility shift assay. When
either HpaII or MspI methylase was employed, the binding
affinity of NPI was reduced several-fold (Fig. 5; lanes
2, 4). However, when the same probes were incubated with
nuclear proteins from SW480 cells expressing tat, the
binding affinity of NPI remained virtually the same, and
a second factor, NPII (Fig. 5; lanes 5-7), appeared. It
is not known whether NPII plays any role in allowing the
binding of NPI, but it is tempting to speculate such a
scenario. Neither nuclear protein is the NFκB protein as
determined by mobility shift studies employing a mutant
NFκB HIV LTR in which both GGG sequences in the two
binding motifs are mutated to CTC (courtesy Drs. G. Nable
and D. Baltimore; data not shown). These observations
are consistent with the obviation of the methylation-
induced transcriptional block by tat (Fig. 4). One might
speculate that tat, either alone or in concert with other
factors, may orchestrate activation of transcription in
trans regardless of the methylation state. In the
absence of a transactivation, cis-level expression would
remain inactive as long as the LTR was methylated,
thereby preserving latency. One may envision the
hypermethylation of HIV LTR CpG sequences perhaps prior
to or after integration into host DNA. In such a model,
the expression of virus would be minimal to non-existent
until cellular activation occurred, or until a
heterologous trans-acting factor was encountered. We
believe that the results described in this study support
DNA methylation as a viable control mechanism for
latency. Further studies, currently in progress, will
elucidate the methylation profile of integrated,
latent/active HIV by genomic sequencing techniques. This
information will contribute to our understanding of the

cellular control of retrovirus expression and to the mechanism of transactivator function.

ACKNOWLEDGEMENTS

We acknowledge Drs. H. P. Saluz and N. B. K. Raj for helpful discussions. We thank J. Simkins for technical assistance, and B. Schneider for excellent preparation of the manuscript.

REFERENCES

1. Lui K-J, Lawrence DN, Morgan WM, Peterman TA, Haverkos HW, Bregman DJ (1986). A model-based approach for estimating the mean incubation period of transfusion-associated acquired immunodeficiency syndrome. Proc Natl Acad Sci USA 83:3051.

2. Medley GF, Anderson RM, Cox DR, Billard L (1987). Incubation period of AIDS in patients infected via blood tranfusion. Nature (London) 328:719.

3. Folks TM, Powell DM, Lightfoote MM, Benn S, Martin MA, Fauci AS (1986). Induction of HTLV-III/LAV from a nonvirus-producing T-cell line: Implications for latency. Science 231:600.

4. Fauci AS (1988). The human immunodeficiency virus: Infectivity and mechanisms of pathogenesis. Science 239:617.

5. Folks TM, Justement J, Kinter A, Dinarello CA, Fauci AS (1987). Cytokine-induced expression of HIV-1 in a chronically infected promonocyte cell line. Science 238:800.

6. Doerfler W (1983). DNA methylation and gene activity. Annu Rev Biochem 52:93.

7. Doerfler W (1984). DNA methylation and its functional significance: Studies on the adenovirus system. Curr Topics Microbiol Immunol 108:79.

8. Keshet I, Leiman-Hurwitz J, Cedar H (1986). DNA

methylation affects the formation of active chromatin. Cell 44:535.

9. Bird AP (1986). CpG-rich islands and the function of DNA methylation. Nature (London) 321:209.

10. Lindsay S, Bird AP (1988). Use of restriction enzymes to detect potential gene sequences in mammalian DNA. Nature (London) 327:336.

11. Keshet I, Yisraeli J, Cedar H (1985). Effect of regional DNA methylation on gene expression. Proc Natl Acad Sci USA 82:2560.

12. Kruczek J, Doerfler W (1983). Expression of chloramphenicol acetyltransferase gene in mammalian cells under control of adenovirus type 12 promoters: Effect of promoter methylation on gene expression. Proc Natl Acad Sci USA 80:7586.

13. Harbers K, Schieke N, Stuhlmann H, Jahner D, Jaenisch (1981). DNA methylation and gene expression: Endogenous retroviral genome becomes infectious after molecular cloning. Proc Natl Acad Sci USA 78:7609.

14. Graessmann M, Graessmann A, Wagner H, Werner E, Simmon D (1983). Complete DNA methylation does not prevent polyoma and simian virus 40 early gene expression. Proc Natl Acad Sci USA 80:6470.

15. Michalowsky LA, Jones PA (1987). Differential nuclear protein binding to 5-azacytosine-containing DNA as a potential mechanism for 5-aza-2' deoxy cytidine resistance. Mol Cell Biol 7:3076.

16. Weissharr B, Langner KD, Juttermann R, Muller U, Zock C, Klimkait T, Doerfler W (1988). Reactivation of the methylation-inactivated late E2A promoter of adenovirus type 2 by EIA (13S) functions. J Mol Biol 202:255.

17. Macnab JC, Adams RL, Rinaldi A, Orr A, Clark L (1988). Hypomethylation of host all DNA synthesized after infection or transformation of cells by herpes simplex virus. Mol Cell Biol

8:1443.

18. Mosca JD, Bednarik DP, Raj NBK, Rosen CA, Sodroski JG, Haseltine WA, Pitha, PM (1987). Herpes simplex virus type-1 can reactivate transcription of latent HIV. Nature (London) 325:67.

19. Mosca JD, Bednarik DP, Raj NBK, Rosen CA, Sodroski JG, Haseltine WA, Hayward GS, Pitha PM (1987). Activation of human immunodeficiency virus by herpesvirus infection: Identification of a region within the long terminal repeat that responds to a trans-acting factor encoded by herpes simplex virus 1. Proc Natl Acad Sci USA 84:7408.

20. Valerie K, Delers A, Bruck C, Thiriart C, Rosenberg H (1988). Activation of human immunodeficiency virus type-1 by DNA damage in human cells. Nature (London) 333:78.

21. Mosca JD, Bednarik DP (in preparation).

22. Bednarik DP, Mosca JD, Raj, NBK (1987). Methylation as a modulator of expression of human immunodeficiency virus. J Virol 61:1253.

23. Higgins JR, Pederson NC, Carlson JR (1986). Detection and differentiation by sandwich enzyme-linked immunosorbent assay of human T-cell lymphotropic virus typeIII/lymphadenopathy-associated virus and acquired immunodeficiency syndrome-associated retrovirus-like clinical isolates. J Clin Microbiol 24:424.

24. Felgner PL, Holm M (1989). Cationic liposome-mediated transfection. Focus 11:21.

25. Raj NBK, Pitha PM (1983). Two levels of regulation of β-interferon gene expression in human cells. Proc Natl Acad Sci USA 80:3923.

26. Baldwin AS, Sharp PA (1987). Binding of a nuclear factor to a regulatory sequence in the promoter of the mouse H-2KB class I major histocompatibility gene. Mol Cell Biol 7:305.

27. Folks TM, Clouse KA, Justement J, Rabson A, Duh E, Kehrl JH, Fauci AS (1989). Tumor necrosis factor α induces expression of human immunodeficiency virus in a chronically infected T-cell clone. Proc Natl Acad Sci USA 86:2365.

THE ROLE OF DNA METHYLATION IN A RETROVIRUS-INDUCED INSERTIONAL MUTATION OF THE MURINE ALPHA 1(I) COLLAGEN GENE

Michael Breindl, Hedy Chan, and Stefan Hartung

Department of Biology and Molecular Biology Institute, San Diego State University, San Diego, California 92182, USA

SUMMARY

We are studying the expression of the murine alpha 1 type I procollagen (COL1A1) gene using the Mov13 mutation as a tool. In Mov13 mice, a retroviral genome inserted into the first intron of the COL1A1 gene prevents its expression and causes an embryonic lethal mutation. We show here that in Mov13 cells the provirus insertion appears to prevent the developmentally regulated demethylation of the COL1A1 gene, confirming and extending earlier observations. Furthermore, demethylation studies with 5-aza-cytidine and DNA transfection and microinjection experiments indicate that DNA methylation plays no causal role in the retrovirus-induced suppression of the COL1A1 gene in Mov13 cells but rather stabilizes a state of transcriptional repression which is determined and maintained by other factors.

INTRODUCTION

A large body of evidence indicates that DNA methylation at CpG dinucleotides plays an important role in the regulation of tissue-specific gene expression (1,2), X chromosome inactivation (3) and parental imprinting (4,5), although its precise function has yet to be determined. Many tissue-specific genes show a basic correlation between their methylation status and transcriptional activity: they are usually hypomethylated in tissues in which they are expressed and hypermethylated in tissues in which they are not expressed. It is not clear, however, whether the methylation status of a gene is cause or consequence of its transcriptional activity. Activation of silent genes by treatment of cells with the demethylating agent 5-azacytidine (5-aza-C, ref. 6,7) and transfection experiments using methylated and unmethylated DNA (2,8) imply a causal role of DNA methylation in gene regulation. It has been suggested, that DNA methylation interferes with transcription by altering DNA-protein interactions and affecting the formation of active chromatin (9,10), although other studies show that the DNA-binding activity of transcription factors is not affected by methylation (11,12). In many cases the expression of tissue-specific genes does not correlate with their methylation status (13,14), suggesting that DNA methylation plays no causative role in the regulation of transcription but is a secondary event which stabilizes and propagates a state of transcriptional repression or competence determined by other factors.

We are analyzing the molecular mechanisms involved in regulating the stage and tissue-specific regulation of expression of the murine alpha 1 type I procollagen (COL1A1) gene and its suppression in Mov13 mutant cells. In Mov13 mice, insertion of a Moloney murine leukemia virus (M-MuLV) proviral genome into the first intron of the COL1A1 gene interferes with expression of the gene and causes an embryonic lethal mutation (15,16). We have previously analyzed the chromatin structure (17) and promoter activity (18) of the wild-type COL1A1 gene and the mutant Mov13 allele. These studies have shown that the provirus insertion in Mov13 does not inactivate COL1A1 gene expression by causing premature termination of transcription, incorrect splicing, or instability of mRNA. Rather, it interferes in *cis* with early developmental events that are necessary for establishing a transcriptionally competent state of the gene, resulting in an activation of the gene at the level of transcriptional initiation. The provirus insertion in Mov13 also changes the methylation pattern of the COL1A1 gene (19). To better understand the role of DNA methylation in the regulation of COL1A1 gene expresseion in wild-type cells and its suppression in Mov13 we have determined the methylation pattern of the COL1A1 promoter region in various mouse cell lines and tissues and in Mov13 cells.

RESULTS

1. The Mov13 provirus prevents the developmentally regulated demethylation of the COL1A1 promoter.

It has previously been observed that the cellular sequences flanking the proviral integration site in Mov13, including the COL1A1 gene promoter region, are hypomethylated in wild-type fibroblasts, but hypermethylated in mutant cells, and it has been suggested that the provirus induces *de-novo* methylation of these sequences (19,21). Our more extensive studies (20) have shown that the wild-type COL1A1 promoter is methylated, and the gene transcriptionally inactive, in P19 embryonal carcinoma (EC) cells, which resemble multipotent stem cells of early mouse embryos. In adult mice the COL1A1 promoter is unmethylated and the gene is transcribed in a variety of tissues, including tissues that contain mostly cells of non-mesenchymal origin (Table 1). This indicates that during normal mouse development the COL1A1 promoter becomes demethylated in cells derived from each embryonic germ layer, and that demethylation of the COL1A1 promoter is necessary, but not sufficient, for gene expression. We further found that in Mov13 fibroblasts the COL1A1 promoter is methylated and the gene transcriptionally inactive, confirming and extending earlier observations (16,18,19). Thus, our results suggest that the provirus insertion prevents the developmentally regulated demethylation of the COL1A1 promoter rather than inducing a *de novo* methylation of these sequences.

TABLE 1
PROMOTER METHYLATION AND TRANSCRIPTION OF COL1A1 GENE IN MOUSE CELLS AND TISSUES

	P19 EC cells	fibroblasts	Mov13 fibroblasts	Mov13 5-aza-C	hepatoma	WEHI 3B	liver	spleen	kidney	brain
methylation	+	–	+	–	+	–	–	–	–	–
transcription	–	+	–	–	–	–	+	+	+	+

2. Demethylation of the promoter region is not sufficient for reactivation of the COL1A1 gene in Mov13 cells.

To study whether DNA methylation plays a causal role in the transcriptional suppression of the COL1A1 gene, we have treated Mov13 fibroblasts with 5-aza-C and analyzed the methylation pattern and transcriptional activity of the COL1A1 promoter. We found that 5-aza-C treatment leads to demethylation of the COL1A1 promoter in Mov13 cells. Furthermore, we have isolated 5-aza-C-treated Mov13 fibroblast clones in which the COL1A1 promoter showed an unmethylated pattern indistiguishable from that of collagen-producing wild type cells (20). Using RNase protection assays, we were unable to detect COL1A1 transcripts in the 5-aza-C-treated, demethylated Mov13 fibroblasts (20). Thus, the suppressed state of the COL1A1 gene in Mov13 cells appears to be stably propagated in the absence of DNA methylation and in the presence of *trans*-acting factors required for efficient COL1A1 gene transcription.

3. Proviral sequences interfere with COL1A1 promoter activity in Mov13 fibroblasts but not in *X. laevis* oocytes.

To further analyze the molecular mechanisms involved in the transcriptional suppression of the COL1A1 promoter in Mov13 cells, we have transfected cloned DNA fragments derived from both the wild-type and the Mov13 COL1A1 alleles into Mov13 fibroblasts. In a second series of experiments we injected these fragments into *Xenopus laevis* oocytes. We have found that in Mov13 fibroblasts stably transfected with a fragment derived from the wild-type COL1A1 gene, the COL1A1 promoter showed a high transcriptional activity comparable to that of the endogenous COL1A1 gene in wild-type fibroblasts (20, Table 2). This is in agreement with results of others (22) and supports the assumption that Mov13 fibroblasts contain the *trans*-acting regulatory factors required for efficient COL1A1 gene expression, i.e., that the lack of COL1A1 gene expression in 5-aza-C-treated Mov13 fibroblasts is not due to the lack of such factors, but rather to the inaccessability of regulatory sequences to these factors.

TABLE 2
TRANSCRIPTIONAL ACTIVITY OF WILD-TYPE AND MOV13
COL1A1 PROMOTERS AFTER TRANSFECTION INTO
FIBROBLASTS AND INJECTION INTO OOCYTES

	wild-type COL1A1	Mov13 COL1A1 partial	Mov13 COL1A1 complete
transfection	++++	+/-	-
oocyte injection	++++	++++	++++

When Mov13 fibroblasts were transfected with fragments derived from the Mov13 COL1A1 allele containing a partial or complete proviral genome, a strong transcriptional suppression of the COL1A1 promoter was observed (20, Table 2). In contrast, after injection into *X. laevis* oocytes the COL1A1 promoter was transcriptionally active independent of whether the injected DNA fragments contained a partial or complete proviral genome or no proviral sequences at all. Moreover, the pattern of protected RNA fragments obtained with the RNAs from injected oocytes was identical to the pattern obtained with RNA from mouse fibroblasts, indicating correct transcriptional initiation at the COL1A1 promoter in the presence of proviral sequences (20). These results clearly showed that the Mov13 COL1A1 promoter is potentially transcriptionally active, and that the presence of the proviral genome in the Mov13 COL1A1 allele does not *per se* preclude its transcriptional activity.

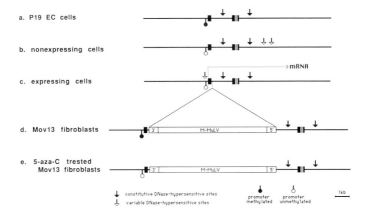

FIGURE 1. Chromatin structure, promoter activity, and DNA methylation of the COL1A1 gene in wild-type and Mov13 cells. The 5'-most exons of the COL1A1 gene are shown as solid boxes and the proviral genome in Mov13 by an open box. For details see the text.

DISCUSSION

We found that the promoter region of the murine COL1A1 gene, which is thought to be predominantly expressed in mesenchyme-derived cells, is methylated in early embryonic cells (P19 EC cells) in which it is not expressed, and that it is unmethylated in all differentiated wild-type tissues and cell types analyzed (with exception of mouse hepatoma cells), including tissues of mainly ectodermal or endodermal origin (brain and liver, respectively, Table 1). This indicates that the developmentally regulated activation of the murine COL1A1 gene is accompanied by a demethylation of its promoter region in cells derived from each embryonic germ layer. We also found that the COL1A1 gene is transcribed in various major tissues of the adult mouse (brain, liver, spleen and kidney). Our studies do not show whether the COL1A1 gene is transcribed in all cells that constitute these tissues or whether subpopulations of mesoderm-derived cells, which may provide supporting frameworks of connective tissue in these organs, are the collagen-producing cells. However, the finding that in all mouse tissues studied the promoter region of the COL1A1 promoter appears to be unmethylated in all cells suggests that the COL1A1 promoter is unmethylated in collagen-producing and non-producing cells, i.e., that its methylation pattern is not correlated with its transcriptional activity. Demethylation of the COL1A1 promoter appears to be necessary, but not sufficient, for transcriptional activity. A similar methylation pattern has been described for the chicken COL1A2 gene in which the DNA around the start site of transcription is not methylated, whether or not the cells synthesize collagen (13).

In sperm, the COL1A1 promoter region was reported to be undermethylated (21). We have shown that it is methylated in multipotent embryonic stem cells and demethylated in differentiated adult cells derived from each embryonic germ layer (20, Table 1). This indicates that a *de novo* methylation of the gene occurs during normal early mouse embryogenesis, followed by an overall demethylation, at least of the promoter region, during further development. In Mov13 fibroblasts we found that the COL1A1 promoter is methylated and the gene transcriptionally inactive, confirming previously reported results (16-18). This suggests that the provirus insertion prevents the developmentally regulated demethylation of the gene rather than inducing a *de novo* methylation. However, this conclusion is based on our study of P19 EC cells only and a final answer to this question will require the analysis of the COL1A1 gene methylation pattern in embryonic tissues.

Treatment of homozygous Mov13 fibroblasts with 5-aza-C can result in a stable demethylation of the promoter region of the mutated COLIA1 gene (20, Table 1). However, even though the methylation pattern of the COL1A1 promoter in some 5-aza-C treated Mov13 fibroblast clones was indistinguishable from that of the wild-type gene in collagen-expressing fibroblasts, demethylation did not result in transcriptional activation of the mutant gene. This indicates that the suppressed state of the COL1A1 gene in Mov13 cells appears to be stably propagated in the absence of DNA methylation and in the presence of *trans*-acting factors required for efficient COL1A1 gene transcription. Although alternative interpretations are possible, the failure to activate the COL1A1 gene by 5-aza-C-treatment in Mov13 fibroblasts is consistent with the notion that COL1A1 gene activation during normal mouse development is a chronologically regulated cascade of events. An early developmental establishment of a transcription-competent chromatin structure in cells derived from each germ layer, accompanied by demethylation, may be a prerequisite for the subsequent transcriptional activation of the gene by specific transcription factors in collagen-expressing cells at later stages of development. In Mov13 mice, the provirus-induced interference with the early event of this cascade precludes the subsequent transcriptional activation, even when the gene is demethylated by 5-aza-C treatment. This suggests that DNA methylation plays no causal role in the transcriptional suppression of the COL1A1 gene but rather may be indicative of and/or stabilize a repressed state caused by other factors.

Our oocyte injection experiments have shown that the COL1A1 promoter can be transcriptionally active despite the presence of proviral sequences and independent of whether a complete or partial proviral genome is present. On the other hand, after transfection into Mov13 fibroblasts the COL1A1 promoter shows only very low activity when a partial proviral genome is present, and no detectable activity in the presence of the complete proviral genome (20, Table 2). This suggests that after transfection and in the 5-aza-C-treated Mov13 fibroblasts transcription from the promoter(s) in the proviral LTR(s) may inhibit the COL1A1 promoter by transcriptional interference or suppression (23,24). Such mechanisms have been suggested to suppress the retroviral promoter in the downstream (3') LTR when the upstream (5') LTR promoter is active (24). Similarly, in retrovirus vectors containing two promoters, expression of one of the promoters is usually suppressed in *cis* by an epigenetic mechanism when the other promoter is transcriptionally active (23). The finding that in constructs containing two alpha-globin genes, transcriptional interference causes inhibition of the downstream gene by transcription of the upstream gene (25) indicates that transcriptional interference may also be a mechanism involved in regulating normal cellular genes. It is conceivable that transcriptional interference by integrated proviruses may not be limited to the proviral promoters, but may also affect the activity of neighboring cellular genes. The molecular events underlying transcriptional interference or suppression are not clear. The fact that the placement of transcriptional termination signals between the two promoters alleviates inhibition of the downstream promoter suggest that readthrough transcripts from the upstream promoter interfere with transcription from the downstream promoter (24,25). Alternatively, local chromatin structure changes induced by the activity of one promoter may inhibit the activity of a second nearby promoter. We have in fact observed that when circular plasmids containing a complete retroviral genome with both LTRs are injected into *X. laevis* oocytes, the 5' LTR promoter is used for initiation of transcription, but not the 3' LTR (26). Because in these circular plasmids each of the two promoters is located "upstream" of the other, our observation suggests a "dominance" of the 5' promoter over the 3' promoter. Furthermore, because both LTRs have identical sequences, the preferential use of the 5' promoter appears to be determined by proviral sequences located outside the LTRs.

It remains enigmatic how the provirus insertion in Mov13 fibroblasts interferes with demethylation and gene activation, specifically in view of the recent surprising finding that the mutant COL1A1 gene promoter is transcriptionally active in odontoblasts from homozygous Mov13 animals, and that stable, correctly initiated and spliced COL1A1 mRNA is produced in these cells despite the presence of the proviral insertion

(27). It appears that alternate mechanisms must exist to overcome the transcriptional repression of the COL1A1 domain in different collagen-expressing cell types, and that the provirus insertion does not inevitably interfere with transcriptional activation of the gene. Further clues may come from a detailed analysis of *cis*-acting regulatory elements and *trans*-acting factors which regulate the stage and tissue-specific expression of the murine COL1A1 gene. We have recently identified two blocks of regulatory sequences located upstream of the promoter and in the first intron of the COL1A1 gene, respectively, which contribute to the transcriptional control of the gene and are sufficient for tissue-specific expression (28). A similar arrangement of regulatory sequences has been identified in the human COL1A1 gene (29-31). The different elements may contribute differentially to the regulation of the gene in different cell types. In fibroblasts, derepression of the COL1A1 gene may require a "communicative" interaction between regulatory factors and both blocks of elements which is made impossible by the provirus insertion. It has in fact recently been suggested that interactions between the 5'-flanking and intron sequences are involved in the transcriptional control of the human COL1A1 gene (32). However, in other collagen-expressing cell types such as odontoblasts the interaction between regulatory factors and only one of the blocks of regulatory sequences may be sufficient, leaving the proviral insertion without consequence. Further studies of the Mov13 mutation will be revealing both for understanding the tissue-specific regulation of COL1A1 gene expression and the molecular mechanisms of insertional mutagenesis.

ACKNOWLEDGEMENTS

We thank S. Bernstein, W. Stumph, K. MacNamara, D.A. Brenner, the San Diego State University Foundation and the Boehringer Ingelheim Fonds for their support.

REFERENCES

1. Yisraeli, J. and M. Szyf. 1984. Gene methylation patterns and expression. In *DNA methylation: Biochemistry and biological significance* (ed. A. Razin, H. Cedar, and A.D. Riggs), pp. 353-378. Springer Verlag, New York.

2. Cedar, H. 1988. DNA methylation and gene activity. Cell 53: 3-4.

3. Monk, M. 1986. Methylation and the X chromosome. BioEssays 4: 204-208.

4. Reik, W., A. Collick, M.L. Norris, S.C. Barton, and M.A. Surani. 1987. Genomic imprinting determines methylation of parental alleles in transgenic mice. Nature 328: 248-251.

5. Swain, J.L., T.A. Stewart, and P. Leder. 1987. Parental legacy determines methylation and expression of an autosomal transgene: a molecular mechanism for parental imprinting. Cell 50: 719-727.

6. Jones, P.A. 1984. Gene activation by 5-azacytidine. In *DNA methylation: Biochemistry and biological significance* (ed. A. Razin, H. Cedar, and A.D. Riggs), pp. 165-188. Springer Verlag, New York.

7. Jaenisch, R., A. Schnieke, and K. Harbers. 1985. Treatment of mice with 5-azacytidine efficiently activates silent retroviral genomes in different tissues. Proc. Natl. Acad. Sci. 82: 1451-1455.

8. Cedar, H. 1984. DNA methylation and gene expression. In *DNA methylation: Biochemistry and biological significance* (ed. A. Razin, H. Cedar, and A.D. Riggs), pp. 147-164. Springer Verlag, New York.

9. Keshet, I., J. Lieman-Hurwitz, and H. Cedar. 1986. DNA methylation affects the formation of active chromatin. Cell 44: 535-543.

10. Buschhausen, G., B. Wittig, M. Graessmann, and A. Graessmann. 1987. Chromatin structure is required to block transcription of the methylated herpes simplex virus thymidine kinase gene. Proc. Natl. Acad. Sci. USA 84: 1177-1181.

11. Harrington, M.A., P.A. Jones, M. Imagawa, and M. Karin. 1988. Cytosine methylation does not affect binding of transcription factor Sp1. Proc. Natl. Acad. Sci. USA 85: 2066-2070.

12. Höller, M., G, Westin, J. Jiricny, and W. Schaffner. 1988. Sp1 transcription factor binds DNA and activates transcription even when the binding site is CpG methylated. Genes Dev. 2: 1127-1135.

13. Enver, T., J. Zhang, T. Papayannopoulou, and G. Stamatoyannopoulos. 1988. DNA methylation: a secondary event in globin gene switching? Genes Dev. 2: 698-706.

14. McKeon, C., H. Ohkubo, I. Pastan, and B. de Crombrugghe. 1982. Unusual methylation pattern of the alpha2(I) collagen gene. Cell 29: 203-210.

15. Jaenisch, R., K. Harbers, A. Schnieke, J. Löhler, I. Chumakov, D. Jähner, D. Grotkopp, and E. Hoffmann. 1983. Germline integration of Moloney murine leukemia virus at the *Mov13* locus leads to recessive lethal mutation and early embryonic death. Cell 32: 209-216.

16. Schnieke, A., K. Harbers, and R. Jaenisch. 1983. Embryonic lethal mutation in mice induced by retrovirus insertion into the alpha1(I) collagen gene. Nature 304: 315-320.

17. Breindl, M., K. Harbers, and R. Jaenisch. 1984. Retrovirus-induced lethal mutation in collagen I gene of mice is associated with an altered chromatin structure. Cell 38: 9-16.

18. Hartung, S., R. Jaenisch, and M. Breindl. 1986. Retrovirus insertion inactivates mouse

alpha1(I) collagen gene by blocking initiation of transcription. Nature 320: 365-367.

19. Jähner, D. and R. Jaenisch. 1985. Retrovirus-induced *de novo* methylation of flanking host sequences correlates with gene inactivity. Nature 315: 594-597.

20. Chan, H., S. Hartung, and M. Breindl. DNA methylation and expression of the murine alpha 1 (I) collagen gene in normal mouse tissues and Mov13 Cells. Genes Dev. (submitted).

21. Jähner, D., and R. Jaenisch. 1985. Chromosomal position and specific demethylation in enhancer sequences of germ line-transmitted retroviral genomes during mouse development. Mol. Cell. Biol. 5: 2212-2220.

22. Schnieke, A., M. Dziadek, J. Bateman, T. Mascara, K. Harbers, R. Gelinas, and R. Jaenisch. 1987. Introduction of the human pro-alpha1(I) collagen gene into pro-alpha1(I)-deficient Mov-13 mouse cells leads to formation of functional mouse-human hybrid type I collagen. Proc. Natl. Acad. Sci. 84: 764-768.

23. Emerman, M., and H.M. Temin. 1984. Genes with promoters in retrovirus vectors can be independently suppressed by an epigenetic mechanism. *Cell* 39: 459-467.

24. Cullen, B.R., P.T. Lomedico, and G. Ju. 1984. Transcriptional interference in avian retroviruses - implications for promoter insertion model of leukaemogenesis. Nature 307: 241-245.

25. Proudfoot, N.J. 1986. Transcriptional interference and termination between duplicated alpha-globin gene constructs suggests a novel mechanism for gene regulation. Nature 322: 562-565.

26. Breindl, M., H. Kalthoff, and R. Jaenisch. 1983. Transcription of cloned Moloney murine

leukemia proviral DNA fragments injected into *Xenopus laevis* oocytes. Nucl. Acids Res. 11: 3989-4006.

27. Kratochwil, K., K. v.d. Mark, E.J. Kollar, R. Jaenisch, K. Mooslehner, M. Schwarz, K. Haase, I. Gmachl, and K. Harbers. 1989. Retrovirus-induced insertional mutation in Mov13 mice affects collagen I expression in tissue-specific manner. Cell (in press).

28. Rippe, R.A., S.-I. Lorenzen, D.A. Brenner, and M. Breindl. 1989. Regulatory elements in the 5'-flanking region and the first intron contribute to transcriptional control of the mouse alpha 1 type I collagen gene. Mol. Cell. Biol. 9: 2224-2227.

29. Bornstein, P., J. McKay, J.K. Morishima, S. Devarayalu, and R.E. Gelinas. 1987. Regulatory elements in the first intron contribute to transcriptional control of the human alpha 1(I) collagen gene. Proc. Natl. Acad. Sci. USA 84: 8869-8873.

30. Bornstein, P., and P. McKay. 1988a. The first intron of the alpha 1(I) collagen gene contains several transcriptional regulatory elements. J. Biol. Chem. 263: 1603-1606.

31. Rossouw, C., W.P. Vergeer, S.J. du Plooy, M.P. Bernard, F. Ramirez, and W. de Wet. 1987. DNA sequences in the first intron of the human pro-alpha 1(I) collagen gene enhance transcription. J. Biol. Chem. 262: 15151-15157.

32. Bornstein, P., J. McKay, D.J. Liska, S. Apone, and S Devarayalu. 1988b. Interactions between the promoter and first intron are involved in transcriptional control of alpha 1(I) collagen gene. Mol. Cell. Biol. 8: 4851-4857.

ROLE OF DNA METHYLATION IN THE REGULATION OF GENE EXPRESSION IN PLANTS[1]

Richard M. Amasino, Manorama C. John, Manfred Klaas, and Dring N. Crowell

Department of Biochemistry, University of Wisconsin-Madison, Madison, Wisconsin 53706

ABSTRACT We have studied the role of DNA methylation in the regulation of T-DNA genes in plant cells. Our results indicate that DNA hypermethylation can suppress T-DNA gene expression. The suppressed state of T-DNA genes is heritable; however, spontaneous or 5-azacytidine-induced activation of suppressed T-DNA genes is always associated with extensive gene demethylation. We have found that DNA in the most DNAse I-sensitive fraction of plant chromatin is depleted in 5-methylcytosine, indicating that the plant genome is organized into methylated and unmethylated compartments that exhibit differences in chromatin structure and protein composition. We also discuss the effect of 5-azaC treatment on the level of methylation of the entire plant genome.

INTRODUCTION

Methylation of cytosine residues at the 5 position of the pyrimidine ring is a post-replication modification of DNA that occurs in many prokaryotic and eukaryotic organisms. In animals, 5-methylcytosine (m^5C) occurs at CpG dinucleotides, and there is much evidence that DNA methylation is involved in the regulation of gene expression (5). For example, there is an inverse correlation between the transcriptional activity of

[1]This work was supported by a grant from the National Science Foundation.

certain genes and the level of methylation of these genes. Moreover, different tissues often exhibit cell type-specific DNA methylation patterns (20), suggesting that DNA methylation may play a role in the establishment of tissue-specific patterns of gene expression throughout development. Further evidence for the involvement of DNA methylation in the regulation of gene expression comes from the observation that silent, methylated genes can often be activated by treating cells with 5-azacytidine (5-azaC), a potent inhibitor of DNA methylation (12). Furthermore, in vitro methylated DNAs, when introduced into the genome of animal cells, adopt an inactive chromatin conformation and are not expressed (13). It has also been shown that DNA methylation may inhibit the binding of specific transcription factors to discrete DNA sequence elements (3,24).

In plants, the effects of DNA methylation have not been as thoroughly studied as in animal systems. However, recent evidence indicates that DNA methylation is involved in the regulation of plant gene expression. For example, levels of ribosomal gene methylation change during plant development (23). The level of methylation of T-DNA genes in crown gall tumor cells, and of transposable elements and storage protein genes in maize, is inversely correlated with expression (2,4,6,10,19,21). Furthermore, silent, hypermethylated T-DNA genes can be activated by treating cells with 5-azaC (2,10,19). DNA methylation has also been shown to reduce the affinity of a DNA binding protein for its target site in a maize transposable element (7).

Plant genomes are often methylated at both CpG dinucleotides and CpNpG trinucleotides (N=any nucleotide) (8), and greater than 30% of the total C residues may be methylated (8,15). The significance of the increased number of target sites and increased amount of C methylation in plant compared to animal genomes is unknown.

In this report, we review our work on i) the role of DNA methylation in *Agrobacterium*-derived T-DNA oncogene expression in plant cells, ii) the response of plant genomes to 5-azaC treatment, and iii) the organization of plant genomes into compartments that contain different levels of DNA methylation.

RESULTS AND DISCUSSION

We have investigated the effect of DNA methylation on

the expression of T-DNA genes in *Nicotiana tabacum* (tobacco) crown gall tumor cells. Crown gall is a tumorous disease of plants that is caused by the transfer of a portion of the *Agrobacterium tumefaciens* tumor-inducing plasmid (T-DNA) to the genome of plant cells (18). The T-DNA contains three oncogenes that encode enzymes of plant hormone (auxin and cytokinin) biosynthesis (25). Two of these genes, *Iaa*H and *Iaa*M, encode enzymes that convert tryptophan into the auxin, indole-3-acetic acid. The other T-DNA oncogene, *ipt*, encodes an isopentenyl transferase that catalyzes the rate limiting step in the production of the cytokinin, zeatin. The plant hormones auxin and cytokinin stimulate the division and differentiation of plant cells in cell culture (22) and presumably in the intact plant. The unregulated production of these hormones due to T-DNA gene expression results in the outgrowth of a tumor on the plant and the ability of tumor cells to proliferate in cell culture on hormone-free medium (1). The T-DNA also contains genes that encode enzymes involved in the biosynthesis of unusual metabolites (opines) that serve as a carbon and nitrogen source for Agrobacteria (9). Crown gall disease is, therefore, a form of genetic parasitism in which the bacterium genetically engineers plant cells for opine production. The conversion of these genetically engineered plant cells into a tumor by T-DNA oncogene expression greatly increases the number of cells serving the bacteria.

We have characterized transformed cell lines in which T-DNA oncogene expression appears to be suppressed by DNA methylation. These lines were first identified as sectors of a crown gall tumor line that had differentiated into normal-appearing plants (2). The normal phenotype of these plants and the observation that cells of these plants required auxin and cytokinin for proliferation in cell culture indicated that T-DNA oncogenes were no longer expressed. (Crown gall tumor cells that express T-DNA oncogenes and proliferate on hormone-free medium cannot differentiate into normal-appearing plants due to unregulated hormone production.) DNA and RNA blot hybridization analyses of nucleic acids isolated from these normal-appearing plants demonstrated that T-DNA genes were present, but not transcribed (2).

To determine if DNA methylation was involved in suppression of T-DNA genes, we treated cells *in vivo* and in cell culture with 5-azaC and measured the effect of this treatment on T-DNA gene expression (Fig. 1). We found that wounding plants containing silent T-DNA genes with a needle dipped in a solution of 5-azaC induced tumorous growth at the wound site (Fig. 1a). For cell culture studies, we established suspension cultures of cells from these plants in medium containing plant hormones and, during a period of active growth, exposed the cells to 5-azaC for 2-3 cell doublings. The treated cells were then plated on medium without hormones; only cells that re-initiate T-DNA gene expression are able to grow on this medium. In several independent trials, 5-azaC treatment always resulted in greater than 50% of the cells exhibiting the tumorous phenotype of hormone-independent growth (14; Fig. 1b). We analyzed T-DNA gene expression and methylation patterns in a number of clones exhibiting the tumorous (i.e., hormone-independent) phenotype and found

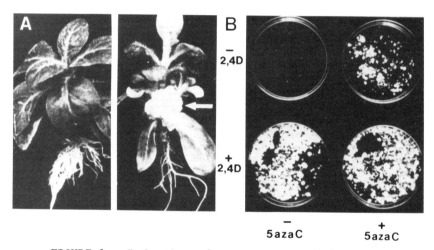

FIGURE 1. Induction of tumorous growth by 5-azaC treatment. A. Whole plants carrying a silent copy of T-DNA were inoculated with 5 μl of water (left) or 100 μM 5-azaC (right) at a needle puncture in the middle of the stem. Plants are shown 4 weeks after inoculation. The site of inoculation and subsequent tumor development is indicated for the plant on the right by the arrow. B. Suspension cultures of cells carrying a silent T-DNA were treated with or without required 5-azaC and plated on medium with or without phytohormone (±2,4D). 5-azaC-treated cells were capable of growth without phytohormones.

that the re-initiation of T-DNA gene expression and
demethylation of T-DNA genes were always associated with
the resumption of tumorous growth in culture (2,11). The
strikingly high rate of re-initiation of tumorous growth
following 5-azaC treatment and the correlation among T-DNA
gene demethylation, T-DNA gene expression and hormone-
independent growth suggest that DNA methylation can
regulate T-DNA gene expression.

To further address the question of whether the
demethylating effects of 5-azaC were responsible for T-DNA
gene activation, we examined the effects of 5-azaC treatment on genomic DNA methylation and T-DNA gene expression
in this system (14). We found that a 2-10 μM dose of
5-azaC present for 2-3 cell doublings was sufficient to
provide a high level of re-initiation of tumorous growth,
and that 5-deoxyazaC and 5-azacytosine were much less
effective at these concentrations. The effectiveness of
these analogs in the tumor re-initiation assay correlates
with their ability to demethylate the plant genome:
5-azaC was more effective at causing genomic demethylation
in this system than 5-azadeoxyC or 5-azacytosine.

5-azaC was the most effective demethylating agent
tested. However, demethylation of the tobacco genome was
incomplete after 5-azaC treatment (14). 5-azaC concentrations that resulted in re-activation of T-DNA genes in
virtually 100% of the treated cells caused only a modest
decrease in genomic cytosine methylation levels from 34%
to 23% m^5C (expressed as $[m^5C/(m^5C + C)] \times 100$). We tried
to further demethylate the genome of this tobacco cell
line by prolonged treatments (6 weeks) with sublethal
doses of 5-azaC (replenished twice weekly), but the level
of genomic methylation never dropped below approximately
20%, the level that was attained after one to two weeks of
treatment. Furthermore, levels of genomic methylation
returned to the control level of 34% by 4-5 weeks (12-15
cell doublings) after removal of 5-azaC. However, T-DNA
genes remained active and demethylated after genomic
methylation levels had returned to normal. Thus, T-DNA
genes and the bulk of the genome respond differently to
5-azaC treatment: i) T-DNA genes undergo extensive
demethylation after 5-azaC treatment, whereas 5-azaC-
induced demethylation of the genome is not extensive, and
ii) T-DNA genes remain active and demethylated upon
removal of 5-azaC, whereas genomic methylation levels
return to normal. One possible explanation for this
behavior is that the T-DNA insert in this cell line is

more sensitive to 5-azaC-induced demethylation than the bulk of the genome, and the extensive demethylation of T-DNA upon 5-azaC treatment blocks remethylation after removal of the drug. Alternatively, all DNA sequences may be equally susceptible to the demethylating activity of 5-azaC, but normal plant genomic sequences contain "signals" for remethylation that are not present on the T-DNA. Thus, normal plant genomic sequences may be in equilibrium between demethylation and remethylation. In this model, specific DNA sequences, DNA configurations (e.g., tandem duplications) or simply lack of transcription might serve as signals for DNA remethylation. In the cell line we have studied, the T-DNA was originally present in a tandem array of approximately 20 copies when DNA hypermethylation occurred, but in derivative lines that had lost tumorous properties, a genomic rearrangement eliminated all but one T-DNA copy (2). Thus, the T-DNA insert that remained demethylated in this line after 5-azaC treatment may have lost the signal that originally led to its hypermethylation.

We wished to examine whether the silent, methylated state of this T-DNA insert was maintained through meiosis and into subsequent generations. Accordingly, we crossed plants containing a silent T-DNA copy with normal tobacco plants and examined the phenotype of 100 progeny. All of the progeny developed into normal-appearing plants. Approximately half of the progeny (55/100) inherited a silent T-DNA, since tissues of these plants resumed tumorous growth when treated with 5-azaC (11). One of the progeny of this cross that contained a silent T-DNA, called CX2, was chosen for further study as described below.

We have isolated derivatives of CX2 cells that spontaneously (i.e., without 5-azaC treatment) re-initiated tumorous growth. These lines were isolated by plating large numbers of CX2 cells on hormone-free medium, which selects for tumorous growth. Spontaneous revertant tumor lines (SRs) arose at a frequency of approximately 10^{-7} (11). In all SRs examined, T-DNA gene expression had resumed and T-DNA methylation levels had decreased (11). The overall level of m^5C in the genome, however, was unchanged, indicating that spontaneous reversion was not due to a lesion in the maintenance methylation system. We have also isolated, at a high frequency by 5-azaC treatment, tumorous revertants of CX2 that express T-DNA genes and contain extensively demethylated T-DNA. That

5-azaC treatment results in a high frequency of T-DNA activation and T-DNA demethylation and that low frequency, spontaneous T-DNA activation is also accompanied by T-DNA demethylation is consistent with a model in which DNA methylation serves as a primary determinant of T-DNA gene expression in this system.

All of the derivatives of line CX2 that re-initiated tumorous growth, either spontaneously or after 5-azaC treatment, expressed the T-DNA *ipt* oncogene (11). We investigated the methylation patterns of the *ipt* gene in line CX2 and in tumorous revertants of this line to determine the extent of demethylation at the *ipt* locus that accompanied tumorous reversion. We were particularly interested in the methylation patterns of the *ipt* gene in spontaneous revertants since these revertants arose at a very low frequency. This low frequency of reversion suggested that perhaps only sites critical for gene expression had undergone demethylation. In this study, we used DNA blot analysis to assay cleavage of restriction sites in the 5', coding and 3' regions of the *ipt* gene after digestion with the cytosine methylation-sensitive restriction enzymes DdeI, HhaI, HpaII, MspI and PstI. The results of this analysis are summarized in Fig. 2. There is a striking difference in DNA methylation patterns at the *ipt* locus between line CX2, where the gene is silent, and lines in which tumorous growth resumed. In line CX2, the *ipt* region is methylated at every site analyzed. In the 5-azaC and spontaneous revertants of CX2, there is extensive demethylation in the region spanning the *ipt* gene. The majority of the sites analyzed lie in the coding region of the gene, and with the degree of demethylation observed, no conclusions can be made regarding whether there exist a small number of specific sites that must be demethylated for gene activation to occur. Perhaps extensive demethylation is a prerequisite for gene expression in this system, and the overall density of methylation controls *ipt* gene expression. Alternatively, demethylation of specific sites in the gene may cause gene expression, and expression may lead to further demethylation. Since we cannot analyze methylation patterns in these tumorous revertants until many weeks after the reversion event (to obtain sufficient tissue for analysis), we are not able to distinguish between these possibilities.

We have also examined the relationship between DNA methylation and DNAse I-sensitivity of DNA in plant

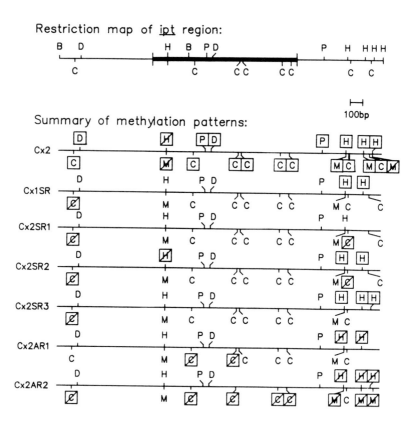

FIGURE 2. Analysis of the DNA methylation pattern of the region surrounding the *ipt* gene. The restriction map of the *ipt* region is derived from the published nucleotide sequence (16). The thick line represents the region found in the mature *ipt* mRNA. The following restriction enzyme site abbreviations are used: B, BamHl; C, HhaI; D, DdeI; H, HpaII; M, MspI; P, PstI. In the summary of methylation patterns, CX2 is the line carrying a silent T-DNA copy, CXSR lines are spontaneous tumorous revertants and CXAR lines are 5-azaC-induced tumorous revertants of cells that had previously carried a silent T-DNA copy. Boxed restriction sites are fully methylated, boxed sites containing a diagonal line are partially methylated and unboxed sites are unmethylated.

chromatin. DNAse I-sensitivity is thought to be a property of genes that exist in a chromatin conformation characteristic of active genes (e.g., DNAse I accessibility may reflect the accessibility of a gene to the transcriptional apparatus). For this analysis, we purified chromatin from pea, barley and maize (corn) and subjected the chromatin to limited DNAse I digestion. The smallest fragments of DNA generated by this treatment represent DNA sequences that are most DNAse I-sensitive. DNA from fractions of chromatin differing in DNAse I-sensitivity were purified by agarose gel electrophoresis followed by electroelution. The m^5C levels in acid hydrolysates of DNA from these fractions of different DNAse I sensitivities were then analyzed by HPLC (15). The results of this analysis are presented in Table I. In all three plant species tested, the m^5C content of DNA

Table 1: Methylation levels of different size fractions of DNaseI-digested chromatin and protein-free DNA of pea, barley, and corn.

plant	$\%m^5C^1$ in total DNA	chromatin			DNA	
		size[2]	$\%m^5C$	% of digest[3]	size	$\%m^5C$
pea	26.2	0.20–0.54	13.2	2	0.20–0.68	24.1
		0.54–1.10	17.3	4	0.68–2.60	25.0
		1.1–2.8	21.4	9	2.6–6.0	26.4
		2.8–9.0	24.7	14		
		9.0–>30	28.0	71		
barley	25.7	0.12–0.48	13.9	2	0.12–0.37	21.2
		0.48–1.05	16.3	3	0.37–1.50	23.4
		1.05–2.2	19.5	6	1.5–4.6	25.3
		2.2–6.7	24.1	11		
		6.7–>30	25.7	78		
corn	27.2	0.12–0.71	16.3	4	0.12–0.58	24.1
		0.71–1.40	22.7	10	0.58–1.80	26.4
		1.4–3.7	26.0	15	1.8–5.0	28.0
		3.7–16	30.2	31		
		16–>30	31.8	39		

[1] The methylation levels are $\%m^5C$ of total C plus m^5C.
[2] Size expressed as number of kilobasepairs (kb) as estimated by agarose gel electrophoresis and comparison to size standards.
[3] The %DNA of the total digest in each fraction was estimated from the intensity of base absorption during HPLC runs and from separation of aliquots on ethidium-bromide stained agarose gels.

from the most nuclease-sensitive fraction was approximately half that of the genome as a whole. However, when purified, protein-free DNA was analyzed in the same way, no significant difference was observed between the m^5C content of nuclease-sensitive DNA and that of total DNA. Therefore, plant genomes, like animal genomes (17), appear to be organized into regions that differ in methylation level. Furthermore, regions that contain relatively low levels of DNA methylation exist in a DNAse I-sensitive chromatin conformation due to the chromosomal proteins associated with these regions of DNA.

ACKNOWLEDGMENTS

We thank members of the lab for critical discussion and Carolyn Kunen for preparing the manuscript. R.M.A. is a scholar of the James D. and Dorothy Shaw Scholars' program.

REFERENCES

1. Amasino RM, Miller CO (1982). Hormonal control of tobacco tumor morphology. Plant Physiol 69:389.
2. Amasino RM, Powell ALT, Gordon MP (1984). Changes in T-DNA methylation and expression are associated with phenotypic variation and plant regeneration in a crown gall tumor line. Mol Gen„Genet 197:437.
3. Becker PB, Ruppert S, Schutz G (1987). Genomic footprinting reveals cell type-specific DNA binding of ubiquitous factors. Cell 5:435.
4. Bianchi MW, Viotti A (1988). DNA methylation and tissue-specific transcription of the storage protein genes of maize. Plant Mol Biology 11:203.
5. Cedar H (1988). DNA methylation and gene activity. Cell 53:3.
6. Chandler VL, Walbot V (1986). DNA modification of a maize transposable element correlates with loss of activity. „Proc Natl Acad Sci USA 83:1767.
7. Gierl A, Lutticke S, Saedler H (1988). *TnpA* product encoded by the transposable element En-1 of *Zea mays* is a DNA binding protein. EMBO J 7:4045.
8. Gruenbaum Y, Naveh-Many T, Cedar H, Razin A (1981). Sequence specificity of methylation in higher plant DNA. Nature 292:860.
9. Guyon P, Chilton M-D, Petit A, Tempe J (1980).

Agropine in "null-type" crown gall tumors: Evidence for generality of the opine concept. Proc Natl Acad Sci USA 77:2693.
10. Hepburn AG, Clarke LE, Pearson L, White J (1983). The role of cytosine methylation in the control of nopaline synthase gene expression in a plant tumor. J Mol Appl Genet 2:315.
11. John MC, Amasino RM (1989). Extensive changes in DNA methylation patterns accompany activation of a silent T-DNA ipt gene in Agrobacterium-transformed plant cells. Mol Cell Biol, in press.
12. Jones PA (1985). Altering gene expression with 5-azacytidine. Cell 40:485.
13. Keshet I, Lieman-Hurwitz J, Cedar H (1986). DNA methylation affects the formation of active chromatin. Cell 44:535.
14. Klaas M, John MC, Crowell DN, Amasino RM (1989a). Rapid induction of genomic demethylation and T-DNA gene expression in plant cells by 5-azacytosine derivatives. Plant Mol Biol, in press.
15. Klaas M, Amasino RM (1989b). DNA methylation is reduced in DNaseI-sensitive regions of plant chromatin. Plant Physiology, in press.
16. Lichtenstein C, Klee H, Montoya A, Garfinkel D, Fuller S, Flores C, Nester EW, Gordon MP (1984). Nucleotide sequence and transcript mapping of the tmr gene of the pTiA6NC octopine Ti plasmid: A bacterial gene involved in plant tumorigenesis. J Mol Appl Genet 2:354.
17. Naveh-Many T, Cedar H (1981). Active gene sequences are undermethylated. Proc Natl Acad Sci USA 78:4246.
18. Nester EW, Gordon MP, Amasino RM, Yanofsky MF (1984). Crown gall: a molecular and physiological analysis. Annu Rev Plant Physiol 35:387.
19. Peerbolte R, Leenhouts K, Hooykaas-van Slogeren GMS, Wullems GJ, Schilperoort RA (1986). Clones from a shooty tobacco crown gall tumor II. Irregular T-DNA structures and organization, T-DNA methylation and conditional expression of opine genes. Plant Mol Biol 7:285.
20. Razin A (1984). DNA methylation patterns: Formation and biological functions. In Razin A, Cedar H, Riggs AD (eds): "DNA methylation. Biochemistry and Biological Significance," New York: Springer-Verlag. p 127.
21. Schwartz D, Dennis E (1986). Transposase activity of

the *Ac* controlling element in maize is regulated by its degree of methylation. Mol Gen Genet 205:476.
22. Skoog G, Miller CO (1957). Chemical regulation of growth and organ formation in plant tissues cultured *in vitro*. Symp Soc Exp Biol XI, p 118.
23. Watson JC, Kaufman LS, Thompson WR (1987). Developmental regulation of cytosine methylation in the nuclear ribosomal RNA genes of *Pisum sativum*. J Mol Biol 193:15.
24. Watt F, Molloy PL (1988). Cytosine methylation prevents binding to DNA of a HeLa cell transcription factor required for optimal expression of the adenovirus major late promoter. Genes & Development 2:1136.
25. Weiler EW, Schroder J (1987). Hormone genes and crown gall disease. Trends Biochem Sci 12:271.

BIOSYNTHESIS AND DISTRIBUTION OF METHYLCYTOSINE IN WHEAT DNA. HOW DIFFERENT ARE PLANT DNA METHYLTRANSFERASES?

Hartmut Follmann[1], Hans-Jörg Balzer, and Roland Schleicher

Fachbereich Chemie (Biochemie) der Philipps-Universität,
D-3550 Marburg (Fed.Rep.Germany)

ABSTRACT One reason for the high methylcytosine content of plant DNA may lie in the specificity and activity of plant DNA methyltransferases. We have analyzed the properties of the DNA methylase system previously purified from wheat embryo and find that it differs markedly enough from the known mammalian enzymes to establish a plant-specific type of DNA methyltransferase. These differences reside, inter alia, in molecular weight, specificity towards DNA substrates of varying methylation, and insensitivity towards inhibition by 5-azacytidine.

INTRODUCTION

Understanding the biosynthesis and functions of methylcytosine (mCyt) in eukaryote DNA requires knowledge of the structure and specificity of DNA methyltransferases (EC 2.1.1.37) which catalyze the postreplicative modification. However, these enzymes have been difficult to purify and characterize for many years, and most previous studies have dealt with rather few DNA methylases isolated from mammalian cells. The enzymes of rat and mouse tissues, human placenta, and HeLa cells are now known in much greater detail, and similarities with bacterial type II DNA methylases begin to emerge from gene sequences (1-6). These advances correlate favorably with our rapidly increasing insight into the role of DNA methylation in gene expression and cell differentiation (7).

[1] To whom correspondence should be directed. New address: Fachbereich Biologie-Chemie der Universität, D-3500 Kassel (Fed.Rep.Germany).

On the contrary, there is a distinct lack of information about the enzymology of DNA methylation in plants. This is surprising, and unsatisfactory, in view of the long-known fact that plants contain on the average much more mCyt than animal DNAs (up to 10 mol% vs. 1 mol%). While the importance of DNA methylation is well recognized in the plant sciences (in particular, in plant genetics) it may be symptomatic that DNA methylase activity could not as yet be obtained in stable form from a prominent plant like maize, despite repeated efforts in our own and in other laboratories.

We have previously studied deoxyribonucleotide synthesis, DNA replication, and methylcytosine formation in germinating wheat (8,9). It was found that the isolated wheat embryo, which represents an almost synchronous tissue during early hours of germination provides a very good experimental system for the study of plant DNA methylation. An active DNA methylase could be purified from this source, and some of its properties have been described (10).

Wheat DNA methyltransferase differs from mammalian DNA methylases in two characteristics: It is much smaller, and its specificity towards different DNA substrates in vitro appears to be special. The molecular weight of the wheat enzyme, established by density gradient centrifugation and gel permeation chromatography is 55,000 whereas the mammalian DNA methylases possess molecular weights around 190,000 (2,3,6). On denaturing electrophoresis gels our preparation exhibits two protein bands corresponding to M_r=35,000 and 55,000. Unfortunately, attempts were unsuccessful to separate these proteins further, or to establish by activity staining (11) whether both are essential for activity. However we have recently observed the same polypeptide pattern in a DNA methylase preparation from soybean root tips (12) suggesting that plant enzymes do comprise both these components. With respect to specificity, wheat DNA methylase is most active towards native plant DNA substrates (which have already high mCyt content), less active towards bacterial or mammalian DNA (which is little methylated), and acts very little on synthetic polydeoxyribonucleotides (10). In contrast, mammalian DNA methylases prefer singlestranded foreign (e.g.,micrococcal) DNA and polydeoxyribonucleotides in vitro, although hemimethylated duplex DNA appears to be the physiological substrate.

Finally an immunological analysis of DNA methylases from different species has recently been performed with an antibody directed against human methyltransferase (13). No crossreactivity could be observed with the wheat enzyme whereas the antibody did precipitate rat and mouse DNA methylases.

In this paper we describe two additional features of DNA methylation in wheat embryo which strengthen the dissimilarities between plants and other organisms. We have completed a more systematic review of the substrate specificity of DNA methylase from wheat, and studied the effect of the methylation inhibitor 5-azacytidine in the plant tissue. It remains undecided, however, whether or not DNA methylation in wheat is representative for plants in general.

DNA SUBSTRATES FOR WHEAT DNA METHYLASE IN VITRO

"Substrate specificity" of a DNA methylase is a multitudinous term: It can refer to the kind of cytosine residues (dinucleotides) that are being methylated, to the type and origin of DNA as a whole, or to more subtle local conditions of the methylatable site within specific flanking regions or DNA fractions. Wheat DNA methylase was considered very unspecific in our previous analysis (10) because in vitro it could modify the different C-N sequences to an equal extent, methylate native bacterial, animal, and plant DNAs with comprable activity (60-100 % of the maximum rate), and react almost equally well with an unmethylated and a hemimethylated plasmid DNA (i.e., show both de novo and maintenance methylase activity). The presence in plants of an active and highly unspecific DNA methylase could per se provide a straightforward explanation for the high amount of methylcytosine in plant DNA. It does not explain, however, why methylated plant DNA is preferred over little methylated DNA substrates in vitro. Moreover the distribution of methylcytosines in plant DNA is not uniform (see below), and a rather specific enzyme apparatus should in fact be expected for the generation and maintenance of specific patterns. It was therefore necessary to characterize the wheat DNA used as methylase substrate in more detail, and to include yet other types of DNA.

A decrease in DNA methylation during the germination of wheat has been reported earlier (14). We have analyzed the methyldeoxycytidine content in the wheat variety used in our experiments and found a rapid and substantial drop in its fraction accompanying the first round of DNA replication (Table 1). Such undermethylation is of course fully compatible with the onset of gene expression and cell differentiation in the germinating embryos but it contrasts with the increase in enzyme activity observed at the same time (9).

TABLE 1
5-METHYL-2'-DEOXYCYTIDINE AND 2'-DEOXYCYTIDINE IN WHEAT DNA[a]

Germination period	5-methyl-deoxycytidine	deoxycytidine	mCyt / (mCyt + Cyt)
hours	mol-%	mol-%	%
0	5.48 ± 0.05	17.66	23.7
5	4.49 ± 0.15	18.55	19.5
10	4.20 ± 0.07	19.21	18.0
15	4.14 ± 0.04	19.50	17.5
20	4.24 ± 0.05	20.69	17.0
25	3.94 ± 0.04	20.02	16.6
30	3.84 ± 0.12	21.34	15.2

[a] DNA was purified from winter wheat (Triticum aestivum, var. "Kormoran"), hydrolyzed enzymatically and the hydrolysate analyzed by HPLC as described (15).

DNA samples from wheat and from various other sources were purified by ribonuclease and pronase treatment and by hydroxylapatite chromatography, and the native DNAs were used as substrates of purified wheat DNA methylase under identical conditions. These samples included bacterial and mammalian DNA containing zero or few mCyt residues, plant DNA with relatively low (green algae), avergae (wheat, maize), or very high mCyt content (Scilla); also included was a recently discovered halophage DNA with fully methylated genome (16). The results are summarized in Figure 1 and Table 2.

It is obvious from these data that the in vitro specificity of wheat DNA methylase is more complex than previously anticipated. The more highly methylated DNA from resting wheat is not a good substrate whereas hypomethylated DNAs from plant seedlings are very good methyl group acceptors. Scilla DNA, in which about half the cytosine residues are already modified, again is a poor substrate. However, the figures in Table 2 do not establish a simple, inverse correlation of mCyt content and substrate activity but suggest that an intermediate, optimum number of preexisting methylcytosines and/or other, species- (plant-)specific DNA structures are required for maximum enzyme activity.

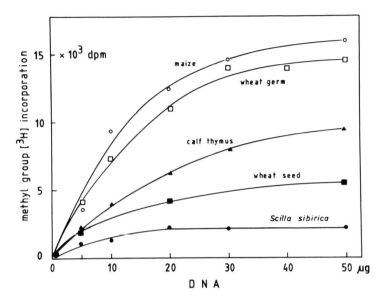

FIGURE 1. Methylation of native DNA samples with purified wheat DNA methylase under standard conditions (1 h at 25°, pH 7.5) (10). DNA was obtained from wheat or maize germ, dry wheat seeds, bulbs and shoots of the bluebell (S.sibirica), and calf thymus, as indicated.

TABLE 2
SUBSTRATE SPECIFICITY OF WHEAT DNA METHYLTRANSFERASE IN VITRO

Source of DNA	Methylation rate (%)	$\frac{mCyt}{mCyt + Cyt}$ (%)
pBR 322 (linear)	73	0
Escherichia coli	63	1
Methanobacterium thermoautotro-phicum	81	1
calf thymus	59	4
Scenedesmus obliquus (green algae)	66	12
germinating wheat embryo	**100**	15
maize seedling	**110**	17
wheat seeds, dry embryo	30	23
Scilla sibirica (bluebell)	16	45
phage φN	~ 50	100

The inclusion of fully methylated phage φN DNA in these assays was meant as a control experiment, in which negligible reactivity was expected. The surprising enzyme-catalyzed incorporation of radioactive methyl groups into the phage DNA (which was free of contaminating bacterial DNA) cannot be explained at this time. Pulse-chase experiments with labeled and unlabeled S-adenosylmethionine have ruled out a reversible, artefactual methylation reaction. It should be noted that another methylcytosine-substituted phage DNA, Xp 12 (17) could also be radioactively methylated in vitro (M.Ehrlich, personal communication).

Most plant DNAs contain a high proportion of repetitive sequences, and an enrichment of mCyt residues in repetitive DNA has been observed (18,19). We have therefore separated sonicated wheat DNA into several fractions of increasing sequence complexity, viz. highly and moderately repetitive, and single-copy DNA, and subjected these fractions to in vitro methylation (Table 3). Base analysis by HPLC revealed highly varying methylcytosine-to-cytosine ratios (10-25 %).

TABLE 3
DIFFERENTIAL METHYLCYTOSINE DISTRIBUTION AND IN VITRO METHYLATION IN WHEAT DNA FRACTIONS

DNA fraction[a]	total DNA	unique sequences	moderately repetitive	highly repetitive
$c_0 t$ value		> 100	0.2	0-0.04
% of total DNA	100	22	46	32
mCyt/mCyt+Cyt (%)	23	10	21	25
enzymatic methylation (cpm/μg DNA)		110	270	330
methylation in nuclei (cpm/μg)[b]	61.400	38.300	54.400	72.500

[a] fractionationation was done by thermal hydroxylapatite chromatography
[b] isolated wheat embryo nuclei were labeled with S-adenosylmethionine for 3 hours under standard conditions (9), the DNA was purified and then fractionated

The figures in Table 3 demonstrate a clear correlation between native methylcytosine levels of the different DNA sequences and their methyl group acceptance in vitro and in vivo. Thus, single-copy DNA, of lowest mCyt content, was modified only one third the amount of repetitive DNA by the isolated enzyme and also in intact nuclei (bottom line). Fractionation of genomic wheat DNA appears to provide a very simple and efficient system in which DNA methylation under physiological conditions can be simulated. These experiments will be extended in the future.

INHIBITION OF DNA METHYLATION BY 5-AZACYTIDINE

The abovedescribed specificity studies could be confirmed and refined with artificially hypomethylated plant DNAs, depleted in methylcytosine to varying degrees. Practicable conditions for undermethylation of genes are also in great demand in plant genetics (20). In mammalian cells, micromolar concentrations of the nucleoside analog 5-azacytidine (azaC) usually induce hypomethylation and gene activation; the analog is incorporated into DNA and then covalently inactivates DNA methylase (21,22). However, azaC has only rarely been used in plant molecular biology and the effects were inconclusive. We have therefore attempted to suppress DNA methylation in the germinating wheat embryo by azacytidine treatment under a variety of conditions.

These experiments have met with very little success. Only after treating embryos on agar plates, or isolated nuclei in a methylation medium with azaC in the millimolar concentration range could some inhibition of methyl group incorporation be observed (Figure 2). The plants deteriorated rapidly in the presence of higher than 5-10 mM azaC concentration. The decrease in methylation could not even be linked to any measurable azacytosine incorporation, or reduced methylcytosine content of the DNA as the base composition of all samples remained unchanged. DNA derived from embryos germinating in presence of 5 mM azaC was a slightly better in vitro DNA methylase substrate (results to be published elsewhere) but again the underlying changes were apparently too small for direct analytical detection, and the system was not of any practical use.

DNA methylation in wheat thus differs fundamentally from the azaC-sensitive process in mammalian cells. The azacytidine resistance of wheat could not be due to limitations of uptake, phosphorylation, or reduction to the 2'-deoxyribotide because these reactions have previously been studied with pyrimidine nucleosides (8) and did function in the wheat embryo. A more likely explanation is that the plant enzyme is not irreversibly inactivated by the inhibitor. Inhibition of thymidylate synthase by 5-fluorodeoxyuridylate (a mechanistically related system) is also reversible in plants (23) but irreversible in other organisms.

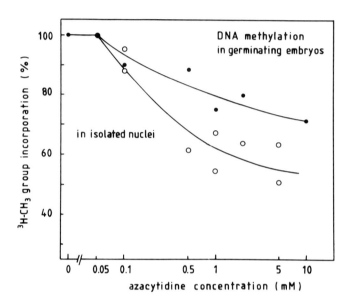

FIGURE 2. Effect of azacytidine on DNA methylation in wheat. Embryos were germinated for 15 h on agar plates (●), or nuclei were prepared from untreated embryos and then incubated with azaC for 3 h. The 100 % value (no azaC) represents 10^5 cpm ^{3}H-CH$_3$ incorporation/mg DNA.

Azacytidine sensitivity may be varying in different plant species. The nucleoside produced phenotypical aberrations but, like in wheat, no distinct molecular changes in maize seedlings (20). Some undermethylation was detected in germinating rice after 3 days of azaC treatment (24), and also in tobacco cells (25). In no plant has azacytosine incorporation into DNA been demonstrated. Thus, the mechanisms of azaC action in plants have yet to be established.

PLANT DNA METHYLTRANSFERASES ARE DIFFERENT

The amount of methylcytosine in average plant DNA is eight times higher than in animal DNA. It is now certain that gene expression is regulated by DNA methylation in the plant kingdom (26-28) like in mammals (7) but it is also obvious that not all the modified cytosines present in plant DNA could be engaged in the control of gene activity. What other functions they fulfill, or which advantage the heavily methylated DNA regions confer to a plant, is still a mystery. In any event it is a priori very reasonable that the enzyme apparatus for DNA methylation should differ substantially in plants and animals. Either there have to be two different DNA methylases in plants, one responsible for the synthesis of regulatory methylcytosines and the other for bulk methylation, respectively; we have searched for such a duplexity in wheat embryo extracts but found none. Or the specificity and other properties of plant DNA methylases (and auxiliary components) must be special enough to emcompass both activities. All present evidence supports the case of one, plant-specific methylation system. The evidence is summarized in Table 4 and will be briefly reviewed here.

TABLE 4
DNA METHYLATION IN MAMMALIAN AND IN PLANT CELLS

Subject	in mammals	in plants	ref.
mCyt content in DNA	< 1 mol%	5-10 mol%	
methylated sequences	predominantly C-G	C-N, C-N-G	(29)
M_r of DNA methylases	(<) 190,000	55-60,000	(2,3,10, 12,30)
crossreactivity with anti-human methylase antibody	yes	no	(13)
preferred substrate	denatured DNA	native DNA	
metal ion effects	Zn^{++} protein?	Zn^{++} inhibits	(12,31)
azacytidine effect	sensitive to to μM azaC	little sensitive to mM azaC	
methylated DNA-binding proteins	sequence-specific	unspecific	(32,33)

Molecular weight of DNA methylases. Mammalian DNA methylases are large proteins (mol. weight, 190,000; one polypeptide chain) which are susceptible to proteolytic breakdown and therefore frequently contain smaller entities (around M_r 100,000-160,000 (2-4)). We have never observed such a large protein in our most active and most rapidly purified DNA methylase preparations of M_r=55,000. Enzymes of the same size have also been found in soybean roots and in the unicellular green algae, Chlamydomonas reinhardii (30), and thus appear to represent the plant-specific type of DNA methylase. Small components (M_r=22,000; 35,000) are always co-purified in the preparations (10,12,30), the nature of which is unknown. This pattern bears a remarkable similarity to frog virus DNA methylase (34) which is also a small enzyme producing highly methylated DNA and requires two dissimilar polypeptides (M_r 26,000 and 18,000) for activity.

Substrate specificity. Both animal and plant DNA methylase are very active towards hemimethylated DNA, in accordance with their physiological role in postreplicative maintenance methylation. Apart from this, they differ strongly in their in vitro activity towards other DNA substrates or polydeoxyribonucleotides. A thorough comparison is not possible, however, because the nature of ssDNA as preferred substrate of the mammalian enzymes has not been varied too much (for example, in combination of a mammalian methylase with plant DNA). It is clear that the specificity of wheat DNA methylase observed in vitro matches all physiological demands in that it modifies all the C-N dinucleotide sequences, can produce a high degree of methylation and at the same time recognize single-copy DNA for low-level methylation (Table 3). From what is known about the mammalian methylases it is unlikely that they could combine all these activities.

Other properties. Rat DNA methylase is stimulated up to tenfold by divalent cobalt ions (35), and it has been hypothesized that the mouse enzyme might be a zinc protein (31). Wheat DNA methylase is inhibited strongly by 2 mM zinc and other transition metal ions (12). The differences in azaC sensitivity have been emphasized above. Finally, methylated DNA binding proteins (MDBP) and other chromatin components (32,33,36,37) interact with methylated DNA and interfere with DNA methylation, and these proteins appear to differ markedly in their sequence specificity, or non-specificity, in animal and plant cells. Such proteins have not yet been included in systematic in vitro enzyme studies.

ACKNOWLEDGEMENTS

These studies have been supported by a grant (to H.F.) from Deutsche Forschungsgemeinschaft, Fo 50/14. We thank Gabriele Schimpff-Weiland, Anke Nebendahl, and Thomas Hast for their assistance in some of the experiments.

REFERENCES

1. Bolden A, War C, Siedlecki JA, Weissbach A (1984) J Biol Chem 259:12437-12443
2. Adams RLP, Gardiner K, Rinalde A, Bryans M, McGarvey M, Burdon RH (1986) Biochim Biophys Acta 868:9-16
3. Pfeifer GP, Drahovsky D (1986) Biochim Biophys Acta 868:238-242
4. Hitt MM, Wu T-L, Cohen G, Linn S (1988) J Biol Chem 263: 4392-4399
5. Bestor T, Laudano A, Mattaliano R, Ingram V (1988) J Mol Biol 203:971-983
6. Spiess E, Tomassetti A, Hernaiz-Driever P, Pfeifer G (1988) Eur J Biochem 177:29-34
7. Cedar H (1988) Cell 53:3-4
8. Schimpff G, Müller H, Follmann H (1978) Biochim Biophys Acta 520:70-81
9. Theiss G, Follmann H (1980) Biochem Biophys Res Comm 94:291-297
10. Theiss G, Schleicher R, Follmann H (1987) Eur J Biochem 167:89-96
11. Hübscher U, Pedrali-Noy G, Knust-Kron B, Doerfler W, Spadari S (1985) Anal Biochem 150:442-448
12. Schleicher R (1989) Ph D thesis, Universität Marburg
13. Pfeifer GP, Kohlmaier L, Tomassetti A, Schleicher R, Follmann H, Pfohl-Leskowicz A, Dirheimer G, Drahovsky D (1989) Arch Biochem Biophys 268:388-392
14. Drozhdenyuk AP, Sulimova GE, Vanyushin BF (1976) Mol Biol (Moscow) 10:1378-1386
15. Gehrke CW, McCune RA, Gama-sosa MA, Ehrlich M, Kuo KC (1984) J Chromatography 301:199-219
16. Vogelsang-Wenke H, Oesterhelt D (1988) Mol Gen Genet 211: 211:407-414
17. Kuo T-T, Tu J (1976) Nature 263:615
18. Shmookler Reis R, Timmis JN, Ingle J (1981) Biochem J 195:723-734
19. Deumling B (1981) Proc Natl Acad Sci USA 78:338-342

20. Brown PTH, Yoneyama K, Lörz H (1989) Theor Appl Genet in the press
21. Taylor SM, Jones PA (1982) J Mol Biol 162:679-692
22. Santi DV, Norment A, Garrett CE (1984) Proc Natl Acad Sci USA 81:6993-6997
23. Bachmann B, Follmann H (1987) Arch Biochem Biophys 256: 244-252
24. Sano H, Kamada I, Youssefian S, Wabiko H (1989) J Cell Biochem Suppl 13D:214
25. Durante M, Cecchini E, Citti L, Geri C, Nuti Ronchi V, Parenti R (1984) Proc Intl Symp Plant Tissue and Cell Culture, Czechoslovak Acad Sci Prague p.327-328
26. Bianchi MW, Viotti A (1988) Plant Mol Biol 11:203-214
27. Kunze R, Starlinger P, Schwartz D (1988) Mol Gen Genet (214:325-327
28. Schwartz D (1989) Proc Natl Acad Sci USA 86:2789-2793
29. Gruenbaum Y, Naveh-Many T, Cedar H, Razin A (1981) Nature 292:860-862
30. Sano H, Sager R (1980) Eur J Biochem 105:471-480
31. Bestor T (1988) Biochem Soc Transact 1988:944-947
32. Supakar PC, Weist D, Zhang D, Inamdar N, Zhang X-Y, Khan R, Ehrlich KC, Ehrlich M (1988) Nucl Acids Res 16:8029
33. Zhang D, Ehrlich KC, Supakar PC, Ehrlich M (1989) Mol Cell Biol 9:1351-1356
34. Essani K, Goorha R, Granoff A (1989) J Cell Biol Suppl 13D:222
35. Pfohl-Leskowicz A, Baldacini O, Keith G, Dirheimer G (1987) Biochimie 69:1235- 1242
36. Kautiainen TL, Jones PA (1985) Biochemistry 24:1193-1196
37. Caiafa P, Mastrantonio S, Attina M, Rispoli M, Reale A, Strom R (1988) Biochem Internat 17:863-875

A NOVEL STRATEGY FOR IDENTIFYING POTENTIAL TARGETS OF ALTERED DNA METHYLATION IN NEOPLASTIC TRANSFORMATION[1]

Andrew P. Feinberg and Shirley Rainier

Howard Hughes Medical Institute, and Departments of Internal Medicine and Human Genetics, University of Michigan Medical School, Ann Arbor, MI 48109

ABSTRACT In order to identify possible targets of altered DNA methylation in carcinogenesis, as well as to define the earliest changes in gene expression in transformation, we have developed a novel strategy to isolate pretransformed cells after treatment with an analogue of 5-azacytidine. We treat single C3H 10T1/2 cells with 5-aza-2'-deoxycytidine, and we capture microcolonies of cells that are determined for transformation, i.e., committed to but prior to neoplastic transformation. From these microcolonies, we have constructed cDNA libraries, and we are identifying genes expressed in pretransformants and transformants, but not in treated but nontransformed microcolonies. This is the first time that phenotypically normal cells, committed to transformation without further intervention, have been captured. Studies of these pretransformed cells will permit examination of the earliest changes

---

[1] This work was supported by the Howard Hughes Medical Institute (A.P.F., S.R.), and the National Institutes of Health (grant CA48932 to A.P.F.).

in gene expression in carcinogenesis and the role of DNA methylation in transformation.

INTRODUCTION

There are three reasons to suspect that altered DNA methylation may play a role in carcinogenesis. First, tumor cells and cells transformed _in vitro_ show substantial alterations in gene expression, affecting as much as 5% of cellular message (1), or about 1,000 genes (2). These alterations in gene expression may be related to the fact that most if not all properties of tumor cells are observed in some normal cells at some time during normal development, a biologic principle first enunciated in the early 19th century by Laennec (3). Since DNA methylation can play an important role in the control of normal gene expression, it has been postulated that altered DNA methylation might be involved in the abnormal gene expression that characterizes cancer (4). While this is a tantalizing hypothesis, there is as yet no experimental evidence to support it.

Second, 5-azacytidine (5-azaCR), which causes hypomethylation of DNA and is nonmutagenic (5), nevertheless transforms cells at high frequency, both _in vitro_ and _in vivo_ (6). Since experiments showing differentiation _in vitro_ caused by 5-azaCR have been considered strong arguments for a role of DNA methylation in differentiation, a similar argument can be made for a role in transformation. Third, an abundant body of circumstantial evidence suggests a role for altered DNA methylation in experimental and human tumors. In particular, we and others have shown that 5-methylcytosine content decreases approximately 10% in human malignancies (7,8). Furthermore, individual genes normally fully methylated in human colonic mucosa, such as γ-crystallin and γ-globin, are substantially hypomethylated in all colorectal carcinomas and premalignant adenomas (8,9,10), indicating that DNA hypomethylation is global and occurs prior to malignancy. Thus, hypomethylation is the earliest defined change in multistep human carcinogenesis, and it affects as many as 1/3 of single copy genes. In human colorectal cancer, we have shown that altered DNA methylation precedes, perhaps by years, development of mutations in cellular oncogenes and losses of tumor suppressor genes on chromosomes 5, 17, and 18 (11). Indeed, all human malignancies that have been systematically examined show alterations in DNA methylation (6).

There are several biological limitations to proving a role for altered DNA methylation in human carcinogenesis. First, tumor cell heterogeneity will confound an analysis of critical alterations in DNA methylation in early carcinogenesis. Heterogeneity in morphological characteristics, elaboration of biochemical markers, and resistance to chemotherapy and radiation, all must involve changes in gene expression and perhaps DNA methylation, as well. Second, human tumors are substantially contaminated with normal cells. Since DNA methylation shows tissue specific differences, even a small fraction of normal contaminating cells will perturb a molecular study. Third, one must identify the exact normal precursor cell in order to compare meaningfully DNA methylation in transformed cells to that in nontransformed cells. However, that is quite difficult to know with certainty, since most tissues include diverse cell types, only one of which is the progenitor of a given type of malignancy. Fourth, and most importantly, since we have shown that altered DNA methylation is already well established even in the smallest premalignant tumors (8), one would need to isolate cells while they still appear normal, in order to determine whether changes in DNA methylation play a causal role in malignant progression.

We therefore sought to develop an in vitro model system, using an agent that causes hypomethylation of DNA, that would enable us to capture cells destined to become transformed, but prior to any apparent morphological or physiological transformation-specific changes. The ideal experimental in vitro system would have the following three attributes: (i) the cells should have a very low background transformation frequency; (ii) they should be easily transformed by analogues of 5-azaCR; and (iii) they should show a long latency of transformation after treatment, so that we might isolate cells before they have become transformed. We chose for these studies the mouse cell line C3H 10T1/2, which is the most widely used cell line for chemical carcinogenesis studies (5,12,13,14,15), largely because it possesses the first two properties listed above. In addition, there is a long latency between treatment with 5-azaCR and appearance of transformed foci (4-6 weeks). We therefore thought that it might be possible to identify cells committed to transformation, but prior to phenotypic transformation.

We used the following strategy to isolate cells at the earliest stages of transformation. We treat individual cells in microwells with 5-aza-2'-deoxycytidine (5-azaCdR) and allow them to grow for several weeks. The cells are harvested prior to confluence, a fraction of the cells are cryopreserved for later characterization, and the rest are replated. In this way, we can isolate pretransformed cells and ultimately identify the genes which are activated in the cells committed to transformation. We can then examine what role perturbations in DNA methylation play in the activation of these genes.

In order to isolate the genes specific for transformation in this system, a cDNA library was constructed from a transformant and differentially screened. Presumably, there are a large number of genes activated following 5-azaCdR treatment. An advantage of our system is that, in addition to cells committed to transformation, we can also isolate those cells which have been treated identically yet do not become transformed. Thus, we can determine which genes are activated nonspecifically by treatment with 5-azaCdR, and distinguish them from genes expressed specifically in transformation-committed cells. By differential screening, we have now isolated more than one hundred clones, which we are currently characterizing.

MATERIALS AND METHODS

Cell Culture and Treatment

We recloned early passage C3H 10T1/2 clone 8 cells (13) (American Type Culture Collection), which were then maintained in basal medium Eagle's with Earle's salt (Sigma), glutamine (2 mM), 10% (vol/vol) fetal calf serum (Gibco), penicillin (50 U/ml), streptomycin (50 µg/ml) and fungizone (0.25 µg/ml).

Clonal transformation was performed as described (16). Briefly, recloned cells were plated at limiting dilution in 24-well dishes to obtain zero to one cell per well by Poisson distribution. The following day, 1 µM 5-azaCdR, dissolved in phosphate buffered saline (PBS), was added to the dishes (control dishes were given an equal volume of PBS). After 24 hours, the cells were washed with PBS and refed. Subsequently, the cells were fed twice weekly.

Four weeks later, the individual microcolonies, which were less than 50% confluent, were washed with Hank's balanced salt solution, trypsinized and subdivided. Three parts were cryopreserved in 10% (vol/vol) dimethyl sulfoxide. The fourth aliquot of each microcolony was replated in a 24-well dish and fed twice weekly. These replated microcolonies were observed at least once a week for transformation.

Transformation Assays

Cells were assayed for growth rate, serum independence, and saturation density (17). The microcolonies were assayed in soft agar as previously described (18), but in 24-well dishes.

Library Construction and Screening

RNA was harvested from confluent dishes (19), and poly A+ RNA was purified by oligo(dT) cellulose chromatography (20,21). A λgt10 library was prepared (22), and the library was packaged (Gigapack Gold, Stratagene) and plated on the C600 Hfl- strain of E. coli. The plaques were transferred to nylon by established methods (23). For library screening, RNA was derived from transformants, pretransformants, and treated but nontransformed cells. Probes were prepared by oligo-dT primed synthesis of first strand cDNA with AMV reverse transcriptase in the presence of 1 mCi ^{32}P-dATP (6000 Ci/mmole, Amersham). The probes were alkaline-treated, precipitated with linear acrylamide, and denatured in water. Inserts from λgt10 cDNA clones were oligolabeled to probe Northern blots as described (24).

RESULTS

Capture and Characterization of Pretransformed C3H 10T1/2 Cells After Treatment With 5-azaCdR

We have developed a novel experimental strategy (16) to isolate cells determined for transformation by 5-azaCdR (Figure 1). Individual cells in 24-well dishes are treated for 24 hours with 1 µM 5-azaCdR, washed and refed. The cells are allowed to grow to subconfluence (approximately 4-6 weeks). Each colony is trypsinized and subdivided for cryopreservation and replating. The replated cells are allowed to grow and are observed for transformation. Typically, approximately 10% of the surviving colonies become morphologically transformed. The ancestral cryopreserved cells of all of the transformed clones, as well as an equal number of treated but nontransformed clones, are replated and characterized for morphology, growth rate, saturation density and growth in soft agar.

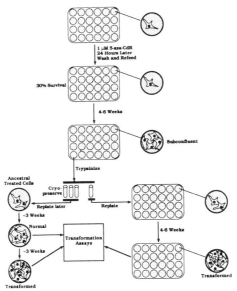

FIGURE 1. <u>Experimental strategy.</u> Individual cells in microwells are treated with 5-azaCdR. At subconfluence, 3/4 of each microcolony is cryopreserved and the remainder is maintained in culture for an additional 4-6 weeks (bottom right). The ancestral cryopreserved cells of transformants and nontransformants are then assayed periodically (bottom left). Reprinted from <u>Proc Natl Acad Sci USA</u> (ref. 16) with permission.

The treated but nontransformed ancestral cells displayed the normal monolayer morphology (Figure 2B), similar to the corresponding descendent cells (Figure 2A). Therefore, cryopreservation had no detectable effect on these cells.

The morphology of the descendent cells of a representative transformant is shown in Figure 2C. The transformed cells showed loss of contact inhibition of growth and appeared as a criss-crossed multilayered array of cells, indicative of transformation. When the ancestral cryopreserved cells of this clone were replated and allowed to grow to confluence, the cells displayed a normal morphology (Figure 2D). However, when the ancestral cells were grown in culture for an additional 20 days before confluence, they recapitulated a transformed morphology. These morphological changes suggest that we were able to isolate pretransformed cells, which we define as cells committed to transformation yet phenotypically normal.

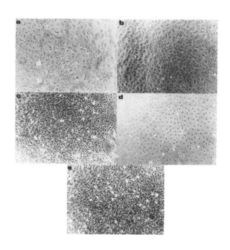

FIGURE 2. Morphology of 5-azaCdR-treated untransformed, pretransformed, and transformed cells. Individual microcolonies, isolated as described in Fig. 1, were grown to confluence on 25 cm^2 dishes for phase contrast photomicrography (x25). (a) 5-azaCdR-treated untransformed descendant microcolony; (b) ancestral cells of untransformed microcolony shown in lane (a), 77 days after replating; (c) 5-azaCdR-treated transformed descendant microcolony; (d) pretransformed ancestral cells of microcolony shown in lane (c), 22 days after replating; and (e) ancestral cells of microcolony shown in lane (c), 50 days after replating. Reprinted from Proc Natl Acad Sci USA (ref. 16) with permission.

We found that while the saturation densities of the transformants were on average two-fold greater than those of the treated but nontransformed microcolonies, the growth rates were not significantly greater in the transformants (16). Since we maintained cells at subconfluence and passaged them as little as possible, we had not selected for growth rate per se, an important advantage of this experimental design.

Growth in soft agar correlated well with the transformed phenotype. All of the transformed microcolonies and an equal number of treated but nontransformed colonies and their pretransformed ancestral cells were examined for growth in soft agar. The time of appearance of morphological transformation and growth in soft agar after cryopreservation varied from 30 to 50 days for the pretransformants (Table 1).

TABLE 1

GROWTH CHARACTERISTICS OF 5-AzaCdR-TREATED CELLS

Cell Type	Morphological Transformation	Growth in in Soft Agar
Transformed descendants	9/9	9/9
ancestral cells-day 14	0/9	0/9
ancestral cells-day 21	3/9	2/9
ancestral cells-day 28	4/9	4/9
ancestral cells-day 56	9/9	9/9
Nontransformed descendants	0/9	0/9
ancestral cells-day 14	0/9	0/9
ancestral cells-day 56	0/9	0/9

In contrast, the ancestral cells of treated but nontransformed colonies did not eventually grow in soft agar (Table 1). Thus, ancestral pretransformed cells were inevitably committed to eventual transformation.

We considered the possibility that the frozen cryopreserved vials of ancestral pretransformed cells could have contained a completely transformed cell that had yet to overgrow the cells within those vials, therefore misleading us in our interpretation. However, sister vials of pretransformants or nonpretransformants always contained cells of the same type. Furthermore, since the growth rates of the pretransformants and nontransformants were roughly the same, the only way a transformed subpopulation of cells could overgrow the other cells on the dish was if the cells were allowed to remain confluent for a substantial period of time. In the growth of the ancestral cells, cultures were permitted only to become subconfluent before they were split, so this problem could not arise. Finally, in three instances, vials of pretransformants were themselves plated at limiting dilution to single cells, to determine whether or not pretransformation was a property of all of the cells within a given cryopreserved vial. In two of the three cases, nearly all of the individually plated ancestral cells eventually became transformed, and in the third, half became transformed. Therefore, we believe that most or all of the cryopreserved ancestral cells in a given vial are in fact committed to transformation. Similarly, single-cell isolates of all ancestral nonpretransformants, observed over the same interval, did not become transformed. Therefore, we have isolated pretransformed cells determined for transformation, after treatment with 5-azaCdR.

Strategies for Identifying Genes Specific for Transformation and Pretransformation

We have adopted several strategies for isolating transformation-specific genes from 5-azaCdR-treated cells. One of these approaches is differential hybridization to screen cDNA libraries derived from transformants or pretransformants, using labeled cDNA derived from transformants, pretransformants, and treated but nontransformed cells. This approach will enable us to determine whether genes are expressed specifically in transformants or pretransformants. The treated but nontransformed cells serve as the ideal control, since they will distinguish those genes activated nonspecifically by 5-azaCdR.

This approach is limited to a degree by the abundance of the message in the probe, and we thus might miss relatively nonabundant transcripts that are differentially expressed. For this reason, we are also attempting two other approaches for isolating transformation-specific genes: subtracted library screening, and DNA-mediated gene transfer. However, those experiments are beyond the scope of this review.

In preliminary experiments using differential hybridization, a λgt10 cDNA library, comprising 2×10^6 recombinant clones, was constructed from a representative transformant. We screened 1.7×10^5 clones on 32 175-cm^2 filters, probing the filters serially with radioactive cDNA prepared from two transformed microcolonies and two treated but nontransformed microcolonies. We picked the plaques which hybridized with the probes derived from transformants, but did not hybridize with probes derived from nontransformed colonies. These differentially hybridizing plaques were replated at very low density for plaque purification, and differentially screened again. In this manner, 140 clones were isolated, which represent less than 0.1% of the plaques. The inserts from several λgt10 cDNA clones were excised from agarose, labeled and hybridized to Northern blots. RNA blot hybridization of a representative clone showed significant expression of two large transcripts (~6-7 kb) in transformants, and extremely low levels of these transcripts in both untreated and treated but nontransformed microcolonies (Figure 3). The mRNA was also expressed in pretransformed cells (Figure 3), indicating that this gene is activated early in the transformation process. Thus, we were able to identify genes expressed specifically and early in cells committed to transformation, and we could distinguish these from genes activated nonspecifically by 5-azaCdR.

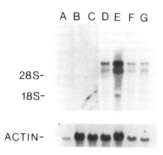

FIGURE 3. RNA from transformants, pretransformants, untreated cells, and treated but nontransformed cells, probed with a cDNA identified by differential screening. Total RNA was extracted from confluent cultures, electrophoresed on a 1% agarose/6% formaldehyde gel, and blot hybridized with the 1.8 kb Eco RI insert of a differentially detected λgt10 clone. The two prominent bands represent ~6 and ~7 kb mRNA. Sources of RNA from which labelled cDNA probes were prepared are as follows: (a) untreated recloned C3H 10T1/2 cells; (b,c) 5-azaCdR-treated but nontransformed microcolonies; (d,f,g) transformed 5-azaCdR-treated microcolonies; (e) ancestral pretransformed 5-azaCdR-treated cells, progenitors of transformants represented in (d).

DISCUSSION

We have captured 5-azaCdR-treated C3H 10T1/2 cells prior to transformation. To our knowledge, this is the first time that cells destined to become transformed without further manipulation have been isolated. In parallel, we have isolated cells treated in the identical manner, but that will not become transformed. Pretransformed cells display a normal morphological and physiological phenotype, but most or all of these cells eventually become transformed in culture without further intervention. Our strategy permits identification of genes expressed specifically in transformed and pretransformed cells, and we have begun to isolate these genes.

We would like to know whether these genes, expressed specifically in pretransformants and transformants, are known genes or genes whose role in carcinogenesis has yet been appreciated. In addition, we would like to determine whether there is a relatively small subset of genes expressed specifically in pretransformants. Our preliminary data suggest that this is so, since only 0.1% of plaques hybridized specifically to cDNA generated from cells committed to transformation.

Importantly, while 140 plaques specific for transformation were isolated, more than 10 times as many hybridized with probes derived from treated but nontransformed cells. Thus, our experiments enable us to distinguish those genes whose expression is specific for transformation from the much larger number activated nonspecifically by 5-azaCdR. We could show, for example, that sequences homologous to the Moloney murine leukemia virus long terminal repeat, which were reportedly expressed early in 5-azaCdR-induced transformation of C3H 10T1/2 cells (17), were in fact expressed nonspecifically as a result of drug treatment (16).

We also wish to determine whether the genes we identify are involved in other pathways of carcinogenesis. This should be relatively easy to do, since C3H 10T1/2 cells transformed by a variety of agents are readily available. Finally, we would like to determine the effect of the genes we identify when expressed experimentally in normal cells.

Even if the genes we identify are known genes, their mechanism of activation should be of considerable interest, given the fact that 5-azaCdR was used to transform the cells. Since the transformation frequency after treatment with 5-azaCdR was several logs higher than the mutation frequency, at least as scored by ouabain resistance (data not shown), point mutation is unlikely to be the mechanism by which 5-azaCdR transforms these cells. While the mechanism could be direct activation of genes by altered DNA methylation, another possibility is suggested by studies of Schmid, et al. (25). They observed that 5-azaCR causes decondensation of centromeric heterochromatin in normal human cells, which leads to chromatid interchanges and chromosomal rearrangements. These rearrangements might alter the structure of transforming genes or cause the loss of tumor suppressor genes.

Therefore, it is possible that altered methylation at a critical gene or genes leads to genetic instability at other loci. While genetic instability has been a subject of considerable speculation for decades, there is as yet no direct experimental proof that it is involved in human carcinogenesis. However, our studies of human colorectal cancer suggest a possible relationship between altered DNA methylation and genetic instability. We have found that hypomethylation is the first and earliest observed alteration in premalignant tumors, and it is followed by genetic alterations at multiple chromosomal loci (11). One of the exciting prospects of the strategy described here is that it may enable us to dissect out a potential cascade of molecular alterations, in which altered DNA methylation at one locus leads to alterations of other genes.

Our studies should also provide useful insights into early changes in gene expression in transformation. Studies of abnormal gene expression in transformed cells suggest that a surprisingly large number of genes are involved. For example, solution hybridization studies of virally-transformed cells suggest the novel expression of over 1,000 genes, even though transformation is presumably due to a single causal event (2,26). Other studies indicate a change in expression of 3-5% of genes after transformation, most of which are nonspecific (1). By isolating cells prior to transformation, we should be able to determine which changes in gene expression in transformation are specific for transformation and might play a causal role in defining the transformed phenotype.

Ultimately, we would like to know whether the differentially expressed genes in transformants and pretransformants are also activated in human tumors. Since we previously found so many genes hypomethylated in human colorectal cancer, we plan to reexamine those tumors, to determine whether conserved genes activated in pretransformants might also play a causal role in human malignancy.

ACKNOWLEDGEMENTS

We thank Anne Brancheau for preparing the manuscript. This work was supported in part by the National Institutes of Health (grant CA48932 to Andrew P. Feinberg, M.D., M.P.H.).

REFERENCES

1. Scott MRD, Westphal KH, Rigby PWJ (1983). Activation of mouse genes in transformed cells. Cell 34:557.
2. Grady LJ, Campbell WP (1973). Nonrepetitive DNA transcripts in nuclei and polysomes of polyoma-transformed and nontransformed mouse cells. Nature New Biol 243:195.
3. Pitot HC (1981). Fundamentals of Oncology, Marcel Dekker.

4. Holliday R (1979). A new theory of carcinogenesis. Br J Cancer 40:513.
5. Landolph JR, Jones PA (1982). Mutagenicity of 5-azacytidine and related nucleosides in C3H 10T1/2 clone 8 and V79 cells. Cancer Res 42:817.
6. Riggs AD, Jones PA (1983). 5-methylcytosine, gene regulation, and cancer. Adv Cancer Res 40:1.
7. Gama-Sosa MA, Slagel VA, Trewyn RW, Ehrlich M (1983). The 5-methylcytosine content of DNA from human tumors. Nucleic Acids Res 11:6883.
8. Feinberg AP, Gehrke CW, Kuo KC, Ehrlich M (1988). Reduced genomic 5-methylcytosine content in human colonic neoplasia. Cancer Res 48:1159.
9. Feinberg AP, Vogelstein B (1983). Hypomethylation distinguishes genes of some human cancer from their normal counterparts. Nature 301:89.
10. Goelz SE, Vogelstein B, Hamilton SR, Feinberg AP (1985). Hypomethylation of DNA from benign and malignant human colon neoplasms. Science 228:187.
11. Law DJ, Olschwang S, Monpezat J-P, Lefrancois D, Jagelman D, Petrelli N, Thomas G, Feinberg AP (1988). Concerted nonsyntenic allelic loss in human colorectal carcinoma. Science 241:961.
12. Reznikoff CA, Bertram JS, Brankow DW, Heidelberger C (1973). Quantitative and qualitative studies of chemical transformation of cloned C3H mouse embryo cells sensitive to postconfluence inhibition of cell division. Cancer Res 33:3239.
13. Reznikoff CA, Brankow DW, Heidelberger C (1973). Establishment and characterization of a cloned line of C3H mouse embryo cells sensitive to postconfluence inhibition of division. Cancer Res 33:3231.
14. Kennedy AR, Cairns J, Little JB (1984). Timing of the steps in transformation of C3H 10T1/2 cells by X-irradiation. Nature 307:85.

15. Male R, Bjerkvig R, Lillehaug JR (1987). Biological and biochemical characterization of cell lines derived from initiation-promotion transformed C3H 10T1/2 cells. Carcinogenesis 8:1375.
16. Rainier S, Feinberg AP (1988). Capture and characterization of 5-aza-2'-deoxycytidine-treated C3H 10T1/2 cells prior to transformation. Proc Natl Acad Sci USA 85:6384.
17. Hsiao WL, Gattoni-Celli S, Weinstein IB (1985). Effects of 5-azacytidine on the progressive nature of cell transformation. Mol Cell Biol 5:1800.
18. Macpherson I, Montagnier L (1964). Agar suspension culture for the selective assay of cells transformed by polyoma virus. Virology 23:291.
19. Strohman RC, Moss PS, Micou-Eastwood J, Spector D, Przybyla A, Paterson B (1977). Messenger RNA for myosin polypeptides: isolation from single myogenic cell cultures. Cell 10:265.
20. Edmonds M, Vaughan MH Jr, Nakazato H (1972). Polyadenylic acid sequences in the heterogeneous nuclear RNA and rapidly-labeled polyribosomal RNA of HeLa cells: possible evidence for a precursor relationship. Proc Natl Acad Sci USA 68:1336.
21. Aviv H, Leder P (1972). Purification of biologically active globin messenger RNA by chromatography on oligothymidylic acid-cellulose. Proc Natl Acad Sci USA 69:1408.
22. Gubler U, Hoffman BJ (1983). A simple and very efficient method for generating cDNA libraries. Gene 25:263.
23. Maniatis T, Fritsch EF, Sambrook J. Molecular Cloning: A Laboratory Manual, p321, 1982.
24. Feinberg AP, Vogelstein B (1984). A technique for radiolabeling DNA restriction endonuclease fragments to high specific activity. Anal Biochem 132:6.
25. Schmid M, Haaf T, Grunert D (1984). 5-azacytidine-induced undercondensations in human chromosomes. Hum Genet 67:257.

26. Groudine M, Weintraub H (1980). Activation of cellular genes by avian RNA tumor viruses. Proc Natl Acad Sci USA 77:5351.

THE ACTIVATION/INACTIVATION-PRONE PROLACTIN GENE IN GH$_3$ RAT PITUITARY CELLS: SILENCING BY EMS AND REACTIVATING BY 5-AZACYTIDINE[1]

Iain K. Farrance[2], Julie Morris, Todd E. Arnold, Iris S. Hall[3], and Robert Ivarie

Department of Genetics, University of Georgia, Athens, Georgia 30602

ABSTRACT GH$_3$ rat pituitary tumor cells express two pituitary-specific polypeptide hormone genes encoding prolactin (PRL) and growth hormone (GH). Epigenetic variants were induced at high frequency by EMS and reverted by 5-azacytidine, a DNA methylation inhibitor. Here, we summarize evidence that EMS induced methylation of a CpG site(s) in the prolactin gene, or a gene regulating its expression, that led to its transcriptional inactivation. PRL-deficient cells are defective in utilizing the PRL promoter efficiently and contain all transcription factors necessary to initiate transcription at the PRL promoter. Thus, the PRL gene itself appears to be the EMS target that is modified to enhance its enzymatic methylation *in vivo*. EMS-modified CpG sites stimulate DNA methylase activity and a molecular model is presented on how a guanine N7 ethyl may create a fraudulent "hemimethyl-

---

[1]This work was supported by NCI grant CA34066 to RI. IF and TA were supported by an NIH training grant GM07103 and IH by a University of Georgia Fellowship.
[2]**Present address:** Department of Anatomy, University of California, San Francisco, CA 94143.
[3]**Present address:** New England Biolabs, 32 Tozer Rd., Beverly, MA 01915

ated" CpG site. An EMS target sequence may be one of all of the six TG-elements found in the PRL gene.

INTRODUCTION

GH_3 cells are an established tumor cell line from rat anterior pituitary that express two pituitary-specific polypeptide hormones (1,2), prolactin (PRL) and growth hormone (GH). Thus, they mimic stem cells that appear transiently during pituitary differentiation that give rise to PRL-producing mammotrophs and GH-producing somatotrophs in the adult gland (3-5).

Several years ago, we set out to isolate GH- and PRL-deficient mutant lines by conventional mutagenesis and made an unusual observation: PRL^- variants were induced at >10% frequency by ethyl methanesulfonate (EMS) and reverted at ~50% frequency by inhibiting DNA methylation with 5-azacytidine (6,7). Both the rate of PRL synthesis and its mRNA levels were reduced by >50-fold in some variants (6,8). The EMS effect was specific to PRL because no other gene in GH_3 cells detected on two-dimensional gels, including the GH gene, was silenced in randomly selected PRL^+ clones after EMS mutagenesis (6). Also, variants did not appear to arise via DNA repair because PRL^- lines were not induced by UV-irradiation (7).

Here, we summarize our findings on the properties of the lines and how they were generated by EMS at the level of DNA methylation. We ask, first, how does EMS modify CpG sites to stimulate methylation of cytosine residues by DNA methylase, and second, are the deficient cells defective in PRL transcription or in some post-transcriptional process?

RESULTS AND DISCUSSION

Genetic Stability of PRL⁻ Lines and the Isolation of GH⁻ Subclones

An unusual property of the prolactin-deficient lines is their genetic instability (Figure 1). One line (B2) reverted to wild-type at ~10% frequency even after several rounds of serial subcloning for the deficient phenotype while another (B3) gave rise to only one wild-type revertant in over 320 clones (6,7). Hence, there was a wide range in the reversion frequencies at least as assayed by these two lines (e.g., 0.3-10%).

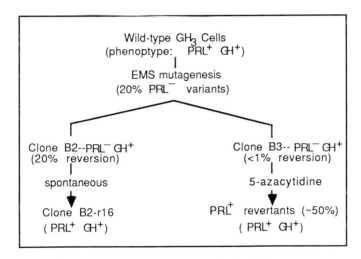

Figure 1. Cells and their lineages.

We also recovered GH-deficient lines from the genetically unstable B2 line in experiments where we asked whether PRL⁺ revertants of the B2 line were as unstable as the PRL⁻ parent (8). When w-t B2 revertants were subcloned, the "on" state of the prolactin gene in several clones was stable, but surprisingly, one gave rise to >90% GH⁻ subclones. Two of these GH⁻ clones were analyzed and found to have suffered a lesion(s) reducing GH synthesis and its mRNA levels by ~25-fold. The lines were slow growing and eventually lost. The nature of their defect(s) is completely unknown.

The EMS Hypothesis

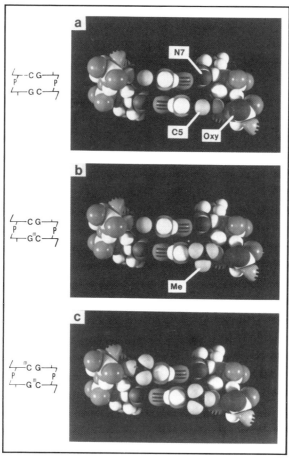

Figure 2. CPK space-filling models of the CpG methylation site.

To explain both the high frequency of variant induction by EMS and spontaneous reversion, we argued that EMS had induced enzymatic methylation of CpG sites. As one test of this idea, the low reverting B3 line was treated with the DNA methylation inhibitor, 5-azacytidine, and then subcloned to measure reversion to wild-type (7). Reversion rose to ~50% at doses of 5-azacytidine giving ~20% survival. Thus EMS induction of at least one PRL⁻ line appeared to be coupled to methylation of cytosines in GH_3 DNA.

A model by which EMS might promote methylation of CpG sites by DNA methylase is illustrated in Figure 2 (12). It is based on known properties of the mammalian DNA methylase. A major site of methylation in vertebrate DNA is CpG (9,10). A methyl from SAM is transferred to the C5 position

Epigenetic Variants in Prolactin Expression 233

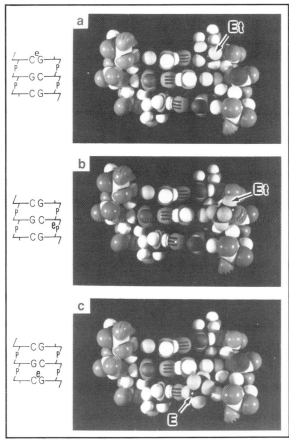

Figure 3. CPK space-filling models of a *hemiethylated* GpCpG site.

of cytosine by the DNA methylase. CpG sites are symmetrically methylated in genomic DNA (Fig.2c). When fully methylated sites are replicated, hemimethylated sites (Fig.2b) are created that are methylated by the enzyme much more efficiently than are unmethylated sites (11). The C5 position of the pyrimidine ring lies in the major groove of B DNA ~4.6Å from guanine N7, the major site of alkylation by EMS (13). An ethyl at guanine N7 might rotate to cytosine C5 and "mimic" a hemimethylated site. If recognized by the enzyme, a methyl would be introduced on cytosine on the opposite strand and subsequently transmitted to daughter cells by maintenance methylation.

As Fig. 3a shows, an ethyl at guanine N7 3' to cytosine cannot reach cytosine C5 because of steric interference with the pyrimidine ring. When located 5' to cytosine, however, the ethyl can rotate near cytosine C5 (Figure 3c). Closest proximity to cytosine C5 is achieved by a major groove proximal ethyl triester (Fig. 3b: an ethyl on a free oxygen of the phosphodiester bond). EMS places ~70% ethyls at guanine N7 and ~20% at the phosphodiester (13).

EMS-Modified Poly(dC-dG)·Poly(dC-dG) Stimulates DNA Methylase Activity

To test the model directly, an alternating dC-dG copolymer, averaging 650 bp in length, was modified with EMS *in vitro* and assayed for its ability to accept methyls from partially purified rat liver DNA methylase (12). The enzyme was over 1,000-fold purified from cell protein but only ~50-fold purified from nuclear proteins. It lacked detectable nuclease and phosphatase activities as well as protein and RNA methylase activities. The enzyme preparation also exhibited a number of properties characteristic of more purified mammalian DNA methylases: (1) G+C-rich DNAs were better substrates than A+T-rich DNAs; (2) enzyme activity was highly salt sensitive; (3) 5-methylcytosine was the only base detectably methylated by the enzyme; and (4) the enzyme was strongly stimulated by hemimethylated poly(dC-dG)·poly(dC-dG).

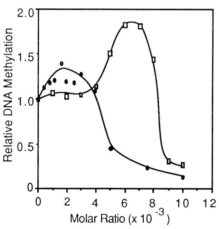

FIGURE 4. Activity of DNA methylase on dC-dG copolymers treat with MMS (●) or EMS (□).

Methylase activity on poly(dC-dG)·poly(dC-dG) at varying molar ratios of EMS:CpG is shown in Figure 4. At ratios of 6-7,000, there was ~2-fold stimulation in methylation of cytosine residues followed by a sharp drop in the capacity of the substrate to accept methyls. This drop largely reflected depurination because of acidification during the modification reaction in relatively weak buffer. If buffer strength was increased, the inhibition was no longer observed.

At the peak of stimulation, the extent of ethylation of guanine N7 averaged 2.5% in the polymer. A crude estimate of the increased affinity of the enzyme for a "hemi-ethylated" CpG site is summarized in Table 1. If no hemimethylated sites occurred in an alternating dC-dG polymer of 100 CpG sites (or 200 residues), ~1 methyl would be transferred per unit time. Using the data of Gruenbaum et al. (11) and assuming all sites are hemimethylated (or 50% modification), then 100 methyls would be transferred per unit time. If only one site were hemimethylated or 0.5% modification, then 2 methyls would be transferred-- one to the 99 unmethylated sites and one to the hemimethylated site. Because 2.5% guanines were ethylated at N7 and the stimulation was 2-fold, we estimate the affinity for a 7-ethylguanine-containing CpG site to be 1/5th (0.5%/2.5%) the affinity for a natural hemimethylated site, or ~20-fold.

Table 1. Estimating the affinity of partially purified DNA methylase for an ethylated CpG

DNA	% C or G modified	Relative affinity	Relative incorp.
(CpG:GpC)	0	1x	1
(^{m}CpG:GpC)	50	100x	100
(^{m}CpG:GpC) N7-ethylated	.5	100x	2
(CpG:GpC)	2.5	~20x	2

Site-Specific Placement of N7-Ethylguanine

The foregoing experiment leaves two questions unanswered. First, is it the alkyl group at N7 itself or the positive charge introduced on the imidazole ring of guanine that stimulates DNA methylase? In the former, the length of the alkyl group would influence activity while in the latter stimulation would largely be independent of length and reflect instead a helix distortion. This latter possibility is not as unlikely as it might seem since Smith et al. (14) have shown that the enzyme can be stimulated by certain mismatches in CpG sites. Second, is N7 guanine the only ethylated site that can stimulate the enzyme or might the ethyltriester or O^6 ethyl also be involved in the stimulation?

To address the first question, we asked whether poly(dC-dG)·poly(dC-dG) modified by *methyl* methanesulfonate (MMS) was as good a substrate for DNA methylase as the ethylated copolymer. As Figure 4 shows, MMS-modified copolymer also stimulated the enzyme but at much lower MMS:CpG ratios. Although the extent of guanine N7 methylation was not determined, the shift in the curve to lower ratios is consistent with the high reactivity of MMS relative to EMS (13). The lower stimulation of methylase activity at what is likely to be higher levels of DNA modification implies that the length of the alkyl group, not the positive charge on the imidazole ring, is more important for stimulation.

We also tested whether ethyl nitrosoguanidine- and ethylnitrosourea-modified copolymers accepted methyls more efficiently than an unmodified substrate. ENU and ENNG produce more ethylphosphotriesters in DNA than does EMS (14) However, ENNG- and ENU-modified poly(dC-dG)·poly(dC-dG) accepted methyls no better and no worse than an unmodified substrate. This result implies that the ethyltriester is not the adduct responsible for stimulation of methylase activity.

Site-Specific Placement of N7-Ethylguanine in a Methylation Site 5' and 5' to Cytosine

To test directly whether the N7 ethyl is responsible for enzyme stimulation, N7-ethyldGTP has been synthesized and used to place N7-ethylguanine 5' and 3' to cytosine in a GCG site by the combined use of terminal transferase and DNA polymerase I (Figure 5; 15,16). N7-ethyldGTP was synthesized by direct alkylation of dGTP with diethylsulfate, purified by TLC, and shown to be N7-ethyldGTP by its chemical and spectrophotometric properties (16). It was tailed onto the 3' end of a 13-mer (lane 1, inset) which abuts the 3' G in the GCG site with an average of ~2 N7-ethyldGMP's per primer (lane 2). The tailed primer was hybridized to the template 35-mer, and the unpaired N7-ethyldGMP residues removed and the primer extended by the Klenow fragment of DNA polymerase I (lane 3).

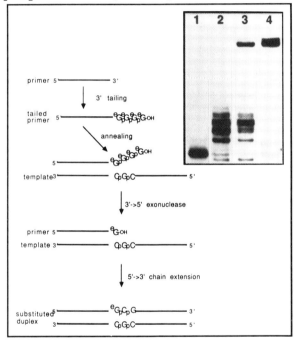

Figure 5. Site-specific placement of N7-ethylguanine in a GCG site.

The yield of purified 35-mer duplex (lane 4) has ranged from 10-12% relative to the initial primer concentration with ~84% N7-ethylguanine at the 3' site as determined by its piperidine sensitivity. Experiments are underway to test the efficiency with which substrates containing N7-ethylguanine 5' and 3' to cytosine can accept methyls relative to unmodified and hemimethylated copolymers.

PRL⁻ Cells Contain Reduced Levels of Properly Initiated, Processed and Terminated PRL Transcripts

We had shown initially that the rate of synthesis of the PRL polypeptide in EMS-induced prolactin-deficient variants was accompanied by a comparable drop in the levels of its cytoplasmic mRNA (6,8). Additional experiments were undertaken to determine whether the defect in the deficient B3 line occurred in the initiation or termination of transcription or in RNA processing.

The PRL gene is 10 kb in length (17,18) and its 1 kb coding sequence is interrupted by four intervening sequences (Figure 6). Primer extension and S1 mapping analysis (19) has shown that the rare PRL transcripts in deficient cells contain normal 5' and 3' termini implying that deficient cells are utilizing the normal PRL promoter and transcriptional terminator. Furthermore, Northern blot analysis has shown that PRL mRNA processing intermediates are the same but substantially reduced in the variant line compared to wild-type and revertant cells. Taken together, these data indicate that the rare transcripts in the deficient line are correctly initiated, processed and terminated.

Although these data are also consistent with an enhanced rate of PRL mRNA turnover in PRL⁻ cells, increased degradation would have to occur at the level of primary nuclear transcripts to explain the Northern blot data. More recent experiments, however, have supported the conclusion that the deficient line is defective in utilizing the endogenous promoter effectively (see below).

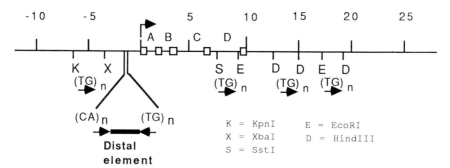

Figure 6. Schematic of the rat prolactin gene with the locations of 6 known TG-repeat elements and the distal promoter element.

Cis vs Trans Mechanisms Inactivating the PRL Gene

The foregoing experiments could not distinguish between two possible gene targets by which EMS could inactivate the PRL gene. In the first (*cis* model), the PRL gene itself is the EMS target: an ethylated CpG site(s) in the PRL gene is enzymatically methylated by the maintenance methylase and silences the gene. In the second (*trans* model), a PRL-specific transcription factor is silenced via EMS-induced methylation of a CpG site(s) in its gene which in turn leads to PRL gene inactivation.

To distinguish between these two possibilities, the PRL-promoter was fused to the bacterial chloramphenicol acetyltransferase gene (CAT) and the chimeric constructs transfected into wild-type and PRL-deficient GH_3 cells. The prolactin gene promoter contains two elements that have been shown to control gene's transcription (Figure 7). The distal element is flanked by TG-repeats and contains four sites which bind pituitary transcription factor pit-1, which is required for both PRL and GH expression (20-23). This region also contains an estrogen response element. The proximal promoter region is just upstream from the TATAA box and contains three pit-1 binding sites and a site binding a "common" transcription factor. Both the proximal and distal promoter elements have been shown to confer "cell-type" specificity to PRL gene transcription when assayed in pituitary and nonpituitary cell types (22).

Both the proximal promoter construct as well as the construct containing both elements exhibited cell-type specificity in that CAT activity has been detected in w-t GH_3 cells but not in rat-2 fibroblasts nor in rat hepatoma cells (HTC). The distal element increased CAT activity by ~8-fold consistent with observations on the effects of upstream element on the proximal promoter. In PRL-deficient B3 cells, both constructs were as active as they were in wild-type cells (24).

These results imply that PRL-deficient B3 cells are fully competent to initiate transcription

from the PRL promoter. They also suggest that the
cis model is a more likely explanation for how B3
cells arose by EMS treatment than the *trans* model.
Hence, B3 cells must contain all the transcription
factors (cell type-specific and common) necessary
to initiate transcription from this region of the
PRL gene. Consistent with this conclusion, DNaseI
footprinting and mobility shift assays have shown
that whole cell and nuclear extracts from wild-type
and PRL-deficient cells do not differ in pit-1
levels nor in other factors binding to the proximal
and distal elements (19). It is possible, however,
that other regions of the gene also influence the
activity of the PRL promoter and these are
currently under analysis (see below).

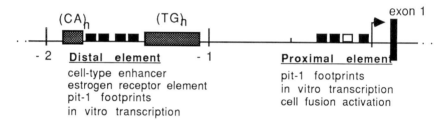

Figure 7. Proximal and distal elements of the rat PRL promoter locating *cis* regulatory sites.

Prolactin Gene Inactivation and the Simple Repeat Poly(dT-dG)·Poly(dC-dA)

If the prolactin gene itself is the EMS
target, some sequence feature of the PRL gene must
make it much more susceptible to EMS-induced
enzymatic methylation than other genes in GH_3
cells. Two-dimensional gel analysis of PRL^+ EMS
survivors (6) showed that other genes active in GH_3
cells were not detectably inactivated. The PRL
gene certainly is not susceptible by virtue of
containing G+C-rich regions. Rather, sequenced
regions of the gene (~9 kb; 18) indicate that it is
a "low" density CpG gene containing only 48 CpGs
(0.5%), somewhat lower than the genomic average
(25). Fifteen of the CpGs lie in 2.5 kb of
sequenced 5' flanking DNA and none occur in pit-1

binding sites (26). Furthermore, no CpG-rich region occurs anywhere in unsequenced regions of the gene as indicated by the rarity of HhaI and HpaII sites (27). There does not appear to be widespread regional methylation of the gene in the deficient line because a CpG site in the first exon/intron and one in the 5' flanking region of the PRL gene have been assayed by Southern blotting after restriction with methylation-sensitive restriction enzymes and did not differ in their methylation state (e.g., the sites were not methylated in either cell line; Morris & Ivarie, unpubl. obs.). Even though the PRL gene does not contain many CpG sites as potential targets for EMS, other *in vivo* characteristics of the prolactin gene (e.g. chromatin structure or DNA secondary structure) could make the CpG sites in the gene more suseptible to EMS modification than CpGs in other genes.

An unusual feature of the gene that might be a candidate for EMS modification involves the simple repeat element, poly(dT-dG)·poly(dC-dA). TG-elements fall into a class of moderately repeated sequences ubiquitous to eukaryotic DNA from fungi to man but absent from prokaryotic genomes (28). In human DNA, there are 50-100,000 copies of the element or one per ~50 kb. This amounts to less than one TG-element per gene. The element readily forms Z DNA (29) and often contains internal CpGs. Deamination of 5-methylcytosine to thymine in a CpG site causes transition mutations to TpG/CpA. An interesting structural feature of a TpG/CpA site is the methyl at the C5 of

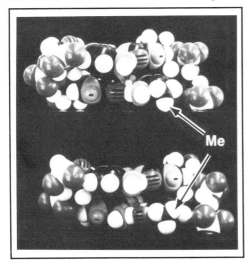

Figure 8. Comparison between a hemimethylated CpG site (upper) vs a TpG/CpA site (lower).

thymine (Figure 8). Although such a site is similar to a hemimethylated CpG site (a pyrimidine 5 methyl, a 3' guanine; a cytosine on the opposite strand, etc.), it is a poor substrate for methylation *in vitro* by partially purified DNA methylase (Farrance & Ivarie, unpubl. obs.).

Compared to the genome as a whole, the prolactin gene contains a very high density of the TG-element with at least six occurring inside and flanking the gene (see Figure 6). The distal promoter element is flanked by an inverted TG-repeat; another occurs ~2 kb upstream, one in the fourth intron and two others over 5 kb downstream from the polyadenylation site (30). This is about a 10-fold higher density of the repeat per gene than known for other mammalian genes.

The high frequency of TG-elements raises two questions. First, why does the gene contain so many repeats and second, might they be the EMS target sequence? The fact that the distal promoter element is flanked by TG repeats raises the possibility that other repeats are associated with elements influencing prolactin expression. TG-elements might be the EMS target if *hemiethylated* TG/CA sites promote enzymatic methylation in a heritable manner. There is no precedent for this, but heritable methylation is not impossible--only the CA-containing strand need undergo "maintenance" methylation. CpA is the second most abundant methylation dinucleotide in mammalian DNA (9,10). The fact that the known DNA methylase does not use TG-repeat DNA as a substrate may suggest a new DNA methylase. If so, the specificity cannot reside solely in TG/CA because the dinucleotide is so abundant and only a small fraction is actually methylated. Thus, specific TG/CA's may be the key sites of modification. With the newly developed method of genomic sequencing (31), we will soon learn whether CA methylation is an important part of vertebrate methylation.

Nonetheless, the biggest problem with poly(dT-dG)·poly(dC-dA) as the EMS target lies with the frequency of the repeat in genomic DNA with about one per 50 kb. If the elements were the target, prolactin TG-elements would have to be more suseptible to EMS than other elements or other genes would have been silenced at high frequency with EMS.

REFERENCES

1. Ivarie RD, Baxter JD, Morris JA (1981) Interaction of thyroid and glucocorticoid hormones in rat pituitary tumor cells: Specificity and diversity of the responses analyzed by two-dimensional gel electrophoresis. J Biol Chem 256:4520-4528.

2. Ivarie RD, Morris JA, Eberhardt NL (1980). Hormonal domains of response: Actions of glucocorticoid and thyroid hormones in regulating pleiotropic responses in cultured cells. Rec Prog Horm Res 36:195-239.

3. Chetelain A, Dupuoy JP, Dubois MP (1979). Ontogeny of cells producing polypeptide hormones in the fetal hypophysis of the rat: Influence of the hypothalamus. Cell Tissue Res 196:409-427.

4. Watanabe YG, Daikoku S (1979). An immunohistochemical study on the cytogenesis of adenohypophysial cells in fetal rats. Dev Biol 68:557-567.

5. Hoeffler JP, Boockfor FT, Frawley LS (1985). Ontogeny of prolactin cells in neonatal rats: Initial prolactin secretors also release growth hormone. Endocrinol 117:187-195.

6. Ivarie RD, Morris JA, Martial JA (1982). Prolactin-deficient variants of GH3 rat pituitary tumor cells: Linked expression of prolactin and another hormonally responsive protein in GH3 cells, Mol Cell Biol 2:179-189.

7. Ivarie RD, Morris JA (1982). Induction of prolactin-deficient variants of GH3 rat pituitary tumor cells by ethyl methanesulfonate: Reversion by the DNA methylation inhibitor, 5-azacytidine, Proc Natl Acad Sci USA 79:2967-2970.

8. Ivarie R, Morris J (1983). Phenotypic switching in GH3 rat pituitary tumor cells: Linked expression of growth hormone and another hormonally responsive protein. DNA 2:113-120.

9. Razin A, Riggs AD (1980). DNA methylation and gene function. Science 210:604-610.

10. Doerfler W (1983). DNA methylation and gene activity. Ann Rev Biochem 52:93-124.
11. Gruenbaum U, Cedar H, Razin, A (1982) Substrate and sequence specificity of a eukaryotic DNA methylase. Nature 295:620-622
12. Farrance, IK and Ivarie R (1985). Ethylation of poly(dC-dG)·poly(dC-dG) by ethyl methanesulfonate stimulates the activity of mammalian DNA methyltransferase *in vitro*. Proc Natl Acad Sci USA 82:1045-1049.
13. Singer B, Kusmierek JT (1982). Chemical mutagenesis. Ann Rev Biochem 52:655-693.
14. Smith SS, Hardy TA, Baker DJ (1987). Human DNA (cytosine-5)methyltransferase selectively methylates duplex DNA containing mispairs. Nucleic Acids Res 15:6899-6916.
15. Farrance, IK, Ivarie R (1989). Syn-thesis of N7-ethyldeoxyguanosine-5'-triphosphate and site-specific placement of N7-ethylguanine in a synthetic 35mer. Anal Biochem, in press.
16. Farrance IK, Eadie JS, Ivarie R (1989). Improved chemistry for oligodeoxyribonucleotide synthesis substantially improves restriction enzyme cleavage of a synthetic 35mer. Nucleic Acids Res 17:1231-1245.
17. Chien Y-H, Thompson EB (1980). Genomic organization of rat prolactin and growth hormone genes. Proc Natl Acad Sci USA 77:4583-4587.
18. Cooke NE, Baxter JD (1982). Structural analysis of the prolactin gene suggests a separate origin for its 5' end. Nature 297:603-606.
19. Farrance IK, Hall IS, Morris J, Ivarie R (1989). Prolactin-deficient GH_3 rat pituitary tumor cells express prolactin transcription factor pit-1 but have reduced levels of primary PRL transcripts that harbor normal 5' and 3' termini. Submitted.
20. Nelson C, Albert VR, Elsholtz HP, Wu LI-W, Rosenfeld MG (1988). Activation of cell-specific expression of rat growth hormone and prolactin genes by a common transcription factor. Science 239:1400-1405.
21. Cao Z, Barrons EA, Carrillo AJ, Sharp ZD (1987). Reconstitution of cell type-specific transcription of the rat prolactin gene *in vitro*. Mol Cell Biol 7:3402-3408.

22. Nelson C, Crenshaw III EB, Franco R, Lira SA, Albert VR, Evans RM, Rosenfeld MG (1986). Discrete cis-active genomic sequences dictate the pituitary cell type-specific expression of rat prolactin and growth hormone genes. Nature 322:557-5562.

23. Gutierrez-Hartmann A, Siddiqui S, Loukin S (1987). Selective transcription and DNase I protection of the rat prolactin gene by GH_3 pituitary cell-free extracts. Proc Natl Acad Sci USA 84:5211-5215.

24. Arnold TA, Ivarie R (1989). Prolactin-deficient GH_3 rat pituitary tumor cells are fully competent to initiate transcription at the prolactin gene promoter. Manuscript in preparation.

25. McClelland M, Ivarie R (1982). Asymmetrical distribution of CpG in an average mammalian gene. Nucleic Acids Res 10:7865-7877.

26. Maurer RA (1985). Selective binding of the estradiol receptor to a region at least one kilobase upstream from the rat prolactin gene. DNA 4:1-9.

27. Durrin LK, Weber JL, Gorski J (1984). Chromatin structure, transcription, and methylation of the prolactin gene domain in pituitary tumors of Fischer 344 rats. J Biol Chem 259:7086-7093.

28. Morris JA, Kushner SR, Ivarie R (1986). The simple repeat poly(dT-dG)·poly(dC-dA) common to eukaryotes is absent from eubacteria and archaebacteria and rare in protozoans. Mol Biol Evol 3:343-355.

29. Galazka G, Palecek E, Wells RD, Klysik J (1886). Site-specific OsO_4 modification of the B-Z junctions formed at the $(dA-dC)_{32}$ region in supercoiled DNA. J Biol Chem 261:7093-7098.

30. McFarlane D, Hall IS, Farrance IK, Morris J, Ivarie R (1986). The rat prolactin gene contains at least six poly(dT-dG)·poly(dC-dA) repeats. Nucleic Acids Res 14:7805.

31. Saluz H, Jost J-P (1989) A simple high-resolution procedure to study DNA methylation and in vivo DNA-protein interactions on a single-copy gene level in higher eukaryotes. Proc Natl Acad Sci USA 86:2602-2606.

ANTIVIRAL ACTIVITIES OF 3-DEAZA NUCLEOSIDES—INDIRECT INHIBITORS OF METHYLATION

Peter K. Chiang

Department of Applied Biochemistry,
Walter Reed Army Institute of Research,
Washington, DC 20307-5100

ABSTRACT 3-Deazaadenosine and 3-deaza-(±)aristeromycin are indirect inhibitors of methylation reactions by inhibiting S-adenosylhomocysteine hydrolase. Both compounds have potent antiviral activities. However, there is no clear evidence that they exert their antiviral activities via the inhibition of the methylation of the 5'cap of viral mRNA.

INTRODUCTION

3-Deazaadenosine (3-deaza-Ado) and 3-deaza-(±)aristeromycin (3-deaza-Ari) (Fig. 1) are potent inhibitors of S-adenosylhomocysteine (AdoHcy) hydrolase (1-4). However, because the equilibrium of the enzymatic reaction (Fig. 2) is much in favor of synthesis of AdoHcy, both 3-deaza nucleosides can also function as alternative substrates, generating novel AdoHcy congeners, 3-deaza-AdoHcy or 3-deaza-AriHcy (1-5). As a consequence of the inhibition of AdoHcy hydrolase by 3-deaza nucleosides, cellular AdoHcy accumulates, and in some instances, the generation of 3-deaza-AdoHcy and 3-deaza-AriHcy can also occur. The attendant accumulation of cellular AdoHcy, 3-deaza-AdoHcy or 3-deaza-AriHcy will lead to the inhibition of methylation reactions (1, 5-10). There are many biological effects associated with the administration of 3-deaza nucleosides to cells or animals, such as cellular differentiation (11-15), the inhibition of aldosterone-stimulated sodium transport (8), the modulation of immune responses (16, 17), and antiviral effects (2, 18-20).

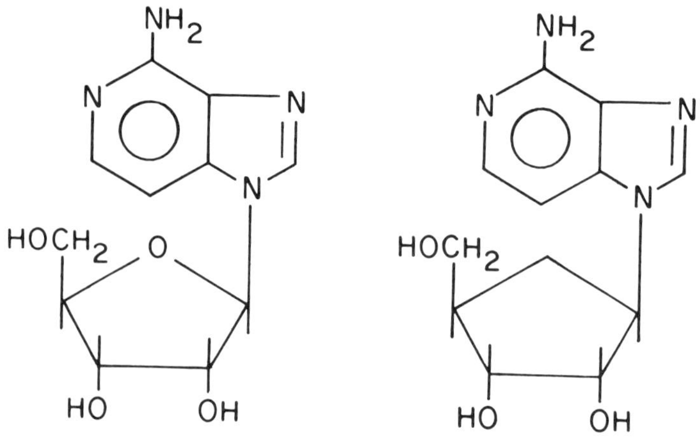

3-DEAZA-
ADENOSINE

3-DEAZA-
(±)ARISTEROMYCIN

FIGURE 1. Chemical structures of 3-deazaadenosine and 3-deaza-(±)aristeromycin.

The potent antiviral activities of the 3-deaza nucleosides have been postulated to be due to the inhibition of the methylation of the 5' cap of viral mRNA. Nonetheless, a detailed study into the mode of action of the antiviral activities of 3-deaza nucleosides is still lacking. The present paper examines the antiviral effects of 3-deaza-Ado and 3-deaza-Ari on the vaccinia virus, which was grown in HeLa cells (21,22), and the plaque formation was then assayed by BSC-1 cells.

RESULTS

Table 1 shows the potencies of these two 3-deaza nucleosides in inhibiting the plaque formation of vaccinia in comparison to sinefungin. 3-Deaza-Ari, at 0.5 μM, inhibited vaccinia plaque formation by 84 %; at 4.0 μM plaque formation was inhibited by 99.8 % without apparent cytotoxicity to the HeLa cells. In contrast, 3-deaza-adenosine showed moderate cytotoxicity and was less potent in inhibiting vaccinia plaque formation. Sinefungin, an analog of AdoHcy lacking the ability to inhibit AdoHcy hydrolase (23), was the least potent and required 200 μM

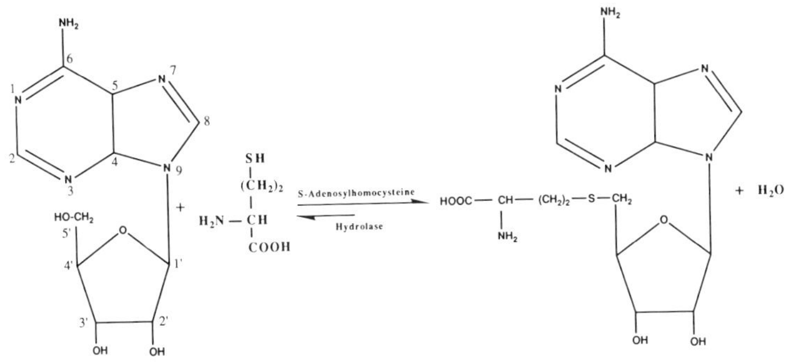

FIGURE 2. Reaction catalyzed by S-adenosyl-homocysteine hydrolase.

TABLE 1
INHIBITION OF PLAQUE FORMATION OF VACCINIA VIRUS

drug addition (µM)	% plaque inhibition	% control HeLa cell number
3-Deaza-(±)aristeromycin:		
0.5	84.0	122
1.0	99.2	144
2.0	99.6	110
4.0	99.8	110
3-Deazaadenosine:		
15.0	88.0	86
30.0	93.4	78
Sinefungin:		
50.0	20.8	100
100.0	84.4	120
200.0	89.6	90

to achieve a 90% inhibition of plaque formation, although there was little cytotoxicity.

Because 3-deaza-Ari was the most potent antiviral in the series, its mode of action was investigated further.

When HeLa cells were incubated with 4 μM 3-deaza-Ari, there was a 1.5-fold increase in [^{35}S]AdoMet and a 10-fold increase in [^{35}S]AdoHcy (Fig. 3). Thus 3-deaza-Ari affected the metabolism of AdoMet and AdoHcy in a manner consistent with its known action as a potent inhibitor of AdoHcy hydrolase, i.e. an accumulation of AdoHcy leading to an increased level of AdoMet due to the inhibition of

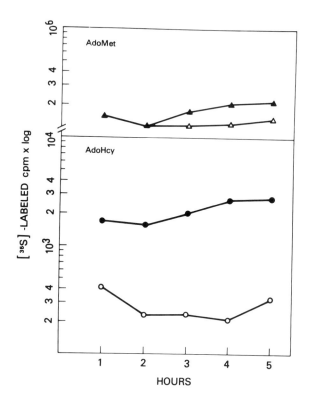

FIGURE 3. Effect of 4 μM 3-deaza-Ari on cellular levels of [^{35}S]AdoMet and [^{35}S]AdoHcy. HeLa cells (6 x 10^5) were first incubated in 80 ml of minimum essential medium (MEM) containing 5% horse serum, and 80 μM [^{35}S]methionine; 2 h later 3-deaza-Ari was added. At the indicated time, the cells were centrifuged at 15,000 x g for 10 min; the cell pellet was extracted with 1.0 ml of 5% sulfosalicylic acid, and then analyzed by high pressure liquid chromatography. Open circles or triangles represent the untreated cells.

methionine before the addition of 3-deaza-Ari, the cellular AdoMet and AdoHcy were most likely in isotopic equilibrium with [^{35}]methionine, because the half-turnover time for AdoMet is between 10-15 min and for AdoHcy about 5-10 min (5).

However, both 3-deaza-Ari and 3-deaza-Ado had no effect on the incorporation of [^{3}H]thymidine into total DNA of vaccinia-infected HeLa cells or cytoplasmic (vaccinia) DNA (Fig. 4). In spite of the apparent

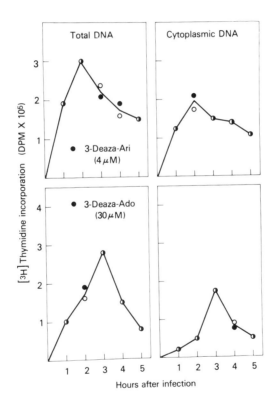

FIGURE 4. Incorporation of [^{3}H]thymidine into total DNA of vaccinia-infected HeLa cells and cytoplasmic (vaccinia) DNA. Experimental details were performed as described in (22). Open circles represent untreated cells.

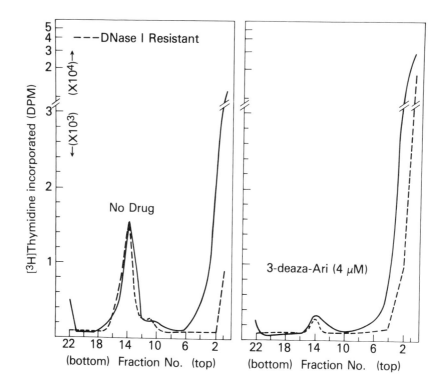

FIGURE 5. Fate of vaccinia DNA in HeLa cells treated with 4 μM 3-deaza-Ari. HeLa cells (5 x 10^7) were first infected with vaccinia virus (30 PFU/cell) in 10 ml of MEM containing 5% horse serum, plus or minus 3-deaza-Ari. After 30 min, 90 ml of the same medium was added to the incubation mixture; 20 μci [^{3}H]thymidine was added 2 h later. After an additional incubation of 1 h, the cells were centrifuged, washed with phosphate-buffered saline, and suspended in 100 ml of fresh MEM, containing 5% horse serum, 100 μM unlabeled thymidine, and 4 μM 3-deaza-Ari. After 17 h, the cells were analyzed for [^{3}H]thymidine-labeled DNA by sucrose gradient (22). Broken line indicates DNA treated with pancreatic deoxyribonuclease for 30 min at 37°.

synthesis of vaccinia DNA, [^{3}H]thymidine incorporation into intact viral particles was drastically reduced in the presence of 3-deaza-Ari (Fig. 5). However, at 4 µM 3-deaza-Ari, there was no observable inhibition in the methylation of the 5' cap of vaccinia mRNA, as measured by the method of Boone and Moss (22) using [methyl-^{3}H]methionine. Nor was there a significant difference in the methylation of cytoplasmic lipids, as assayed by methyl-^{3}H incorporation from [^{3}H-methyl]methionine relative to [1,2,-^{14}C]choline (5, 24) into the organic phase (total lipid fraction) of vaccinia-infected or uninfected HeLa cells (not shown).

DISCUSSION

It has been shown previously that 3-deaza-Ado has no effect on DNA or RNA synthesis in chick embryo fibroblasts with or without infection with Rous sarcoma virus (18). Nor does 3-deaza-Ari inhibit the incorporation of [^{14}C]uridine into the RNA of HL-23 (c-type RNA) virus (2). Thus, the results reported here are consistent with the above in that 3-deaza-Ari did not affect DNA or RNA synthesis of vaccinia virus.

The lack of inhibition of methylation of the 5' cap of vaccinia mRNA by 3-deaza-Ari is in agreement with other findings that the cap structure is relatively insensitive to inhibition by 3-deaza nucleosides (9), S-tubercidinyl-homocysteine, or other AdoHcy analogs (25, 26). This observation may reflect the inherent resistance of the 5' cap 7-methylguanosine to perturbation, which is imperative for mRNA function and stability. A plausible conclusion for now is that the 3-deaza nucleosides exert their antiviral effects by inhibiting other vital methylation reaction(s) essential for the packaging of functional vaccinia virus. The present findings should also be viewed with respect to the fact that some viruses lack the 5'cap for their mRNAs (27-29).

ACKNOWLEDGEMENTS

I thank Bernard Moss of NIH, in whose laboratory the experiments were performed.

REFERENCES

1. Chiang, PK, Richards, HH, Cantoni, GL (1977). Mol Pharmacol 13: 939.
2. Montgomery, JA, Clayton SJ, Thomas, HJ, Shannon, WM, Arnett, G., Bodner, AJ, Kim, IK, Cantoni, GL, Chiang PK (1982). J Med Chem 25: 626.
3. Kim, IK, Zhang, CY, Chiang, PK, Cantoni GL (1985). Arch Biochem Biophys 226: 65.
4. Chiang PK (1985). In Paton DM (ed): "Methods in Pharmacology: Methods Used in Adenosine Research," New York: Plenum Press, p 127.
5. Shattil, SJ, Montgomery, JA, Chiang, PK (1982). Blood 59, 906.
6. Chiang, PK, Cantoni GL (1979). Biochem Pharmacol 28: 1897.
7. Garcia-Castro, I, Mato JM., Vasanthakumar, G, Weismann, WP, Schiffmann, E, Chiang, PK (1983). J Biol Chem 258: 4345.
8. Wiesmann, WP, Johnson, JP, Miura, GA, Chiang, PK (1985). Am J Physiol 248: f43.
9. Backlund PS Jr, Carotti D, Cantoni, GL (1986). Eur. J. Biochem 160: 245.
10. Ueland PM (1983). Pharmacol Rev 34:223.
11. Chiang PK (1981). Science 211: 1164.
12. Chiang PK, Brown ND, Padilla, FN, Gordon RK (1987). In Aarbakke, J, Chiang PK, Koeffler HP (eds): "Tumor Cell Differentiation: Biology and Pharmacology," NJ: Humana Press, p 231.
13. Harris, M (1982). Cell 29: 483.
14. Scarpa, S, Strom R, Bozzi, A, Aksamit, RR, Backlund PS Jr, Chen, J, Cantoni GL (1984). Proc Natl Acad Sci USA 81: 3064.
15. Aarbakke, J, Miura, GA, Prytz, PS, Bessesen, A., Sløordal, L, Gordon, RK, Chiang PK (1986). Cancer Res 46: 5469.
16. Morita, Y, Chiang PK, Siraganian, RP (1981). Biochem Pharmacol 30: 785.
17. Medzihradsky, JL, Zimmerman, TP, Wolberg, G, Elion, GB (1982). J Immunopharmacol 4: 29.
18. Bader, JP, Brown NR, Chiang, PK, Cantoni, GL (1978). Virology 89: 494.
19. De Clercq, E, Montgomery, JA (1983). Antiviral Res 3: 17.
20. De Clercq, E, Berstrom DE, Holy, A, Montgomery, JA (1984). Antiviral Res 4: 119.

21. Boone, RF, Moss, B (1977). Virology 79: 67.
22. Moss, B, Rosenblum, EN, Katz, E, Grimley PM (1969). Nature 224: 1280.
23. Im, YS, Chiang, PK, Cantoni, GL (1979). J Biol Chem 254: 11047.
24. Chiang, PK, Im, YS, Cantoni, GL (1980). Biochim Biophys Res Commun 94: 174.
25. Camper, SA, Albers, RJ, Coward, JK, Rottman, FM (1984). Mol Cell Biol 4: 538.
26. Pugh, CSG, Borchardt, RT (1982). Biochemistry 21: 1535.
27. Pelletier, J, Kaplan, G, Racaniello, VR, Sonenberg N (1988). Mol Cell Biol 8: 1103.
28. Shih, DS, Park, IW, Evans, CL, Jaynes, JM, Palmenberg, AC (1987). J Virol 61: 2033.
29. Leung, DW, Browning, KS, Heckman, JE, RajBhandary, UL, Clark, JM (1979). Biochemistry 18: 1361.

CHANGING METHYLATION PATTERNS DURING DEVELOPMENT

Aharon Razin, Dale Frank, Michal Lichtenstein[2], Zeev Paroush, Yehudit Bergman[2], Moshe Shani[3], and Howard Cedar.

Department of Cellular Biochemistry
The Hebrew University-Hadassah Medical School
Jerusalem, Israel

ABSTRACT Although the pattern of gene methylation in eukaryotic organisms is relatively fixed, changes in this pattern can occur during development. Tissue specific genes undergo demethylation in their cell type of expression and this process is controlled by flanking cis acting sequences. Another type of demethylation is specific for methylated CpG islands and its activity is restricted to early stages of development. This demodification could be responsible for recycling the methylation changes which occur on the X-chromosome and in satellite DNA.

---

[1]This work was supported by grants from NIH (to H.C. and A. R.) by the Israel Cancer Research Fund (to H.C.) and by the Israel U.S. Binational Science Foundation (to A.R., H.C. and Y.B.).
[2]Department of Experimental Medicine, The Hebrew University-Hadassah Medical School, Jerusalem, Israel.
[3] Institute of Animal Sciences, ARD, The Volcani Center, Bet Dagon, Israel.

INTRODUCTION

Differential gene expression in animal cells is regulated by many varied molecular mechanisms, including DNA methylation. Since the level of this modification on any given gene is inversely proportional to its activity, methylation has been implicated as an inhibitor of RNA transcription *in vivo* (1). These conclusions are backed up by more direct experiments using DNA transfection with methylated gene constructs. In almost all cases studied, it can be shown that DNA methylation strongly inhibits gene expression in cultured fibroblasts (2,3). Further evidence for a role of DNA modification comes from experiments with the drug 5-Azacytidine, which can activate specific genes in various cell lines presumably by local demethylation (4).

Despite these convincing data obtained from studies in tissue culture, the role of DNA methylation during normal development *in vivo* has not been well characterized. Tissue specific genes are frequently found to be highly methylated at most CpG residues in germ line DNA. Even though genomic sequencing has not been used to assay all CpG sites, most methyl sensitive sites assayed in tissue specific genes in sperm DNA were in a modified state (5). Presumably, a similar methylation pattern is present in the female germ line, but this has not as yet been confirmed experimentally due to the difficulty in obtaining sufficient quantities of oocyte DNA. According to this picture, the vast majority of tissue specific genes begin their life in the organism in the methylated state and presumably remain modified throughout embryogenesis. Only in the tissue of expression each individual gene undergoes specific demethylation; in the rest of the organism these genes remain highly methylated. In contrast to tissue specific genes, housekeeping genes have a unique sequence structure characterized by the presence of CpG islands at the 5' ends (6). These islands are always unmethylated, both in somatic tissues and in germ line DNA (7).

ACTIVE AND INACTIVE GENE CONFORMATIONS

With this methylation profile in mind, one can now propose a model of how this modification affects gene expression *in vivo*. The fact that almost all genes are inhibited by DNA methylation in a transfection assay, strongly suggests that methyl moieties are recognized as a general signal of gene activity. Thus, already in the germ line and early embryo, the DNA methylation profile plays a role in the determination of gene activity. Tissue specific genes are methylated and thus kept inactive, while housekeeping genes are unmodified and are thus recognized as active sequences. Although the precise mechanism by which gene transcription is inhibited by methylation is unknown, it is clear that this modification mediates its action by affecting protein DNA interactions. In one experiment, methylated and unmethylated DNA sequences were introduced into L-cells by DNA mediated gene transfer. Unmethylated DNA of all types were found to be integrated in a DNaseI sensitive conformation, while methylated DNA always showed an insensitive closed chromatin conformation (8, see figure 1). This suggests that DNA methylation influences a large variety of structural proteins and that this signal alone can determine the activity fate of the inserted DNA, regardless of the nature of the sequence.

It should be noted that accoring to this model, DNA methylation plays a special role in controlling gene activity in the early embryo and in somatic cells. Whereas other regulatory mechanism are based on gene specific interactions, DNA methylation provides a general control signal. Almost all tissue specific genes are globally placed in an inactive conformation simply by virtue of the prescence of its methyl moieties, without the need for specific inhibitory protein factors.

If DNA methylation serves as a general signal of gene repression, how do methylated tissue specific genes get turned on in the correct cell type? In some manner these cells must have the ability to overcome the DNA methylation and

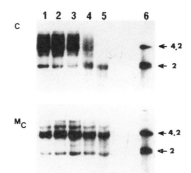

Figure 1. *DNAase sensitivity of methylated and unmethylated transfected DNA.* L cell nuclei (10^8/ml) containing an unmethylated (C) or methylated (^{M}C) M13 α-actin construct (9) were digested with increasing concentrations of DNAase I (lane 1- no enzyme; lane 2-0.3µg/ml; lane 3-1 µg/ml; lane 4-3 µg/ml; lane 5-10µg/ml) for 20 min. at 37° in a volume of 300 µl. Deproteinized DNA was subjected to restriction enzyme analysis, gel electrophoresis, and blot hybridization with the appropriate probe. Digestion of total DNA for both cell types was determined to be identical for each DNAase I concentration by visual analysis of the ethidium bromide stained gels. DNA was digested with PstI and hybridized with probe A. The marker shown in lane 6 represents a mixture of pαCAT DNA (9) digested with PstI and PstI - Sac I. The map indicates the locations of cleavage of PstI (P), HinfI(H) and SacI(S) and the various probes used in the blot hybridizations of the α-actin construct. The large triangle marks the location of a strong hypersensitive site in the actin promotor region (figure taken from reference 8).

activate the appropriate genes. That this is inded the case, was shown by experiments using the muscle specific rat α actin gene (9). When an α CAT chimeric gene was inserted into fibroblast cells by DNA mediated gene transfer its activity was severely inhibited by DNA methylation. On the other hand, both the methylated and non-methylated α actin contructs have identical basal level transcriptional activity when inserted into myoblast cells, and undergo induction following myogenesis with similar kinetics (Figure 2). Thus, although this gene can be repressed by modification, myoblasts have the ability to activate the gene despite its methylation pattern.

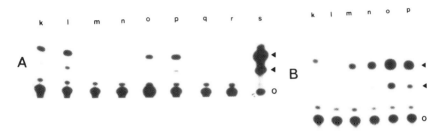

Figure 2. *Cell-type-specific effect of methylation on expression of α-CAT* LTK fibroblasts and L8 myoblasts were contransfected with the methylated a-CAT constructs and either HSV tk or the neomycin resistance-bearing plasmid, pTM respectively. After 2-3 weeks in appropriate selection, a few individual colonies were isolated and the rest (over 100) were pooled together and analyzed. CAT activity was assayed using identical numbers of cells within a given experiment. The pointers at the side of the plates mark the position of the acetylated chloramphericol. The letter "O" marks the position of unacetylated substrate. (A) L cell transfections. Lane s, reaction with purified CAT enzyme (Pharmacia); lanes k-r, independently isolated colonies of unmethylated α-CAT (k and l), HpaII/HhaI methylated α CAT (m and n), unmethylated M13 α-CAT (o and p) and methylated M13 α-CAT (q and r).(B) Myoblast transfections of unmethylated and HPAII/HhaI methylated α-CAT, Lanes k, m, and o methylated α-CAT at 0,96 and 104 hrs. after induction; lanes l, n, and p, unmethylated α-CAT at 0,96 and 104 hrs. after induction (figure taken from reference 9).

When the integrated gene sequences were assayed for methylation, it could be shown that α actin underwent demethylation in the myoblast which mimics the pattern of the endogenous gene. On the other hand, the methylated gene remained fully modified following transfection into the fibroblast cell type.

Gene activation could have occurred in one of two ways. One possibility is that α actin is initially turned on by protein factors which override the DNA methylation, and this process is followed by a secondary demethylation. Support for such a model can be derived from the behavior of several rodent liver specific genes. (10, 11, 12). These genes are activated late in embryonic development while they are still methylated and this is followed by a gradual demethylation process which eventually leads to the adult liver pattern. A second possibility is that α-actin is recognized by protein factors in the myoblast and rapidly undergoes demethylation - which is essential to put the gene in an active conformation.

DEMETHYLATION

In order to learn about the relationship between demethylation and gene activity, we studied the kinetics of the demethylation reaction in myoblast cells using transient transfection with methylated substrates. Chimeric molecules containing the α actin promoter linked to the CAT gene were methylated at HpaII sites *in vitro* and introduced into myoblasts. At various times following transfection, these plasmid constructs were isolated from the nucleus and subjected to analysis with the methyl sensitive restriction enzymes HpaII, SmaI and NciI (see map in Figure 3). At very early times, the α actin promoter can be cut at one specific site with the enzyme NciI, but not HpaII or SmaI. This pattern is due to the fact that NciI recognizes and cuts hemimethylated DNA as well as fully unmethylated DNA (Paroush and Cedar, unpublished results). Thus, even at very early times after the entrance of α actin into the cell, this specific site is recognized and demethylated on one strand of the DNA. This suggests that hemidemethylation may indeed be a necessary step in the activation of the gene, although full demethylation may actually take place at a later time. Using genomic sequencing it has been shown that the vitellogenin gene also undergoes hemidemethylation concomitant with the onset of its transcription (13).

It should be noted that the hemidemethylation seen in the α actin promoter occurs in the absence of any DNA synthesis, since these plasmid constructs all remain cleaveable by DpnI, an enzyme which requires the presence of the bacterial 6mA residues in its recognition sequence. Even minimal DNA replication would have removed these modifications. This is consistent with what is known about demethylation in other systems. When mouse erythroleukemia cells, for example, are induced to diffferentiate, this process is accompanied by a massive genome wide demethylation (14). The kinetics of this reaction and its proposed mechanism clearly indicate that the removal of these methyl groups must occur without DNA replication (15). This is also true for the *in vivo* demethylation of the vitellogenin gene in the liver and the δ - crystallin gene in post-mitotic lens cells (13, 16).

In order to determine whether cell specific demethylation is a general phenomenon, we have begun to study additional genes in different cell types. Igκ is a good example of a gene which is under methylation control. In almost all cell types this gene is methylated, but is completely unmodified in plasmacytoma cells in which it is expressed (17). To study the details of this

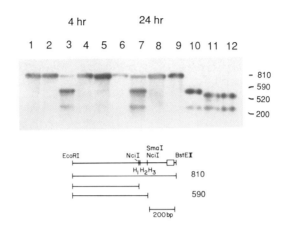

Figure 3. *α-Actin demethylation following transient transfection.* A plasmid containing the EcoRI/ BstEII fragment from the promoter region of the rat α-actin gene was methylated *in vitro* with HpaII and HhaI and transfected into the rat myoblast line L8 (9). At either 4 or 24 hours after the addition of the DNA, cells were harvested and DNA was prepared from nuclei and digested with EcoRI/BstEII (lanes 1 and 5) alone or together with SmaI (lanes 2 and 6), NciI (lanes 3 and 7) or HpaII (lanes 4 and 8). Marker plasmid cleaved with the same order of enzymes is shown in lanes 9 to 12. Following blotting these filters were hybridized with a probe which covers the entire region shown on the map. Note that EcoRI/BstEII produces an 810 bp fragment. When this is cut with Nci I at site H2, approximately 600 bp and 200 bp fragments are produced. Cleavage at both H1 and H3 would yield a 540 bp fragment together with the same 200 bp piece. The map (9) also shows the position of the first gene exon (open box).

demethylation reaction, we introduced a methylated Igκ gene into both fibroblasts as well as a plasmacytoma line. As shown in Figure 4, this gene remained fully modified in the fibroblast cells, but underwent thorough demethylation in the κ specific cell type. Control experiments using other genes indicate that this is a gene specific process. Thus, plasmacytoma cells, but not others, have the ability to recognize this gene and cause its demethylation. Using reverse genetics it should now be possible to dissect out the cis acting elements required for this important step in molecular differentiation.

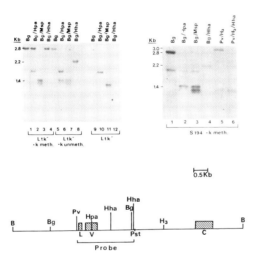

Figure 4. *Methylation analysis of Igκ following its transfection to cells.* A plasmid containing the $V_\kappa 19J_2C_\kappa$ was methylated *in vitro* using HpaII and HhaI methylases and co-transfected into cell line S194 by electroporation or into mouse L-cells either by the calcium phosphate precipitation technique or by electroporation. In both cases colonies containing the plasmid were isolated by selection for neomycin and these were pooled and expanded and then used to prepare DNA for Southern blot analysis. DNA was digested with BglII and in addition with either HpaII, MspI or HhaI in order to assay methylated sites near the V region. In

the case of S194, the 3' HhaI site was analyzed by digestion with PvuII-HindIII. In all cases, 20 µg of DNA were loaded on each lane and the final filters were probed using a nick translated PvuII/PstI fragment. In the case of the L-cell transfection, gel analysis is shown for both the methylated Igκ construct as well as for a non methylated control. In addition to the expected 2.8kb band and its cleavage products, these autoradiograms also show hybridization to the endogenous equivalent of the Igκ gene as is demonstrated by the blot of the untransfected Ltk⁻ control.

A similar demethylation can be observed for the rat insulin I gene when it is transfected into insulinoma cells in cultlure (figure 5). In this system, only specific sites in the endogneous insulin gene are undermethylated *in vivo* (18), and this specificity is mimicked in the exogenous gene constructs. Preliminary experiments indicate that demethylation is not controlled by the same sequence elements involved in gene expression, since the removal of a strong enhancer element about 100 bp upstream of the promoter (19), had no effect on the demethylation of the gene. From both of the above examples, it is clear that each specific cell type must retain the machinery for the demodification reactions. This knowledge should pave the way toward an understanding of its mechanism and of the cis and trans-acting elements which are involved in the process.

CpG ISLANDS

Unlike tissue specific genes, housekeeping genes are usually characterized by the presence of a 5' CpG island which is unmethylated in every somatic tissue and in the germ line. Since the methylation status of these genes is static during development, we assumed that the organism probably lacks the facilities for changing its methylation

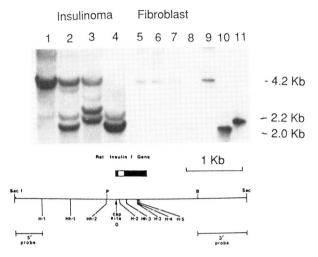

Figure 5. *Demethylation of the rat insulin I gene.* The cloned SacI fragment of the insulin gene was methylated *in vitro* with HpaII and HhaI and co-transfected using a neomycin resistance vector into hamster insulinoma cells or fibroblasts (CHO). DNA from pools of colonies were digested with SacI (lanes 1 and 5) and in addition HpaII (lanes 2 and 6), HhaI (lanes 3 and 7) or MspI (lanes 4 and 8). Marker plasmid DNA has also been cleaved with SacI alone (lane 9) or together with HpaII (lane 10) or HhaI (land 11). The resulting blot filters were hybridized with the 3' probe shown on the map. The accompanying map (18) also pictures the actual location of the insulin I gene and the positions of the HpaII sites (H) and the HhaI sites (Hh). Note that the entire SacI fragment is 4.2kb in length. Digestion of this inserted DNA with HpaII shows that in the insulinoma the cluster of sites H3-H5 is largely unmethylated. The blot also shows that HhaI sites Hh-2 and Hh-3 are also unmodified. All of these sites remain totally methylated in fibroblast cells.

pattern. In order to test this hypothesis, we introduced a methylated APRT gene into mouse zygote cells with the intention of producing a transgenic mouse with a methylated housekeeping gene. We obtained three founder mice containing this construct, but when the APRT sequences were analyzed by restriction enzymes, it became clear that these genes became unmodified *in vivo*. Tail DNA and DNA from several mouse tissues all showed a total demethylation in the island region, while the 3' end of the gene remained fully methylated at the Hpa II and Hha I sites and even underwent *de novo* methylation at additional CpG sites in these non island regions. These results suggest that island sequences are somehow recognized in the early embryo and that these cells have the capability of demethylating such regions. In this case, it was possible that embryonic cell culture lines, such as the F9 teratocarcinoma retain this property. To test this, we introdcuced a methylated APRT gene construct into the F9 cells and studied the resulting DNA methylation pattern after over 30 generations of growth (Figure 6). In a manner similar to the transgenic mice, this methylated CpG island underwent total demethylation. In contrast, these same sequences remained almost fully methylated both in mouse fibroblasts and in rat myoblast cells. Thus, this demethylation process appears to be stage specific and is presumably not present in normal somatic cells.

In order to learn something about the recognition of CpG island sequences, we prepared various constructs for transfection into F9. In one experiment, the APRT promoter was removed from the gene, leaving a truncated island which can not undergo transcription. Surprisingly, the remaining island sequences also became unmethylated, suggesting that this process is not in any way related to gene activity. If an island is simply recognized because of the presence of an unusually high concentration of CpG residues, then any bacterial DNA sequence should also behave as an island. In keeping with this prediction, when a methylated pBR molecule was stably transfected

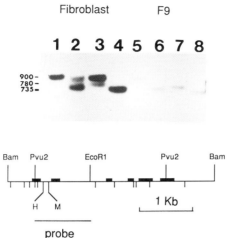

Figure 6. *Methylation state of the APRT gene in transfected cells*. The cloned BamHI fragment containing the APRT gene (dark regions) was methylated *in vitro* with the HpaII and HhaI methylases and transfected into mouse L-cells or F9, teratocarcinoma. DNA from pooled colonies was digested with PvuII/EcoRI (lanes 1 and 5) and, in addition, with the enzymes HpaII (lanes 2 and 6), HhaI (lanes 3 and 7) or MspI (lanes 4 and 8). After electrophoresis and blotting the filters were hybridized to a nick taanslated PvuII/ EcoRI probe. Shown on the map are the HpaII site (M) and the HhaI site (H) which were analyzed in this experiment. The other HpaII and HhaI sites in this region are indicated by vertical lines, and these were assayed using different blotting strategies (data not shown).

into F9, it too underwent efficient demodification. Consistent with the stage specificity of this process, these sequences did not become demethylated in somatic cell types.

What might be the function of the island demethylation observed *in vivo* and in embryonic cell lines? Although most island sequences are indeed constitutively unmethylated, there are special classes of sequences which undergo methylation changes. One important category of methylated islands includes the housekeeping genes present on the X chromosome. These genes become inactivated on one X at the blastocyst stage of development, and this is followed by a massive *de novo* methylation of island sequences, which presumably acts to maintain their inactive state (20). This chromosome specific methylation has been well documented for several constitutive genes, including HPRT, PGK and G6PD, (21, 22, 23). On the other hand, these same sequences, must be kept in an unmethylated state in the later stages of female germ line development as well as in the embryo, where both X chromosomes are active. The demethylation activity described in our study, clearly represents a possible mechanism for controlling X chromosome methylation. In the embryo, it provides the means for keeping islands unmethylated, but its absence in somatic cells insures the continued presence of methylated islands in all cells in which the X remains inactive.

A further DNA species which probably employs a stage specific demethylation process is satellite DNA. In the mouse, these sequences represent over 10% of the genome and over 35% of all CpG residues of the cell (24). A careful look at the 234 bp repeat sequence reveals the presence of 10 CpG sites, representing about 5% of the total unit. This concentration of CpG's is certainly atypical of the rest of the genome, and actually may represent a large tandem domain of island DNA. These sequences are fully methylated in all somatic cells, but are dramatically undermethylated in the male and female germ line and in embryonic cells (25, 26). We propose that the activity described in this paper may be partly responsible for this developmental pattern. In the embryo, it provides a means for demethylating islands and maintaining this undermethylated state. Its absence in somatic cells allows the sequences to remain fully methylated following the *de novo* methylation which occurs at around the blastocyst stage. Other organisms have clear cut mechanisms for physically removing satellite sequences from somatic cells (27) and DNA methylation may accomplish the functional equivalent of this in animal cells.

It should be noted that this island demodification system may represent a mechanism for generally controlling DNA methylation in the embryo. There is strong evidence that gene sequences inserted into mouse zygotes all undergo *de novo* methylation (28). At the same time, island sequences appear to be protected from this general modification (29). This selectivity could be provided by the demethylation activity described in this paper, since it would clearly act to prevent or reverse island methylation. In one sense, the discovery of an island demethylation activity somewhat alters our picture of DNA methylation and its role during normal development. It is still clear that DNA modification represents and efficient and important mechanism for repressing gene activity on a global basis. On the other hand, our results suggest that changes in this methylation pattern may be possible in the embryo. Thus, even though we introduced a methylated APRT gene, the organism was able to recognize it and reestablish a normal undemethylated pattern. Once this pattern is fixed, at the blastocyst stage, however, it cannotbe changed, and remains static in somatic cells.

REFERENCES

1. Cedar, H.(1988) DNA methylation and gene activity.Cell 53: 3.
2. Stein, R., Razin, A. and Cedar, H. (1982) *In vitro* methylation of thehamster APRT gene inhibits its expression in mouse L-cells. Proc. Natl Acad.Sci. USA 79: 3418.
3. Busslinger, M., Hurst, J and Flavell, R.A. (1983) DNA methylation and the regulation of globin gene expression. Cell 34: 197.
4. Jones, P.A. (1984) Altering gene expression with 5- Azacytidine. Cell 40: 485.

5. Yisraeli, J. and Szyf, M. (1984) Gene methylation patterns and expression. In "DNA methylation and its biological significance" (A. Razin, H. Cedar and A. D. Riggs, eds.) Springer-Verlag Inc. New York, p. 353.
6. Bird, A., Taggart, M., Frommer, M., Miller, O.J. and Macleod, D. (1985) A fraction of the mouse genome that is derived from islands of non-methylated, CpG-rich DNA. Cell 40: 91.
7. Stein, R., Sciaky-Gallili, N., Razin, A. and Cedar, H.(1983) The pattern of methylation of two genes coding forhousekeeping functions. Proc. Natl. Acad. Sci. USA 80: 2422.
8. Keshet, I., Lieman-Hurwitz, J and Cedar, H. (1986) DNA methylation affects the formation of active chromatin. Cell 44:535.
9. Yisraeli, J., Adelstein, R., Melloul, D., Nudel, U., Yaffe, D.and Cedar, H.(1986) Muscle specific activation of a methylated chimeric actin gene. Cell 46: 409.
10. Wilks, A.F., Cozens, P.J., Mattaj, J.W. and Jost, J.P. (1982) Estrogen induces a demethylation at the 5' end region of the chicken vitellogenin gene. Proc. Natl. Acad. Sci. USA 79: 4252.
11. Vedel, M., Gomez-Garcia, M., Sala, M. and Sala-Trepat, J. M.(1983) Changes in methylation pattern of albumin and α-fetoprotein genes in developing rat liver and neoplasia. Nucl. Acids Res. 11: 4335.
12. Benvenisty, N., Mencher, D. Meyuhas, O., Razin, A. and Reshef, L.(1984) Sequential changes in DNA methylation patterns of the rat phosphoenol pyruvate carboxykinase gene during development. Proc. Natl. Acad. Sci. USA 82: 267.
13. Saluz, H.P., Jiricny, J. and Jost, J. P. (1986) Genomic sequencing reveals a positive correlation between the kinetics of strand-specific DNA demethylation of the overlapping estradiol/glucocorticoid-receptor binding sites and the rate of avian vitellogenin mRNA synthesis. Proc. Natl. Acad. Sci.USA 83: 7167.
14. Razin, A., Feldmesser, E., Kafri,T. and Szyf, M. (1985) Cell specific DNA methylation

patterns: formation and a nucleosomal locking model for their function. In "Biochemistry and Biology of DNA Methylation". (G.L. Cantoni and A. Razin, eds.) New York Alan R.Liss p.234.
15. Razin, A., Szyf, M., Kafri, T., Roll, M., Giloh, H., Scarpa,S.Carotti, D. and Cantoni, G.L. (1986) Replacement of 5-methylcytosine - A possible mechanism for transient demethylation during differentiation. Proc. Natl. Acad. Sci. USA 83: 2827.
16. Sullivan, C.H. and Grainger, R.M. (1987) δ-crystallin genes become hypomethylated in post mitotic lens cells during chicken development. Proc. Natl. Acad. Sci. USA 84: 329.
17. Mather, L. and Perry, R.P. (1983) Methylation status and DNase sensitivity of immunoglobulin genes: changes associated with rearrangement Proc. Natl. Acad. Sci. USA 80: 4689.
18. Cate, R.L., Chick, W. and Gilbert, W. (1983) Comparison of methylation patterns of the two rat insulin genes. J. Biol.Chem. 258: 6645.
19. Karlsson, O.,, Edlund, T., Barnett Moss, J., Rutter, W.J. and Walker, M.D. (1987). A mutational analysis of the insulin gene transcription control region: Expression in beta cells is dependent on two related sequences within the enhancer. Proc. Natl. Acad. Sci. USA 84: 8819.
20. Lock, L.F., Takagi, N. and Martin, G.R. (1987) Methylation of the Hprt gene on the inactive X occurs after chromosome inactivation. Cell 48: 39.
21. Wolf, S.F., Dintzis, S., Toniolo, D., Persico, G., Lunnen, K.D., Axelman, J. and Migeon, B.R. (1984) Complete concordance between glucose-6-phosphate dehydrogenase activity and hypomethylation of 3' CpG clusters:implications for X chromosome dosage compensation. Nucl. Acids Res. 12: 9333.
22. Wolf, S.F., Jolly, D.J., Lunnen, K.D., Friedmann, T. and Migeon, B.R.(1984) Methylation of the hypoxanthine phosphoribosyl transferase locus on the human X

chromosome: implications for X-chromosome inactivation. Proc. Natl. Acad.Sci. USA 81:2806.
23. Keith, D.H., Singer-Sam, J. and Riggs, A.D. (1986) Active X chromosome DNA is unmethylated at eight CCGG sites clustered in a guanine-plus-cytosine-rich island at the 5' end of the gene for phosphoglycerate kinase. Mol. Cell. Biol.6: 4122.
24. Solage, A. and Cedar, H. (1978) The organization of 5-methylcytosine in chromosomal DNA. Biochemistry 17: 2934.
25. Chapman, V., Forrester, L., Sanford, J., Hastie, N. and Rossant, J.(1984) Cell lineage-specific undermethylation of mouse repetative DNA. Nature 307: 284.
26. Ponzetto-Zimmerman, C. and Wolgemuth, D.J. (1984) Methylation of satellite sequences in mouse spermatogenic and somatic DNAs. Nucl. Acids. Res. 12: 2807.
27. John, B. and Gabor-Miklos, G.L. (1979) Functional aspects of satellite DNA and heterochromatin. Int. Rev. Cytol. 58: 1.
28. Jahner, D. and Jaenisch, R. (1984) DNA methylation in early mammalian development. In "DNA methylation and its biological significance" (A. Razin, H. Cedar, and A.D. Riggs eds.) Springer-Verlag, Inc. New York, p.189.
29. Kolsto, A. B., Kollias, G., Giguere, V., Isobe K.I., Prydz, H. and Grosveld, F. (1986) The maintenance of methylation-free islands in trangenic mice. Nucl. Acids. Res. 14: 9667.

REGULATION OF TRANSCRIPTION BY DNA METHYLATION IN SV40 AND THE HSV TK GENE IN OOCYTES[1]

F. Götz, H. Wagner, H. Kröger, D. Simon

Robert Koch-Institut, Nordufer 20, D-1000 Berlin 65

ABSTRACT SV40 DNA and HSVI TK DNA were methylated in vitro, either with prokaryotic or eukaryotic DNA-methyltransferases; their transcription was studied in Xenopus oocytes. Prokaryotic methylation with Hpa II or Hha I methyltransferase did not effect transcription of SV40 and TK DNA. Methylation with a eukaryotic methyltransferase from rat liver inhibited the SV40 early genes only slightly. The SV40 late genes were inhibited completely, when both, the regulatory region and the 5'end of the gene, were methylated.

Inactivation of the TK gene could be achieved either by methylation of the promoter or of the structural gene. The SV40 enhancer, inserted into the TK upstream region, was not able to reduce the transcription inhibition of the TK gene by complete or partial eukaryotic methylation.

INTRODUCTION

As we have shown previously, SV40 and polyoma early genes are transcribed normally in permissive and nonpermissive cells after in vitro methylation with a eukaryotic DNA methyltransferase (1). They have, so far, remained the only genes desribed which are essentially not influenced by eukaryotic methylation. One possible explanation for this behaviour was that the SV40 enhancer, containing no methylable CpGs in its sequence, was able to override the inhibition of transcription by methylation. W. Dörfler's group recently showed that inactivation of the adeno E2A promoter by methylation with Hpa II methyltransferase was reverted by insertion of a cytomegalovirus enhancer (2).

[1]This work was supported by the Deutsche Forschungsgemeinschaft

We have analysed the inhibition of transcription by methylation of SV40 early and late genes and of the HSVI TK gene in Xenopus oocytes. We found that in oocytes, too, only a slight inhibition of the SV40 early transcription was achieved by extensive methylation with a eukaryotic methyltransferase. In contrast, the SV40 late genes were completely shut off under these conditions.

Transcription of the TK gene in oocytes was inhibited by methylation of the promoter, as well as by methylation of the structural gene. To determine whether the SV40 enhancer, at least in its early orientation, was able to overcome transcription inhibition by methylation, we inserted two copies of the 72-bp SV 40 enhancer into the regulatory region of the TK gene and studied their influence in oocytes.

METHODS

DNAs and in vitro methylation. SV40 DNA and HSVI TK DNA (excised from pATTK) were methylated in vitro with the rat liver methyltransferase and purified as described (3). For partial methylation of the TK gene, the Bam HI insert of pATTK was recloned into pBR 322 after the addition of Hind III linkers. The two Hind III-Bgl II pieces of the TK gene were isolated, methylated separately and religated.

The SV40 enhancer from pAO (donated by P. Chambon) was provided with blunt ends and inserted into the upstream Pvu II site of the Hind III-cloned TK DNA after partial digestion with Pvu II. The primary structure of the enhancer insert was confirmed by sequencing.

Isolation and detection of RNAs. The DNAs (0.5 - 2 ng in 25 nl) were microinjected into stage 6 oocytes. 24 hrs later the oocytes were homogenized by pipetting and treated with SDS (1 %) and proteinase K (500 µg/ml) for 2 hrs at 37°C in a volume of 25 µl/oocyte. After 3 extractions with phenol/chloroform and ethanol precipitation, the DNA was digested with RNase-free DNase I (BRL) at 50 µg/ml in 50 ul/20 oocytes. S1 mapping was performed with the RNA from 5 oocytes. A 395-bp Pvu II-Ava II probe, 5'endlabeled with ^{32}P at the Ava II site, was used for the early SV40 genes and a 354-bp Bgl I-Hpa II probe, labeled at the Hpa II site, for the late SV40 genes. TK RNA was measured by primer extension with a ^{32}P-endlabeled 23-bp primer from nucleotides -58 to -80. The products of S1 mapping and primer extension were separated on 8 % acrylamid-urea gels. Standard procedures were used for the cloning work, S1 mapping and primer extension (4).

RESULTS

For a better understanding, Fig. 1 shows the main regulatory elements and the CpG sites in the promoter regions of SV40 and TK. Early transcription of SV40 depends on the enhancer (located in the 72-bp repeats) and on the 21-bp repeats to which the Sp1-protein binds. Late transcription of SV40 also needs the enhancer and is activated by the large T-antigen, Sp1 and LSF (late transcription factor) which as Sp1 binds to the 21-bp repeats.

The essential stimulating elements in the TK promoter are two Sp1 sites and a CTF binding site (CCAAT).

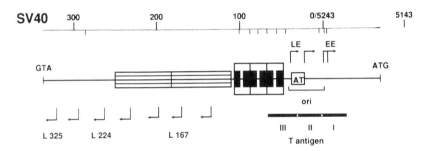

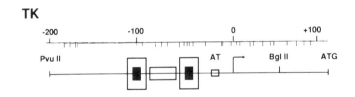

FIGURE 1: Regulatory Regions of SV40 and HSV TK.
SV40: The major late and early transcription starts are designated by arrows. The 72-bp repeats with the enhancers are represented by the striped boxes, the 21-bp repeats containing six Sp1 binding sites (CCGCCC) by the open boxes with the inserts.
TK: The cap site is indicated by an arrow. The filled box represents a CTF binding site (CCAAT) and the open boxes two Sp1 binding sites.
Methylation sites are marked below the upper scale.

Götz et al.

Effect of DNA Methylation on Transcription of SV40 in Oocytes.

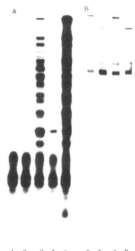

FIGURE 2: S1 Mapping of the SV40 Gene Products after in Vitro DNA Methylation with the Rat Methyltransferase.
A: SV40 late genes. Lanes 1 and 2: methylated (1 ng DNA/oocyte), lane 3: control, lane 4: methylated (1.7 ng DNA/oocyte), lane 5: control.
B: SV40 early genes. Lanes 6 and 8: methylated (1 ng DNA/oocyte), lanes 7 and 9: control.

It had been reported that transcription of the late SV40 genes in oocytes could be inhibited by methylation of the only Hpa II site, 25 bp downstream from the main late cap site (5). In our hands, however, neither methylation of the Hpa II site nor additional methylation of the two Hha I sites in the late genes had any effect on SV40 late transcription (data not shown).

For further experiments, we therefore used a purified DNA methyltransferase from rat liver which methylates every CpG and a small amount of CpA and CpT in addition (6).

There are a total of 27 CpG pairs in SV40 DNA, most of them in the regulatory region comprising replication origin and promoters. The enhancer is free of CpGs. Maxam-Gilbert sequencing (7) of the SV40 regulatory region after in vitro methylation with the rat liver enzyme showed all CpGs completely methylated.

After injection of the methylated SV40 DNA into Xenopus oocytes, late transcription was found completely suppressed. Early transcription was comparatively slightly inhibited (Fig. 2), the average inhibition being about 50 % in different experiments.

Partial Methylation of SV40 Late Genes.

To narrow down the region responsible for transcription inhibition of the late genes by methylation, we isolated the regulatory region (Taq I-Kpn I) or the first 500 nucleotides of the late genes (Kpn I-Hae II) and after methylation ligated them back into the unmethylated rest of the SV40 molecule. As can be seen in Fig. 3, neither methylation of the

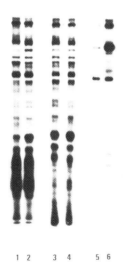

FIGURE 3: S1 Mapping of the SV40 Late Gene Products after Partial Eukaryotic DNA Methylation.
Lane 1: regulatory region (Taq I-Kpn I) methylated; lane 2: control; lane 3: 5'-structural gene (Kpn I-Hae II) methylated; lane 4: control; lane 5: regulatory region + 5'-structural gene methylated; lane 6: control.

280   Götz et al.

promoter nor of the 5'structural gene alone led to a significant inhibition of transcription. Methylation of both, promoter and 5'structural gene, was neccessary for a pronounced effect.

Effect of DNA Methylation on TK Gene Transcription in Oocytes.

Methylation of the TK gene with the Hpa II methyltransferase or with Hpa II- and Hha I methyltransferase did not inhibit transcription in oocytes (data not shown). In contrast, Buschhausen et al. (7) reported the loss of TK activity after microinjection of Hpa II methylated TK gene into cultured cells. A transcription stop at some methylated site in the structural gene would be a possible explanation for that

FIGURE 4: Primer Extension Mapping of the TK Gene Products after Partial or Complete Eukaryotic DNA Methylation. Lane 1: structural gene (Bgl II-Hind III) methylated; lane 2: control; lane 3: regulatory region (Hind III-Bgl II) methylated; lane 4: control; lane 5: regulatory region + structural gene methylated; lane 6: control; lane 7: pBR 322 cut with Hha I, 5'endlabeled with ^{32}P.

difference. As we determined the transcription start by S1 mapping, we would not have observed a stop further downstream. On the other hand, they would not have found TK enzyme activity in case of a transcription stop.

Methylation of the TK gene with the rat methyltransferase did not only shut off transcription when both, promoter and structural gene were methylated, but also when either promotor or structural gene was methylated alone (Fig. 4). For these experiments the TK gene without the vector was cut at the Bgl II site, 50 bp downstream from the cap, and religated after methylation of the promoter or the structural gene. Similar results had beeen obtained by transfection of Ltk- cells with a TK gene methylated by DNA synthesis in the presence of 5-methyl dCTP (9).

Effect of the SV40 Enhancer on Inhibition of TK Transcription by Methylation.

As pointed out above, our results with the early SV40 genes (1) and the results obtained with a cytomegalovirus enhancer inserted into an adeno gene (2) suggested that the SV40 enhancer in its early orientation could be able to override the transcription inhibition by methylation. We therefore introduced two copies of the 72-bp enhancer of SV40 in both orientations into the Pvu II site of the TK gene, 200 bp upstream from the cap site. The enhancement of TK transcription in oocytes with the unmethylated construct was 4 to 5 fold.

Methylation with the rat liver methyltransferase was performed of 1) promoter and structural gene, or 2) promoter only, or 3) structural gene only. After injection into oocytes the transcription products were measured by primer extension (Fig. 5: enhancer in the early orientation; Fig. 6: enhancer in the late orientation). Apparently the SV40 enhancer did not influence the transcription inhibition of TK by methylation under these conditions.

DISCUSSION

We have methylated SV40 DNA and HSV TK DNA in vitro with a DNA methyltransferase from rat liver and studied the inhibition of transcription by methylation in frog oocytes. Whereas SV40 early genes were only slightly inhibited by methylation, SV40 late genes were completely shut off from

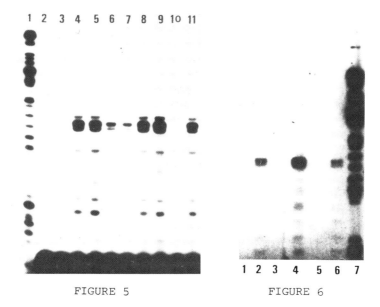

FIGURE 5 (left): Primer Extension Mapping of TK Gene Products after Insertion of an SV40 Enhancer in the Early Orientation and Partial or Complete Eukaryotic Methylation. Lane 1: pBR 322 cut with Hha I, 5'endlabeled with ^{32}P; lanes 2 and 3: regulatory region methylated; lanes 4 and 5: control; lanes 6 and 7: structural gene methylated; lanes 8 and 9: control; lane 10: regulatory region + structural gene methylated; lane 11: control.

FIGURE 6 (right): Primer Extension Mapping of TK Gene Products after Insertion of an SV40 Enhancer in the Late Orientation and Complete Eukaryotic Methylation.
Lane 1: TK gene without enhancer, methylated; lane 2: control; lane 3: TK gene with enhancer in the early orientation, methylated; lane 4: control; lane 5: TK gene with enhancer in the late orientation, methylated; lane 6: control; lane 7: pBR 322 cut with Hha I, 5'endlabeled with ^{32}P.

transcription. Partial methylation of the late genes showed that methylation of the regulatory region as well as of the 5'end of the gene was neccessary for transcription inhibition. Methylation of either piece alone had almost no effect.

Inactivation of the TK gene could be achieved either by methylation of the promoter and the first 50 bp of the gene or by methylation of the rest of the gene.

We propose that for individual genes, the way in which DNA methylation leads to transcription inhibition may be different. In the case of the TK promoter, which was recently shown to be inactivated by 1 to 4 methylgroups only (10), a local perturbation could result in a change of the binding of an activating or repressing protein. With the SV40 late genes the more extended region of methylation neccessary for transcription inhibition could more probably effect the overall structure of the SV40 chromatin.

The reason for the early SV40 genes not being inhibited by methylation remains obscure. The hypothesis that the SV40 enhancer in the early orientation could override the methylation effect, as had been established for the cytomegalo enhancer (2), was not confirmed. The SV40 enhancer inserted into the TK upstream region was not able to reduce the transcription inhibition of the TK gene by complete or partial eukaryotic methylation. Similar results were obtained when the SV40 enhancer was inserted into the Pvu II site 1840 bp downstream from the TK cap site or with a plasmid in which the TK promoter region had been repeaced by the SV40 early promoter and enhancer. In these cases, too, eukaryotic methylation inhibited transcription of the TK gene completely (results unpublished). In the adenovirus model with the inserted cytomegalo enhancer, only 3 methylgroups had been introduced into the promoter by Hpa II methylation, whereas every CpG was methylated in the TK promoter with the SV40 enhancer. It would be of general interest whether the cytomegalo enhancer or any enhancer could also override a full equivalent of eukaryotic methylation.

ACKNOWLEDGMENTS

We thank P. Chambon, Strassbourg, for kindly providing the plasmid pAO.

REFERENCES

1. Graessmann M, Graessmann A, Wagner H, Werner E, and Simon D (1983). Complete DNA methylation does not prevent polyoma and simian virus 40 early gene expression. Proc Natl Acad Sci USA 80:6470
2. Knebel-Mörsdorf D, Achten S, Langner K-D, Rüger R, Fleckenstein B, and Dörfler W (1988). Reactivation of the methylation-inhibited late E2A promoter of adenovirus type 2 by a strong enhancer of human cytomegalovirus. Virology 166:166.
3. Simon D, Stuhlmann H, Jähner D, Wagner H, Werner E, and Jaenisch R (1983). Retrovirus genomes methylated by mammalian but not bacterial mthylase are non-infectious. Nature 304:275.
4. Maniatis F, Fritsch EE, and Sambrook J (1982). "Molecular Cloning - A Laboratory Manual." Cold Spring Harbor, New York: Cold Spring Harbor Lab.
5. Fradin A, Manley JL, and Prives CL (1982). Methylation of simian virus Hpa II site affects late, but not early, viral gene expression. Proc Natl Acad Sci USA 79:5142.
6. Hubrich-Kühner K, Buhk H-J, Wagner H, Kröger H, and Simon D (1989). Non-C-G-Recognition sequences of cytosine-5-methyltransferase from rat liver. Biochem Biophys Res Commun, in the press.
7. Maxam AM, and Gilbert W (1977). A new method for sequencing DNA. Proc Natl Acad Sci USA 74:560.
8. Buschhausen G, Wittig B, Graessmann M, and Graessmann A (1987). Chromatinstructure is required to block transcription of the methylated herpes simplex virus thymidine kinase gene. Proc Natl Acad Sci USA 84:1177.
9. Keshet I, Yisraeli J, and Cedar H (1985). Effect of regional DNA methylation on gene expression. Proc Natl Acad Sci USA 82:2560.
10. Ben-Hattar J and Jiricny J (1988). Methylation of single CpG dinucleotides within a promoter element of the Herpes simplex virus tk gene reduces its transcription in vivo. Gene 65:219.

DNA METHYLATION IN THE 5' REGION OF THE MOUSE PGK-1 GENE AND A QUANTITATIVE PCR ASSAY FOR METHYLATION

J. Singer-Sam, T. P. Yang[1], N. Mori, R. L. Tanguay, J. M. Le Bon, J. C. Flores, and A. D. Riggs

Biology Division, Beckman Research Institute of the City of Hope, Duarte, California, 91010

ABSTRACT We report here the methylation pattern for Hpa II sites (CCGG) in the CpG-rich island at the 5' region of the mouse X-linked phosphoglycerate kinase gene (PGK-1). The data indicate that the active X chromosome is unmethylated at every Hpa II site in this region, whereas the inactive X is partially methylated at several sites and fully methylated at site #7, which is 30 bp upstream of the translation start point. Mouse site #7 corresponds to the Hpa II site in the human PGK-1 gene whose methylation is best correlated with inactivity. Northern blots and measurement of mouse PGK-1 specific activity show tissue-specific differences in expression, with muscle being ten-fold greater than spleen. Other than for X inactivation, no clear correlation between methylation and expression is seen.

We also report a Hpa II-PCR assay for DNA methylation. Genomic DNA is first digested with Hpa II. Then PCR is performed with primers which flank the Hpa II sites(s) being assayed. Only intact DNA will serve as a template for exponential amplification. The assay is sensitive and quantitative; we have used it to determine that site #7 in female DNA is methylated 50.7 ± 1.9 %.

[1]Present address: Dept. of Biochemistry and Mol. Biol., Univ. of Florida, Gainesville, Fla. 32610.

INTRODUCTION

Much evidence has accumulated implicating DNA methylation in the maintenance of X inactivation (reviewed in 1,2). The most convincing of these experiments show that 5 azacytidine, a demethylating agent, can cause the reactivation of genes on the inactive X chromosome in somatic cell hybrids, and increase the transfection efficiency of HPRT, an X-linked gene carried on the inactive X chromosome in the hybrid cells (3,4). In addition, the CpG-rich islands of the X-linked housekeeping genes HPRT, G6PD, and PGK are unusual in that they are hypermethylated on the inactive X chromosome (5,6,7,8), in contrast to the characteristically unmethylated state of autosomal CpG islands (9). We found all 8 Hpa II sites in the CpG island at the 5' end of X-linked human phosphoglycerate kinase (PGK-1) to be unmethylated in male leukocytes, which contain only an active X chromosome; female leukocytes, however, which have an inactive X chromosome as well, contain additional DNA which is completely methylated at these 8 sites (5). Hansen et al (10) have recently found that demethylation at one or more of the Hpa II sites between the transcription and translation initiation points of the human PGK-1 gene correlates with PGK expression in nonselected reactivants.

For comparative purposes, and because the mouse is preferable for many studies, we have characterized the mouse CpG island. We also were particularly interested in the possibility of developing a PCR based assay to measure DNA methylation. Previously, in elegant but laborious experiments, Lock and Martin (11), studying mouse embryos and teratocarcinoma cells, demonstrated that DNA methylation at a site in the first intron of HPRT occurred after X inactivation. Studies with embryos and small numbers of cells would clearly benefit from a more sensitive assay for methylation. The polymerase chain reaction (PCR) method (12) has the required sensitivity, but the amplified product does not retain the 5-methylcytosine pattern of the genomic template DNA. We reasoned that if we chose PCR primers flanking a Hpa II site or sites, prior restriction of the DNA by Hpa II would allow the genomic DNA to serve as a PCR template only if the Hpa II site or sites were methylated, and therefore refractory to cutting by Hpa II. We have found that such an assay can be used quantitatively to measure methylation of specific sites.

METHODS

Cloning of the 5' Region of the Mouse PGK-1 Gene

Our mouse PGK-1 cDNA clone has been described (13). The 1.7 kb EcoR1 insert of this clone was used to screen a mouse genomic library in Charon 4A containing DNA of strain C57/B6 partially digested by EcoR1 (kindly supplied by R.B. Wallace). One genomic clone was isolated containing an 11.5 kb insert including the 5' region of the X-linked PGK gene, as confirmed by comparison with the published sequence (14). We subcloned a 2 kb Xba I fragment containing the translation initiation site and 540 bases upstream of it into Bluescript (Stratagene, Inc.) to obtain the plasmid pMPX2, which was subsequently used for isolation of the 704 bp Pvu II-Xba I probe shown in Fig. 1. Probes A and B were isolated on low melt agarose, prior to labeling by random priming (15). Probe B was synthesized by PCR by use of primers 1 and 2 (see Fig. 1) and pMPX2 as the template.

Methylation Analysis by PCR

<u>Hpa II Treatment</u>. 0.5-1 µg total DNA (carrier added if necessary), was digested by at least a 3-fold excess of Hpa II (60 U/µl, NEBL) for 3 hr at 37 °C, in 0.5X KGB (50 mM potassium glutamate, 12.5 mM magnesium acetate, 0.25% gelatin)(18), in a final volume of 25 µl. Completenesss of cutting was ascertained by the addition of M13 RF DNA to either the entire reaction mix or to an aliquot of each sample. After digestion, the Hpa II was inactivated by placing the samples in a heat block at 100 °C for 10 min. Since Hpa II is relatively resistant to heat inactivation, we recommend that its inactivation be carefully monitored.

The Polymerase Chain Reaction.

The entire Hpa II reaction mix was diluted directly into 1X PCR buffer (10 mM Tris-HCl, pH 8.3, 1.5 mM MgCl2, 50 mM KCl, 0.01% gelatin) and 200 µM dNTPs, in a final reaction volume of 100 µl. (It is necessary to dilute the Hpa II reaction mix at least 4-fold to avoid artifacts due to the KGB buffer during subsequent gel electrophoresis.). In addition, the primers P1, CACGCTTCAAAAGCGCACGTCT, and P3, CTTGAGGGCAGCAGTACGGAAT, shown in Fig.1, were present at a final concentration of 0.2 µM.

After the addition of Taq I polymerase (2.5 U), PCR was carried out at 95 °C for 1 min, 60 °C for 2 min and 73 °C for 2.5 min, for the number of cycles indicated in the figure legends. The reagents, protocol, and thermal cycler were from Cetus-Perkin Elmer.

Gel electrophoresis and quantitation.

After the amplification reaction, a 15 µl aliquot was used for gel electrophoresis in 2% agarose containing 0.4 µg/ml ethidium bromide. The gel was then photographed onto technical film (Kodak, 4415). A video densitometer (Biorad Model 620) together with programs for peak integration (The 1-D Analyst programs, Biorad) were used to determine the intensity of each band, with care taken to be in the linear response range of the film. Alternatively, where increased sensitivity was necessary, the DNA was transferred to Genetrans 45 (Plasco, Inc.) by vacuum-blotting in 0.4 N NaOH as previously described (16). After hybridization with ^{32}P-labelled oligomer P2 (CTTGAGGGCAGCAGTACGGAAT, see Fig. 1), at 60 °C for at least 3 hr, the filter was washed three times for four min each at 60 °C. After autoradiography, the intensity of each band was determined by densitometry as above.

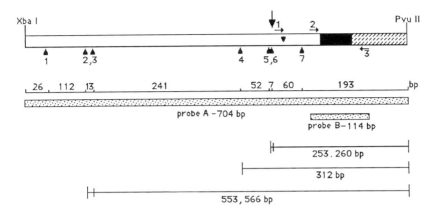

Fig. 1. The 5' region of the mouse PGK-1 gene. The upper portion of the figure shows the 704 bp Xba I-Pvu II restriction fragment, including the first translated region (solid bar), and part of the first intron (hatched bar). The triangles show the positions of the 7 Hpa II sites in this fragment (14). The inverted triangle shows the position in human PGK-1 of the methylated Hpa II site best correlated with inactivity (10). The inverted arrow shows the distance of the transcription start points from the translation initiation site, as determined for human DNA (17). The horizontal arrows show the position of the oligonucleotides P1, P2, and P3 (see METHODS). The lower part of the figure shows the size of restriction fragments expected after cutting by Pvu II, Xba I, and Hpa II. Probes A and B are also shown, as well as the partially methylated fragments found in female DNA (see Text and Fig. 2).

RESULTS

DNA Methylation in the Mouse PGK-1 5' region: Sex and Tissue-Specificity

The mouse 5' region is part of a CpG-rich island, showing 68% homology with its human counterpart (5,14). Fig. 1 schematically shows a 704 bp portion of the mouse sequence, bounded by the restriction enzymes Xba I and Pvu II. Also shown are the 7 Hpa II sites in this region, and the location, relative to the start site of translation, of the Hpa II site in human DNA which shows a perfect correlation between methylation and inactivity (10).

To determine the methylation pattern at these Hpa II sites, DNA was purified from various tissues of female and male Balb C mice, digested with Xba I, Pvu II and Hpa II, and analyzed by Southern blotting (Fig. 2). With probe A, male and female DNA both show bands of 112, 241, and 193 bp, as expected for unmethylated DNA. These are the only bands seen for the male DNAs. The female DNAs show only a faint band at 704 bp, which would be seen if the DNA were fully methylated, as is the case for human female DNA (5). However, in contrast to males, the four female tissues show bands corresponding to partially methylated DNA. Particularly prominent are bands of 260 and 312 bp, suggesting that Hpa II site #7 is methylated but not site 5 and/or 6 in the case of the 260 bp band or site 4 in the case of the 312 bp band. Fig. 2 also shows the same blot after

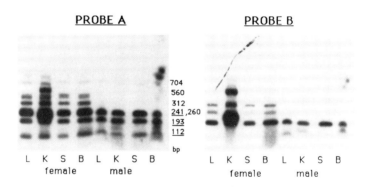

Fig.2. Southern blots showing tissue and sex-specificity of methylation of mouse DNA. DNA from Balb C male and female mice was digested with Pvu II, Xba I, and Hpa II. Completeness of digestion was verified by incubation of an aliquot of each sample with plasmid DNA. Gel electrophoresis, blotting, and hybridization were performed as described (15). The left panel was obtained by hybridization with probe A shown in Fig. 1. The right panel was obtained by rehybridizing the same filter with probe B shown in Fig. 1. L,liver; K, kidney; S, spleen; B, brain.

rehybridization with a 114 bp probe for the 3'end (probe B). Bands of 260 and 312 bp are seen, corresponding to methylation of site #7 for the 260 bp band, and methylation of sites 7 and 5 and/or 6 for the 312 bp band. The other bands seen in female DNA probably result from partial methylation of sites 2,3, and 4.

Kidney DNA is seen to have an additional band at 565 bp, corresponding to sites 4,5,6 and 7 being methylated; this unique band is apparently not related to expression, since kidney DNA is moderately expressed (Fig. 3; Table 1). Female spleen DNA hybridized with probe B shows a higher ratio of the intensity of the 312 bp and 260 bp bands, indicating more complete methylation of site 5, and/or 6.

Tissue Specificity of PGK mRNA Levels and Enzyme Specific Activity

Little information is available about tissue-specific expression of PGK, and still less about transcription of the gene. We therefore analyzed several mouse tissues for mRNA content as shown in Fig. 3. PGK mRNA levels are seen to vary: RNA from abdominal wall and leg muscle gives the greatest signal, and RNA from from spleen and liver the least

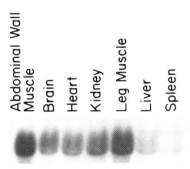

Fig.3. Northern blot analysis showing tissue -specificity of PGK-1 mRNA levels. Total RNA was isolated from Balb C mice and Northern blots were done as described (15). Hybridization was with the entire 1.7 kb mouse cDNA probe (13).

Table I
PGK SPECIFIC ACTIVITY IN VARIOUS MOUSE TISSUES

Tissue	PGK Specific Activity (units/mg protein)[a]
Leg muscle	2.2
Abdominal muscle	2.0
Brain	1.1
Heart	0.7
Kidney	0.5
Liver	0.28
Spleen	0.23

[a] Each tissue was homogenized in 2 ml KPO_4 buffer (0.1 M, pH 6.5) and centrifuged for 20 min at 10,000 rpm; 2-10 µl was then used in a spectrophotometric assay for PGK (19). The initial rate of formation of 1,3-diphosphoglycerate was measured. Assay mixtures contained 5 mM ATP, 10 mM 3-phosphoglycerate, 5 mM magnesium acetate, 0.2 mM NADH, 20 mM triethanolamine-acetate buffer (pH 7.5), and glyceraldehyde-3-phosphate dehydrogenase (150 µg/ml) in a total volume of 1 ml. Protein concentration was determined by the method of Bradford (20).

signal, while heart, kidney, and brain are intermediate. Some values for PGK activity per gram of tissue had previously been reported for various mouse tissues (21), but no systematic comparison of PGK mRNA and PGK specific activity has been done. We therefore measured PGK enzyme activity for several mouse tissues, and found the values shown in Table I. In agreement with the mRNA data, a 10-fold range is seen for the specific activity of PGK, with muscle giving the maximum value, and liver and spleen the minimum.

A Hpa II-PCR Assay for DNA Methylation

The data obtained by Southern blot analysis indicate that Hpa II site #7 is of interest for developmental studies of mouse embryos, for which increased sensitivity would be necessary. We therefore used site #7 as a test system for development of a PCR based assay. We synthesized a set of primers flanking Hpa II site #7 (primers 1 and 3 shown in Fig. 1). These primers direct the synthesis of a 170 bp product only if the template DNA is not cut by Hpa II at site #7. DNA from mouse male or female spleen and muscle was treated simultaneously with Xba I, Pvu II and Hpa II. The DNA was then used as a template for PCR.

Fig. 4 shows that for male spleen DNA, prior digestion with Hpa II totally eliminates the PCR product, whereas for female spleen DNA the PCR product is only reduced by 50% as determined by densitometry. A very similar result is seen for muscle DNA (last 4 lanes). As expected since there is only a single X chromosome in males, the signal seen for uncut male DNA (lanes 4 and 8) is one-half that seen for female DNA cut by Hpa II (lanes 1 and 5). The barely visible band seen in lane 7 exemplifies a potential pitfall and the need for a cautionary note; the assay is so sensitive that even

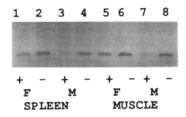

Fig. 4. Hpa II-PCR analysis showing differences between male and female DNA at Hpa II site #7. DNA (700 ng) from male and female spleen or muscle DNA was digested with Hpa II and assayed by PCR as described in METHODS PCR was done for 22 cycles. F, female DNA; M, male DNA; (+), digested with Hpa II; (-), not digested with Hpa II.

a small percentage of uncut or contaminating DNA will give a visible band if enough cycles are done. For the experiment shown in Fig. 4, a few more cycles would have obliterated the differences between male and female. The signal is proportional to template DNA only as long as nucleotides and primers remain in great excess. It is best to stop the PCR reaction while bands are very faint by ethidium bromide staining. Identical temperature cycling is also essential; however our results indicate that adjacent wells in the Perkin Elmer-Cetus thermal cycler were similar enough to detect 50 % differences in the amount of target DNA.

The amount of DNA used in the above experiment (700 ng) represents about 10^5 cells. Table II, below, shows that the Hpa II-PCR assay can determine a two-fold difference when as little as 0.8 ng of genomic DNA is used, corresponding to about 100 cells.

TABLE II
HPA II-PCR ASSAY OF SITE 7 IN SMALL AMOUNTS OF GENOMIC DNA

Amount DNA (ng)	Ratio of PCR Signals[a] (+ Hpa II)/(- Hpa II)
500	1.516/2.724 (0.56)
100	0.948/1.919 (0.49)
20	0.275/0.554 (0.50)
4	0.814/1.657 (0.49)
0.8	0.598/1.207 (0.50)

[a]Female spleen DNA was serially diluted in Tris-HCl 10 mM, EDTA, 1 mM, pH 8, and the amounts of DNA indicated were incubated together with 0.75 µg M13 RF carrier DNA in the presence or absence of Hpa II, as described in METHODS. After heat inactivation of the Hpa II, and 25 cycles of PCR with oligomers P1 and P3 (see Fig. 1), the gel was hybridized with ^{32}P-labelled P2 used as a probe (7×10^6 cpm/ml). After washing and autoradiography, densitometry was done. See METHODS for details.

DISCUSSION

We have developed a PCR assay to measure DNA methylation at specific sites with much smaller DNA amounts than previously possible. Sensitivity is clearly adequate to assay DNA from about 100 cells. The Hpa II-PCR method has the potential for assaying methylation in only a few cells when combined with procedures currently being developed to assay efficiently DNA from a small number of cells (22 and our unpublished results). Hpa II site #7 in the mouse PGK-1 has been a good test system for the method because complete cutting, and thus no signal is expected for male DNA, whereas only a two-fold reduction is expected for female DNA. We find that the two-fold difference can be reproducibly detected; averaging data from 10 experiments, methylation of site # 7 is 50.7 ± 1.9 %.

Our finding that site #7 is methylated in all female mouse tissues assayed is consistent with the finding of Hansen et al. (10) that hypomethylation of this site is correlated with reactivation of PGK-1 on a previously inactive human X chromosome in Chinese hamster cells. These results and our finding that the upstream promoter region of the mouse PGK-1 is not nearly as heavily methylated as is the human promoter region suggests that a likely critical site for methylation silencing of PGK-1 is near and somewhat downstream of the start of transcription. Hpa II sites represent only one-sixteenth of the total CpGs; presumably the methylation of site #7 reflects methylation of a larger region, but verification of this assumption will require genomic sequencing. These studies are currently in progress, but will never match the sensitivity of the Hpa II-PCR assay.

If one examines the Southern blot seen in Fig. 2, the bands due to methylation of Hpa II site # 7 are lighter than might be expected if this site is fully methylated. However, this apparent discrepancy between the Southern blot and the PCR results is due to partial methylation at other Hpa II sites, resulting in a decrease in signal at 312 or 260 bp. Thus for measurement of methylation of a particular site, the PCR assay has an advantage, in that methylation at other sites outside the region assayed do not complicate interpretation of the results.

Spleen, which poorly expresses PGK mRNA (see Fig. 3), has sites 5 and/or 6 methylated as well. Aside from site 5/6, we find that tissue-specific variations in the Hpa II sites methylated are not correlated with gene activity (i.e. kidney DNA, while highly methylated, is intermediate in expression of PGK mRNA), again consistent with the idea that hypomethylation of DNA is necessary but not sufficient for the active X chromosome to function.

The 10-fold difference we see in expression of PGK-1 in different tissues, taken together with the results of Toniolo et al.(23) for G6PD and the increased expression of HPRT in brain (24), show that the distinction between "housekeeping" and "tissue-specific" genes is blurred. Tissue-specific factors must be involved in the control of PGK-1 expression from the active X chromosome.

ACKNOWLEDGEMENTS

We thank Cheryl C. Clark for fine technical assistance, and Drs.Jerome Bailey, Gerry Forrest and Paul Salvaterra for their kind advice and help. This work was supported by NIH grant AG08196.

REFERENCES

1. Monk, M (1986). Methylation and the X chromosome. BioEssays 4:204.
2. Riggs, AD (1989). DNA methylation and cell memory. Cell Biophysics, in press.
3. Mohandas T (1981). Reactivation of an inactive human X chromosome: evidence for X inactivation by DNA methylation. Science 211:393.
4. Venolia L, Gartler SM, Wassman ER, Yen P, Mohandas T, Shapiro LJ (1982). Transformation with DNA from 5-azacytidine-reactivated X chromosomes. Proc Natl Acad Sci USA 79:2352.

5. Keith DH, Singer-Sam J, Riggs AD (1986). Active X-chromosome DNA is unmethylated at eight CCGG sites clustered in a CpG-rich island at the 5' end of the gene for phosphoglycerate kinase. Mol Cell Biol 6:4122.
6. Toniolo D, Martini G, Migeon BR, Dono R (1988). Expression of the G6PD locus on the human X chromosome is associated with demethylation of three CpG islands within 100 kb of DNA. EMBO J. 7:401.
7. Yen PH, Patel P, Chinault AC, Mohandas T, Shapiro LJ (1984). Differential methylation of hypoxanthine phosphoribosyltransferase genes on active and inactive human X chromosomes. Proc Natl Acad Sci USA 81:1759.
8. Wolf SF, Jolly DJ, Lunnen KD, Friedmann T, Migeon BR (1984). Methylation of the hypoxanthine phosphoribosyltransferase locus on the human X chromosome: Implications for X-chromosome inactivation. Proc Natl Acad Sci USA 81:2806.
9. Bird AP (1986). CpG-rich islands and the function of DNA methylation. Nature 321:209.
10. Hansen RS, Ellis N, Gartler SM (1989). Demethylation of specific sites in the 5' region of the inactive X-linked human phosphoglycerate kinase gene correlates with the appearance of nuclease sensitivity and gene expression. Mol Cell Biol 8:4692
11. Lock LF, Takagi N, Martin GR (1987). Methylation of the Hprt gene on the inactive X occurs after chromosome inactivation. Cell 48:39.
12. Saiki RK, Gelfand DH, Stoffel S, Scharf SJ, Higuchi R, Horn GT, Mullis KM and Erlich HA (1988). Primer-directed enzymatic amplification of DNA with a thermostable DNA polymerase. Science. 239:487.
13. Mori N, Singer-Sam J, Lee CY, Riggs AD (1986). The nucleotide sequence of a cDNA clone containing the entire coding region for mouse X-chromosome-linked phosphoglycerate kinase. Gene 45:275.
14. Adra CN, Boer PH, McBurney MW (1987). Cloning and expression of the mouse pgk-1 gene and the nucleotide sequence of its promoter. Gene 60:65.

15. Maniatis T, Fritsch EF, Sambrook J (1982). "Molecular Cloning". New York: Cold Spring Harbor Laboratory.
16. Murakawa GJ, Zaia JA, Spallone PA, Stephens DA, Kaplan BE, Wallace RB, Rossi JJ (1988). Direct detection of HIV-1 RNA from AIDS and ARC patient samples. DNA 7:287.
17. Singer-Sam J, Keith DH, Tani K, Simmer RL, Shively L, Lindsay S, Yoshida A, Riggs AD (1984). Sequence of the promoter region of the gene for human X-linked 3-phosphoglycerate kinase. Gene 32:409.
18. McClelland M, Hanish J, Nelson M, Patel Y (1988). KGB: a single buffer for all restriction endonucleases. Nucleic Acids Res 16:364.
19. Bücher T (1955). In Colowick SP, Kaplan NO (eds): "Methods in Enzymology " New York: Academic Press, p 415.
20. Bradford M (1976). A rapid and sensitive method for the quantitation of microgram quantities of protein utilizing the principle of protein-dye binding. Analyt Biochem 72:248.
21. VandeBerg JL (1985). The phosphoglycerate kinase isozyme system in mammals: biochemical, genetic, developmental, and evolutionary aspects. Isozymes: Current Topics in Biologcial and Medical Research. 12:133.
22. Li H, Gyllensten UB, Cui X, Saiki RK, Erlich HA, Arnheim N (1988). Amplification and analysis of DNA sequences in single human sperm and diploid cells. Nature 335:414.
23. Battistuzzi G, d'Urso M, Toniolo D, Presico GM, Lussatto L (1985). Tissue-specific levels of human glucose-6-phosphate dehydrogenase correlate with methylation of specific sites at the 3' end of the gene. Proc Natl Acad Sci USA 82:1465.
24. Stout, TJ, and Caskey CT (1985). HPRT: gene structure, expression, and mutation. Ann Rev Biochem 19:127-148.

STUDYING DNA METHYLATION AND PROTEIN-DNA INTERACTION SITES IN VIVO

Hanspeter Saluz[°], Stefan Wölfl[*], Antonio Milici[+] and Jean-Pierre Jost[°]

[°] Friedrich Miescher-Institut. P.O. Box 2543, CH-4002 BASEL, Switzerland
[*] Inst. für Molekularbiologie und Biochemie, Freie Universität Berlin, D-1000 Berlin 33, FRG
[+] M.D.Anderson Tumor Institute, Houston, Texas 77030, USA

ABSTRACT Genomic sequencing allows the study of the native state of a genome such as gene specific DNA methylation in a strand-specific manner, protein-DNA interactions to single-nucleotide resolution, mutations and genetic polymorphisms, even in a heterozygote state. Several procedures are available at present, each of which has its own advantages. The classical procedure of genomic sequencing according to Church and Gilbert is compared with a new approach which radically simplifies the method while retaining all the benefits of the original technique. None of these methods enable the study of the native condition of a single DNA molecule in a single cell. Therefore new procedures, based on chain-termination, have to be invented.

INTRODUCTION

The cloning of genes in bacterial hosts results in the loss of relevant information, firstly about the interactions of regulatory proteins with their respective recognition sequences <u>in vivo</u> and secondly, the state of DNA methylation. Furthermore, the study of mutations had until recently required that each mutant DNA be cloned before being sequenced and in this context it must be noted that the actual cloning procedures themselves may introduce artifacts into the sequences of interest, e.g. deletion mutations.

DNA methylation has to date been studied by means of

methylation sensitive restriction endonucleases and their isoschizomers, e.g. HpaII/MspI. Unfortunately all these studies are restricted to certain recognition sequences within the DNA. About 90% of all potential methylation sites cannot be studied by such an approach and from the remaining 10% only fully methylated or fully unmethylated sites can be investigated. Hemimethylated sites are usually undetectable by this means. Although the use of methylation sensitive restriction enzymes has the advantage that knowledge about the DNA primary structure is not required the former considerations clearly show the disadvantage of using such an approach to study DNA methylation. Other methods such as HPLC, TLC, high-voltage electrophoresis or mass-spectrometry do not provide sequence-specific information and are often not sensitive enough. Genomic sequencing (1,2,3) overcomes all these limitations: The existence of hemimethylated DNA in the upstream region of the avian vitellogenin gene was sucessfully demonstrated using this method (4), protein-DNA interaction studies in vivo revealed correlations between gene activation, methylation and protein-DNA interaction patterns in vivo (5).

The procedure of genomic sequencing is essentially as follows (Fig.1): Total genomic DNA is isolated from organs or cells and digested to completion with a restriction endonuclease. The restriction enzyme is chosen such that the target sequence, i.e. the sequence of interest remains intact. The restriction fragments are randomly and base specifically cleaved when subjected to the different chemical sequencing reactions (6,7,8). Billions of these subfragments are separated on a high resolution sequence gel, electrotransferred to a nylon membrane, and covalently crosslinked by uv-irradiation. In an independent experiment a piece of one flanking area is cloned in M13 or similar vectors in order to produce a single-stranded probe, labeled to a very high specific radioactivity, for subsequent hybridization. After processing and exposure of the membrane to an X-ray film, the sequence of interest becomes visible.

DNA Methylation and Protein-DNA Interaction Sites 301

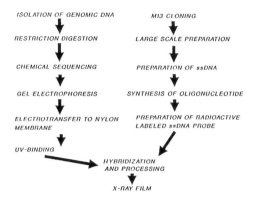

FIGURE 1. Scheme of the classical genomic sequencing procedure.

Why can such a procedure be used to study cytosine methylation ? The reason is very simple: 5-mC reacts poorly with hydrazine and will therefore not be cleaved by piperidine. This results in a gap within the Cytosine specific sequencing ladder (i.e. fragments of the appropriate length are missing). Such gaps could also be due to deamination events of 5-methyl cytosine to thymidine (9) and therefore T-specific sequencing reactions (3) should be performed to ensure detection of a methylated base and not a deamination event.

Unfortunately the genomic sequencing procedure according to Church and Gilbert (1) is rather tricky and time consuming. Therefore other ways of studying the in vivo state of a genome had to be invented (10,11). The pathway of this procedure is shown in Figure 2.

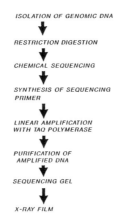

FIGURE 2. Scheme of genomic sequencing with Taq polymerase.

The first three steps are common to both methods: isolation of the genomic DNA, digestion with a restriction enzyme and chemical sequencing of the restriction fragments. In the latter procedure this is followed by selective amplification of the target sequence fragments with Taq polymerase, using a primer labeled to a very high specific activity. The products of linear amplification are applied to a sequencing gel and the sequence information is directly obtained by exposing the gel to an X-ray film. Difficult steps such as electroblotting of the fragments to nylon membranes or time consuming hybridizations or cloning steps are circumvented without losing the benefits of the original procedure. Moreover, as the restriction digest of the genomic DNA is only used to reduce the viscosity of the DNA in the reaction mixture, its choice is not critical as long as the enzyme does not cleave within the target sequence. This allows the use of less expensive restriction enzymes.

THE POWER OF GENOMIC SEQUENCING AS DEMONSTRATED IN THREE EXAMPLES USING THE CHICKEN VITELLOGENIN II GENE AS A MODEL SYSTEM

The methylation state of cytosines has to date been studied with methylation-sensitive restriction endonucleases and their isoschizomers (12) or genomic sequencing (1). An example of the first method is represented in the publication of Wilks et al. (13) where they employed the isoschizomer pair MspI/HpaII to study the methylation state of a hormone-responsive CpG site situated within the upstream area of the avian vitellogenin gene II during the activation of the gene. The results indicated that, upon estrogen injection, the gene was activated before the 5-methylcytosine became demethylated. With the development of genomic sequencing by Church and Gilbert (1) it was possible to study all cytosines in a strand-specific manner. This procedure was applied to reexamine Wilks's results (4). It was found that the demethylation occured first in the upper strand with kinetics similar to the rate of message production whereas in the lower strand it lagged approximately 24 hours behind. Why this discrepancy? The reason is simple: methylation-sensitive restriction endonucleases can usually not be applied to study such hemimethylated sites (14). This CpG site is the only one of this upstream area situated within the recognition sequence of methylation-sensitive restriction enzymes. All the other 11 CpG dinuclcotides of the upstream area had to be studied by means of genomic sequencing and about 50% were not methylated at all so as far as we know they do not play any role in gene regulation. This indicates how dangerous the use of methylation sensitive restriction endonucleases may be, because the probability is low that such a site is situated within the restriction enzyme recognition sequence and general conclusions would not be representative.

As indicated in Fig. 3 (*) genomic sequencing is also a powerful tool to study mutations. When DNA from different hen and rooster tissues was used to study the methylation pattern of the upstream area a site was detected (here upper strand; indicated with a star) whereas in the DNA from all male tissues a Cytosine and in all female tissues tested an Adenosine was found.

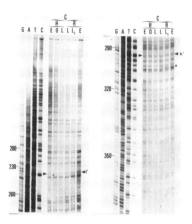

FIGURE 3. Genomic sequencing experiment of the upper strand of the avian vitellogenin gene upstream area (approximative positions <-200>-<-400>). The control sequences G,A,T,C were performed with cloned diluted DNA. The arrow heads indicate unmethylated CpG sites in different genomic DNA (C-specific reactions). H stands for hen and R for rooster, E for erythrocytes, O for oviduct and L for liver. The star indicates the position of a mutation: all rooster tissues show a cytosine whereas all hen tissues showed an adenosine.

GENERAL ANALYTICAL PROBLEMS FOR GENOMIC SEQUENCING PROCEDURES

THE GENOMIC DNA: Any contaminations of the genomic DNA with proteins, carbohydrates, phenol etc. influences the quality of "genomics" negatively (15). Proteins bound to DNA result in improper restriction digestions, unclean modifications of the bases during chemical sequencing or bad transcription efficiencies for genomic sequencing with DNA polymerases. The degree of degradation of the genomic DNA is positively correlated with the increase of background noise (3). It is therefore very important to start genomic sequencing experiments with undegraded and highly purified DNA. Good results were obtained by first isolating nuclei from tissues and a subsequent digestion overnight with proteinase K, several phenol/chloroform extractions followed by dialysis of the DNA against EDTA over 2-3 days as described previously (3, 16).

CHEMICAL SEQUENCING REACTIONS: All genomic sequencing procedures to study DNA methylation and protein-DNA interactions published so far (1,3,11,17) are based on chemical sequencing reactions as described previously (6,7,8). One of the major differences between in vitro and in vivo sequencing is the amount of DNA used for each reaction. In genomic sequencing a much higher concentration of unlabeled DNA is required to guarantee strong signals. This is either achieved by starting with a relatively high amount of genomic DNA or by amplifying the chemically sequenced DNA with Taq polymerase as described by Saluz and Jost (10,11) but using a high number of amplification cycles (>>30 cycles) and several additions of Taq polymerase or as proposed by Mueller and Wold (18) using a combination of Maxam and Gilbert sequencing, with a ligation step of a synthetic linker to the sequencing fragments followed by polymerase chain reaction using two additional synthetic oligonucleotides.

For genomic footprinting experiments, the accepted procedure has until now been to use a fixed concentration of 0.5% dimethyl sulfate (19-21,22-24). In our hands such a concentration resulted in almost no protection, e.g. nearly all bands corresponding to guanosine residues were visible on the X-ray films. Therefore we used several concentrations of dimethyl sulfate (0.5%, 0.05%, 0.005% 0.0005% and 0.00005%) for each trial to obtain fine resolution footprints (5,16).

GEL ELECTROPHORESIS: For genomic sequencing of single copy genes in higher eucaryotes it is usually necessary to use relatively large amounts of DNA. Therefore, to avoid overloading, spacers of up to 1 mm thickness are employed. To guarantee that a good resolution of the same amount of sequence information is to be retained, an increase in gel running time and also in gel length is required (up to 1 m length). Long gels also have the advantage that the distance between bands is increased resulting in a dilution of background due to nonspecific degradation products or hybridization mismatches if the classical genomic sequencing procedure according to Church and Gilbert (1) is used.

ANALYTICAL PROBLEMS FOR THE CHURCH AND GILBERT PROCEDURE

RESTRICTION DIGEST OF GENOMIC DNA: In the classical procedure, pioneered by Church and Gilbert (1) the restriction digestion of the genomic DNA plays a crucial role. The enzyme is used to digest the DNA to create a population of DNA fragments containing the target sequence. Good results will be obtained only if this restriction digest is complete and clean, e.g. no star activity or similar events should occur, to guarantee the creation of specific subpopulations of fragments of identical length and base composition when chemically sequenced. Incomplete digests drastically impair the resolution of the sequencing ladders. To avoid superimposed sequencing ladders the length of the target restriction fragment should be equal to or greater than the length of the hybridization probe in nucleotides plus the size exclusion of the gel in nucleotides (3). The amount of DNA to be digested depends on the size of the genome and the copy number of the gene to be studied. To study single-copy genes in mammals (5 x 10^6 kb per haploid genome) between 25-50 µg of genomic DNA should be used per sequencing reaction to obtain a reasonable signal strength after hybridization.

ELECTROTRANSFER TO NYLON MEMBRANES: One of the technically most difficult aspects of the classical genomic sequencing procedure is the electrotransfer of the DNA fragments from the sequencing gel to the nylon membrane. We found that "GeneScreen plus" membranes were unsuitable, due to their high electrical resistance. A high resolution transfer requires an absolutely tight contact between the gel and the membrane. Gas bubbles trapped in the transfer system result in a distortion or loss of DNA bands in the blot. Additionally, to obtain a quantitative and diffusion- free transfer, several other parameters have to be taken into consideration: electrical resistance, strength of the electrical field, Joule heating, stability of the electrolyte, etc. Good results were obtained with a horizontal system (3) consisting of a lower sieve plate electrode, a lower layer of Whatman 17 paper, the sequencing gel, a nylon membrane (Gene Screen), an upper layer of Whatman 17 paper, an upper sieve plate electrode and and a weight on top guaranting a tight contact between gel and membrane. Since the distance between the electrodes is only 3 cm, a current of approximately 1.5 A (voltage appr. 30-35 V) was fully sufficient to transfer the DNA within 30-35 minutes (room temperature) in 1 x Tris-borate-EDTA buffer (3).

THE HYBRIDIZATION PROBE: In genomic sequencing single-stranded DNA or RNA probes have been used (19-28). Several investigators have claimed that the use of RNA probes causes an increase in background (S. Wölfl, unpublished results). A review on the different methods of labeling hybridization probes is available (15). It is important that the probes are of an adequate length, i.e. 100-180 nucleotides long (3), a high specific radioactivity and sequence homogeneity (3,29).

HYBRIDIZATION AND PROCESSING: In this classical procedure, best results were obtained with the buffer system described by Church and Gilbert (1). For hybridization probes with (G + C) content between 30 and 47% a hybridization temperature of 58°C was sufficient. The hybridization was usually carried out between 2 silicon treated glass plates using a fishing line (diameter: 3.5 mm) as spacer (3) or in rotating cylinders obtained from Bachofer Reutlingen, FRG (S. Wölfl, unpublished results) for a duration of 16 to 24 hours. After hybridization the membranes were washed using the two-buffer system described by Church and Gilbert (1) at an initial temperature of 48°C. The washing itself was performed by letting the buffer cool down to room temperature. Five to seven washing steps (each 5 minutes) were usually sufficient. The buffer volumes were kept as small as possible, e.g. the membrane in the washing box was just covered. Additional details have been provided elsewhere (3).

ANALYTICAL PROBLEMS FOR GENOMIC SEQUENCING WITH TAQ POLYMERASE (10,11):

RESTRICTION DIGESTION: In contrast to the classical genomic sequencing (1), in this method (10,11) the restriction digest is used only to reduce the viscosity of the genomic DNA and is not critical as long as the enzyme does not cut within the target sequence.

LABELING OF THE SEQUENCING PRIMER: For studying the native state of single copy genes in higher eucaryotes, a primer labeled to a very high specific radioactivity is required. A direct end-labeling by kinasing is usually not sufficient. Labeling to a very high specific radioactivity may be obtained by the "filling in" reaction with polymerase I Klenow fragment (11) as demonstrated in Figure 4.

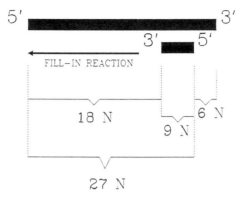

FIGURE 4. Scheme of the labeling of the sequencing primers.

A synthetic 6- to 9-mer segment of the sequencing primer is annealed to the 3' end of a 33-mer oligonucleotide complementary to the primer. The 6 to 9-mer oligonucleotide hybridizes 6 base pairs away from the 3' end of the 33-mer. This allows, upon elongation with one or more radioactively labeled deoxynucleotides, an easy separation of the sequencing primer from the 33-mer oligonucleotide. For the "filling in" reaction high quality deoxynucleotides (sequencing grade) should be used and deoxyinosinetriphosphate or deaza-deoxyguanosine-triphosphate should be avoided, since they decrease the melting temperature (T_m). Primers of this length allow an accurate radioactive labelling and are still sufficiently short to guarantee a rapid annealing.

THE MELTING TEMPERATURE: Since all conditions for genomic sequencing with Taq polymerase have to be chosen stringently it is also necessary to determine the melting temperature (T_m) of the primer oligonucleotide. This has to be done experimentally since estimations based on the formula proposed by Suggs et al. (30) are valid only for oligonucleotides with a maximum length of 21 nucleotides. Twenty-seven-mer duplexes of different base compositions (30-60% G+C content) were tested and all showed a similar melting temperature of approximately 64°C. The annealing to the genomic DNA was therefore performed at a temperature 2°C to 4°C below the T_m-value, e.g. 60°C.

LINEAR AMPLIFICATION OF DNA: The linear amplification is performed as previously described (11). Additionally it should be mentioned that the quality of the dNTPs used should be as good as possible (sequencing grade). We noticed that for the linear amplification (in contrast to the "filling in" reaction of the primer) the use of deaza-dGTP yields an excellent result (deaza-deoxyguanosine reduces the stability of secondary structures. This seems to be important especially in the case of G/C rich areas).

PURIFICATION OF THE REACTION PRODUCTS: It seems that certain enzyme preparations contain additives, such as stabilizers etc., which may negatively influence the separation of the amplification products on the sequencing gel (smears within the sequencing lane). Since common ethanol precipitations or phenol/chloroform extractions do not improve the result, we recommend precipitation of the amplified DNA in 2.74 mM (0.1% final) N-Cetyl-N,N,N-trimethyl-ammonium bromide and a final concentration of approximately 17 mM $(NH_4)_2SO_4$ (usually present in amplification buffer) at 0°C for 20 min. The DNA is recovered by a centrifugation at 30000 x g for 10-15 minutes. After removing the supernatant with a drawn-out capillary, the pellet is redissolved in 100 µl of 0.5 M (NH_4)acetate and precipitated with ethanol. This step should be repeated before adding the sample buffer and loading the gel.

CONCLUDING REMARKS:

Many investigators would be interested in studying the modifications of one single copy gene in one cell, i.e. in an egg. Such an approach is at the present absolutely impossible since all sequencing procedures to study DNA-methylation or protein-DNA interactions in vivo published so far (3,11,17) are based on chemical sequencing reactions. Therefore a certain minimal number of target copies is required in order that sequence ladders can be produced. Although techniques involving exponential amplifications (polymerase chain reactions) cannot be used to study such a problem as long as cleavage steps are involved since only one single band would appear on the sequencing gel, regardless of the number of amplification cycles. This is demonstrated in Figure 5.

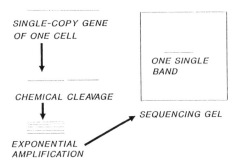

FIGURE 5. Sequencing procedures where cleavage is involved can never be used to study one single DNA molecule, since this would result in only one single sequence-specific band regardless of amplification steps etc.

Therefore to study this and similar problems where only a very few target copies are available, new techniques, based on chain-termination reactions have to be developed. The future availability of techniques which employ radically simplified procedures of this kind should make genomic sequencing methods as commonplace as dideoxy-sequencing, rendering them accessible to every molecular biologist.

ACKNOWLEDGEMENTS

We would like to thank Drs. Andrew Wallace and Karin Wiebauer for their careful pre-submission review of the manuscript.

REFERENCES

1. Church GM, Gilbert W (1984). Proc Natl Acad Sci USA 81: 1991-1995.
2. Saluz HP, Jost JP (1986). Gene 42: 151-157.
3. Saluz HP, Jost JP (1987)."A laboratory Guide to Genomic Sequencing." Basel, Boston:Birkhaeuser.
4. Saluz HP, Jiricny J, Jost JP (1986). Proc Natl Acad Sci USA 83:7167-7171.
5. Saluz HP, Feavers I, Jiricny J, Jost JP (1988). Proc Natl Acad Sci USA 85:6697-6700.
6. Maxam AM, Gilbert W (1980). Meth Enzymol 65:499-560.
7. Fritzsche E, Hayatsu H, Igloi GL, Iida S, Kössel H (1987). Nucl Acids Res 15:1517-5528.

8. Rubin CM, Schmid CW (1980). Nucl Acids Res 8:4613-4619.
9. Razin A, Cedar H, Riggs AD (1984). "DNA Methylation." New York:Springer.
10. Saluz HP, Jost JP (1989). Nature 338:277.
11. Saluz HP, Jost JP (1989). Proc Natl Acad Sci USA 86:2602-2606.
12. Bird AP, Southern EM (1978). J Mol Biol 118:27-47.
13. Wilks AF, Seldran M, Jost JP (1984) Nucl Acids Res 12:1163-1177.
14. Grünbaum Y, Cedar H, Razin A (1981) Nucl Acids Res 9:2509-2515
15. Mattews JA, Kricka LJ (1988). Anal Biochem 169:1-25.
16. Saluz HP, Jost JP (1989). Anal Biochem 176:201-208.
17. Müller P, Salser SJ, Wold B (1988). Genes & Development 2:412-427.
18. Müller P, Wold B (1989). J Cell Biol 13E:299.
19. Nick H, Gilbert W (1985). Nature 313:795-798.
20. Ephrussi A, Church GM, Tonegawa S, Gilbert W (1985). Science 227:134-140.
21. Becker PB, Gloss B, Schmid W, Straehle U, Schütz G (1986). Nature 324:686-688.
22. Gimble JM, Max EE (1987). Mol Cell Biol 7:15-25.
23. Pauli U, Chrysogelos S, Stein G, Stein J, Nick H (1987). Science 236:1308-1311.
24. Becker PB, Ruppert S, Schütz G (1987). Cell 51:435-443.
25. Becker MM, Wang JC (1984). Nature 309:682-687.
26. Giniger E, Ptashne M (1988). Proc Natl Acad Sci USA 85:382-386.
27. Selek SB, Majors J (1987). Nature 325:173-177.
28. Sellek SB, Majors J (1987). Mol Cell Biol 7:3260-3267.
29. Jackson PD, Felsenfeld G (1987). Meth Enzymol 152:735-755.
30. Suggs SV, Hirose T, Magake T, Kawashima EH, Johnson MJ, Itakura K, Wallace RB (1981). "Developmental Biology Using Purified Genes." New York:Academic Press.

HORMONE-DEPENDENT DNA DEMETHYLATION AND PROTEIN DNA INTERACTIONS IN THE PROMOTER REGION AND ESTRADIOL RESPONSE ELEMENT OF AVIAN VITELLOGENIN GENE

Jean-Pierre Jost, Hanspeter Saluz, Iain McEwan, Melya Hughes, Hai-Min Liang, and Marylin Vaccaro

Friedrich Miescher-Institut, P.O. Box 2543
CH-4002 Basel, Switzerland

ABSTRACT Genomic sequencing has been used to study the state of cytosine methylation in the upstream region of the avian vitellogenin gene. In the promoter region a ^mCpG at nucleotide position +10 becomes demethylated in an expression specific manner. The demethylation occurs on both strands simultaneously, correlating well with the rate of vitellogenin mRNA synthesis, the appearance of a DNAse I hypersensitivity site and changes in the specific protein-DNA interaction. On the other hand, two methylated CpGs present in the estrogen response element (ERE) become demethylated in a hormone dependent way, i.e. they are demethylated in liver and oviduct under the influence of estradiol. The demethylation in the upper strand is parallel to the onset of vitellogenin mRNA synthesis whereas for the lower strand there is a lag of about 20 hours. In vivo footprinting of the ERE indicates that the inactive and active genes are always protected by proteins. Besides the estrogen receptor there is another protein binding to the ERE with a high affinity.

INTRODUCTION

The methylation pattern of many genes has been established by using various methylation sensitive restriction enzymes (1,2). The study of the pattern of DNA methylation in different tissues shows a strict correlation between gene activation and hypomethylation within the gene and/or in their 3' and 5' end flanking regions (1,2). The change in the methylation pattern for a given gene depends on the stage of differentiation of the tissue and later depends on the activation of the gene. For example during the embryonic development of the chicken liver two restriction sites (MspI and XhoI) located in the 5' and 3' end flanking region of the very low density lipoprotein gene becomes demethylated between the 7-9th day of embryonic development, time during which liver acquires competence to express the gene (3). A similar observation has been made in the embryonic vitellogenin gene (Saluz et Jost, unpublished observations). Other CpGs become demethylated only upon stimulation of the gene with estradiol (4,5). In the present review we show that under the influence of estradiol changes in the methylation pattern of specific CpGs in the promoter region and the estradiol response element of avian vitellogenin gene occur and this is correlated with changes in the in vivo and in vitro protein-DNA interactions.

THE STUDY OF THE METHYLATION PATTERN BY GENOMIC SEQUENCING AND A MORE RECENT PROCEDURE

The CpGs present in the upstream region of avian vitellogenin gene are shown in Figure 1.

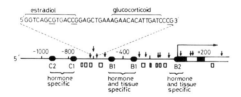

FIGURE 1. Organization of the upstream region of avian vitellogenin II gene. B1, B2, C1 and C2 are the DNAse I hypersensitive sites appearing upon estradiol treatment (20.21). The open boxes below the horizontal line are A + T rich regions (80% A + T). The arrows pointing downwards represent the CpGs.

Only one out of the 13 CpGs shown can be tested for its methylation using a restriction endonuclease (Hpa II site, at nucleotide position -600). To address the question of the methylation state of the other CpGs it was therefore necessary to use the technique of genomic sequencing first described by Church and Gilbert (6) and improved in our laboratory (7,8).

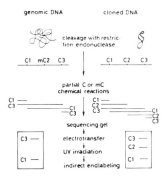

FIGURE 2. Genomic sequencing scheme. The indirect endlabeling is carried out by hybridizing the filters with a labeled single stranded DNA or RNA probe complementary to one end of the restriction fragment.

Figure 2 shows a short outline of the basic principle of genomic sequencing. Total genomic DNA is cut with a suitable restriction enzyme and upon partial reaction with hydrazine and cleavage with piperidine the reaction product is separated on a sequencing gel. Following electrotransfer of the DNA fragments onto a nylon membrane, the DNA is covalently attached to the membrane by UV irradiation and heat treatment. The sequence is then visualized by indirect endlabeling (hybridization with a labeled single stranded DNA fragment complementary to one end of the target sequence) and by subsequent autoradiography. Since methylated cytosine does not react with hydrazine it is not cleaved by piperidine and thus generates a gap in the C sequencing ladder. Such a gap in the sequence could also represent a base transition caused by deamination of the 5 methylcytosine to thymidine. Therfore, it is necessary to perform a T specific control reaction (4) Genomic sequencing as presented above is rather complex and time consuming, therefore we sought to simplify the technique. Total genomic DNA is digested with suitable restriction enzyme, and chemically sequenced as described by Maxam and Gilbert (9), Fritzsche et al (10) and Rubin and Schmid (11). This is followed by the selective amplification with Taq polymerase using a synthetic primer labeled to a very high specific radioactivity. The product of the linear amplification is separated on a sequencing gel and then directly exposed on a X ray film(12).

THE METHYLATION IN THE PROMOTER REGION
OF AVIAN VITELLOGENIN GENE

Of the 2 CpGs situated in the promoter region the CpG at nucleotide position +10 is fully methylated in the inactive gene and becomes demethylated in the active gene. In this case the demethylation is strictly expression specific since it is only observed in the induced liver and not in the other tissues (Fig. 3). The demethylation of this CpG is parallel to the onset of transcription and the appearance of the DNAse I hypersensitive site B2. It is interesting to note that upon cessation of the transcription of the gene both the DNAse hypersensitivity and the demethylation remain and persist for a long time. This has been interpreted as one of the explanations for the so-called memory effect observed during the secondary stimulation of the avian vitellogenin gene. For the secondary stimulation of the gene by estradiol the CpG at position +10 is already demethylated and thus there is no lag between the injection of estradiol and the first appearance of vitellogenin mRNA. However, recently the group of Burch (13) has shown that we may possibly have in addition to the demethylation of CpG other factors responsible for the memory effect. The methylation of the CpG at position +10 may play a crucial role in keeping the gene silent in the liver of immature chicks and roosters. Preliminary in vitro transcription experiments using the synthetic gene fully methylated at this site indicated that ^{m}CpG inhibited the transcription. However, these experiments were carried out using HeLa cell lysate, therefore they need to be confirmed with the homologous transcription system.

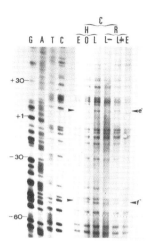

FIGURE 3. The demethylation of mCpG in the promoter region of vitellogenin gene is expression specific. Genomic sequencing was carried out as outlined by Saluz and Jost (7,8). Cytosine specific reactions were carried out with total DNA from hen liver (LH), rooster liver (LR) and hen oviduct (O). Arrows e+ and f+ indicate the position of CpGs.

DOES A CHANGE IN THE STATE OF METHYLATION OF A CpG IN THE PROMOTER REGION BRING ALSO A CHANGE IN THE IN VIVO AND IN VITRO PROTEIN-DNA INTERACTIONS?

To address this important question we used the approcah of in vivo genomic footprinting. A suspension of living cells was treated with various concentrations of dimethylsulfate. Depending on the degree of protection due to proteins, the guanosine residues react to various degree with the alkylating agent. Upon extraction of total DNA and cleavage with suitable restriction enzyme the sites of the modified bases are quantitatively cleaved with piperidine. The rest of the procedure is essentially as for the sequencing (5,14-17).

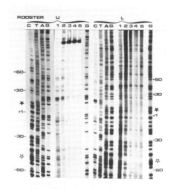

FIGURE 4. Example of an in vivo DMS footprint of rooster hepatocytes. Only the lower strand of the promoter region of vitellogenin gene is shown. Decreasing concentrations of dimethylsulfate (lane 1-5: 0.5%, 0.05%, 0.005%, 0.0005%, 0.00005% respectively) were used to reveal the details cf the footprint. The full star represents the position of the methylated CpG.

By using decreasing amounts of DMS it is possible to determine the lowest effective concentration at which only the unprotected guanosine residues react. Such an experiment is shown in figure 4 (5). At the concentration of 0.5% of dimethylsulfate almost all guanosine residues react very strongly. However, by decreasing the concentration of DMS by as much as 10,000 fold "windows" of protection became apparent. A graphic interpretation of such results is shown in Figure 5 (5).

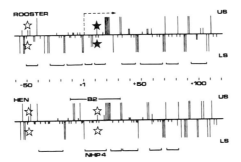

FIGURE 5. Graphic representation of the relative extent of the reaction of DMS with guanosine residues. Vertical bars above each horizontal line represent the upper DNA strand (US) and lines below represent the lower DNA strand (LS). Length of the vertical bars indicates the relative strength of the autoradiogram signals (the longest the strongest). B2 is the tissue specific and expression specific DNAse I hypersensitive site. The full star represents the position of the methylated CpG, and the open star the unmethylated CpG.

At the methylated CpG (full star) in the repressed gene there is a protected window of about 10 nucleotides. Upon activation of the gene this protected area expands to about 20 bp, this is a clear indication that we have a change in the in vivo protein-DNA interaction but it does not tell us which proteins are binding to the DNA. This problem was approached by studying the protein DNA interaction in vitro, using synthetic labeled oligonucleotides and fractionated nuclear proteins from hens and roosters.

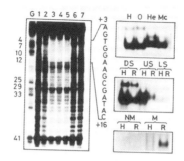

FIGURE 6. DNAse protection in the region of the transcriptional start of the vitellogenin gene and gel shift assays. (Left panel) DNAse I protection experiment was performed on the lower strand of a oligonucleotide duplex (nucleotide position -12 to +16) using a 0.5 M Heparin Sepharose fraction of HeLa cell lysate. (Right) Gel shift assay (only the protein-DNA complexes are shown). (Top panel) complexes between labeled oligonucleotide (position +3 to +16) with nuclear proteins from hen liver (H), oviduct (O), HeLa cell (He) and MCF 7 cells (Mc). (Middle panel) Complex formed between the same labeled oligonucleotide as above with hen liver (H) or rooster liver (R) nuclear extracts. DS, double-stranded DNA. (US and LS) upper and lower single stranded DNA respectively. (Bottom panel) Complex between the labeled oligonucleotide duplex (positions +3 to +29) with hen (H) or rooster liver (R) nuclear fraction (eluted from Heparin Sepharose with 0.5 M KCl). NM and M, unmethylated and methylated oligonucleotides respectively (5).

Figure 6 shows an in vitro DNAse I footprint of the promoter region and it can be shown that there is indeed a strong protein DNA interaction over the methylation site between the nucleotide positions +3 to +16 (5). A labeled synthetic oligonucleotide covering this region was synthesized and used as a substrate for gel shift assays with fractionated nuclear extracts. Figure 6 (right hand bottom panel) shows that a nuclear factor contained in the 0.5 M KCl Heparin Sepharose fraction from the rooster liver binds to the methylated oligonucleotide only. This protein is only present in the rooster where the CpG is methylated and the gene inactive and it could not be detected in egg laying hens. The right hand upper and middle panels show that in a different nuclear fraction (fractionation on heparin Sepharose) there is a ubiquitous protein binding to the same oligonucleotide whereas a third fraction eluted from Heparin Sepharose with different concentration of KCl showed single-strand binding activity. The presence of the single-strand binding activity is estrogen dependent and the protein binds only to the upper DNA strand (mRNA like strand) of the oligonucleotide. What role these different proteins play in the regulation of avian vitellogenin gene is not yet known.

DEMETHYLATION OF THE ESTROGEN RESPONSE ELEMENT IS ESTROGEN DEPENDENT BUT NOT ORGAN SPECIFIC

Experiments in which restriction enzymes Hpa II/Msp I were used indicated that a treatment of immature chicks with estradiol resulted in the demethylation of a CpG situated in the estrogen response element of vitellogenin gene (18). Kinetics of the demethylation as measured by using the same restriction enzyme suggested that demethylation of the Hpa II site occurred long after the onset of vitellogenin mRNA synthesis (19) implying that the demethylation is a consequence and not the cause of gene activation. This question was later reinvestigated by means of genomic sequencing, Saluz et al (4) found that demethylation of the CpG in the ERE occured in a strand specific manner. The demethylation was taking place at first on the upper DNA strand and was parallel to the onset of vitellogenin mRNA synthesis. Then approximately 20 hours later demethylation of the opposite strand occurred. Figure 7 shows a summary of the kinetic of demethylation of the CpG located in the ERE and neighbouring sequences. In contrast to the promoter region demethylation of the ERE is not tissue specific. As for the DNAse I hypersensitive sites C1 and C2 (Figure 1 and Ref. 20, 21) the demethylation of the ERE is only hormone specific and can be observed in the liver and oviduct of egg laying hens (4). Therefore, it seems unlikely that the demethylation of this site is directly responsible for the gene activation.

322   Jost et al.

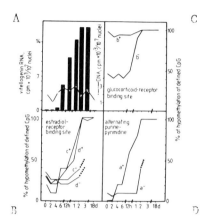

FIGURE 7.   Kinetics of transcription and demethylation of the upstream region of avian vitellogenin gene.
A) The relative rate of vitellogenin mRNA transcription in isolated liver nuclei (solid bars) and the relative rate of incorporation of H^3 dTTP into DNA (upper left panel).
B) Relative amounts of demethylation (%) of the ^mCpG situated in the estrogen response element. The + and - signs are the upper and lower DNA strand respectively.
C) The kinetics of demethylation of CpG, situated within the glucocorticoid response element.
D) The kinetics of demethylation of CpG, present on a short stretch of alternating purine-pyrimidine bases situated at nucleotide position -526 (4).

IN VIVO AND IN VITRO PROTEIN DNA INTERACTIONS ON THE ESTROGEN RESPONSE ELEMENT

As for the promoter region the in vivo interaction of proteins with the ERE were tested by means of dimethylsulfate (DMS) footprinting. Throughout the experiments we used various concentrations of DMS (0.5 to 0.0005%). A summary of the results obtained with hepatocytes of egg laying hens and roosters is given in Fig. 8.

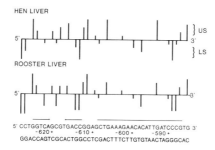

FIGURE 8. Interpretation diagram of the in vivo footprinting of the estrogen response element of vitellogenin gene in egg laying hens (active gene) and rooster (inactive gene) hepatocytes. US and LS stand for upper and lower DNA strand respectively. The length of the vertical lines gives an estimate of the degree of reactivity of each guanosine with dimethylsulfate, the smaller the bar the more protected is the base.

In both cases the ERE is protected by proteins, however, some major differences in the reactivity of the guanosine residues were observed. For example the lower strand of the ERE in hens was better protected than in the roosters and vice versa for the upper strand the protection was better in roosters than in hens. When adult roosters were injected with estradiol the pattern of protection became identical to that of the hens. As for the promoter region these changes indicate that the complex interaction of proteins with DNA changes during gene activation but again it does not tell anything about the nature of the proteins involved in the binding. It is quite certain that the estrogen receptor binds to the ERE as a homodimer (23,24) but it cannot explain by itself the changes in the pattern of protection observed between hen and roosters. In the chromatin of rooster liver we could not detect any tight estrogen receptor binding as this is the case for the egg laying hens. One has to conclude that some other proteins, different from the estrogen receptor, bind the ERE. Recent work from our laboratory indicates that another protein, different from the estrogen receptor binds with high affinity (Kd 10^{-11}M) to the ERE of the avian vitellogenin gene. This protein called the NHP1 is present in all tissues so far tested. It has been purified 45,000 fold from HeLa cells. The protein has a native molecular weight of 170,000 KDa and is composed of two polypeptides of MW 75 and 85 KDa. Titrations made with nuclear extracts from rooster and hen liver indicate that NHP1 is about 4.5 times more abundant than estrogen receptor. The contact points of NHP1 with the ERE are different from the ones shown for estrogen receptor.

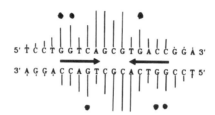

FIGURE 9. Summary of contact points of NHP1 with the estrogen response element. The length of the vertical lines indicates the importance of the corresponding base for the binding of NHP1. The horizontal arrows emphasize the dyad symmetry structure of the estrogen response element. The black dots represent strong contact points of the estrogen receptor (24).

Fig. 9 shows a summary of the results obtained by partial depurination and depyrimidation of the synthetic labeled ERE. Following protein-DNA interaction, the reaction product was separated on a native agarose gel. The protein-DNA complex and the free DNA were extracted, purified, subjected to the piperidine reaction and analyzed on a sequencing gel. As shown in Fig. 9 the strongest contact points between NHP1 and the ERE are located in the middle of the palindrome of the ERE where the demethylation site is situated. The importance of these bases for the binding of NHP1 was also demonstrated by DNA competition assays. The suppression of the 3 bases GCG in the center of the palindrome decreased the binding of NHP1 by about 90% (25).

DISCUSSION

A positive correlation between the kinetics of gene activation and demethylation suggests that there is a cause-effect relationship between the two parameters. This important question could be addressed either by genetical approaches and/or by reconstitution experiments. To date the best evidence for such a relationship has been provided by several groups (26-36). Doerfler and associates

(29,30,35) have shown that the methylation of 3 Hpa II sites in the late E2A promoter of adenovirus type 2 DNA inactivates or reduces the activity of this promoter when tested by transient expression experiments carried out in Xenopus leavis oocytes and mammalian cells. Using the same system Watt & Molloy (36) have shown that the methylation of a specific CpG on the major late promoter alters the binding of a HeLa protein factor and the in vitro transcription of the synthetic DNA in a HeLa lysate system. In both cases however, it has not yet been demonstrated whether the CpGs studied were biologically relevant or not. Also correlated with the demethylation is the appearance of DNAse I hypersensitive sites. Keshet et al (37) have shown that in vitro methylated genes transfected into mouse L cells did not cause DNAse I hypersensitive sites, whereas when the corresponding unmethylated genes were transfected into L cells they were present. Similarly nuclease hypersensitive sites in hypomethylated regions of active X chromosome linked house keeping genes have been described by several groups (38 and references therein). In agreement we find that in the promoter of the avian vitellogenin II gene there is a positive correlation between its hormone specific activation, the demethylation of one CpG and the appearance of the DNAse I hypersensitive site B2. Such a correlation between demethylation and the presence of a DNAse I hypersensitivity site however does not exist in the estrogen response element present in the upstream region of avian vitellogenin gene.

It is generally assumed that the demethylation in higher eucaryotes is a passive phenomenon where the repair methylase is inhibited during replication. In the case of the chicken vitellogenin the inhibition of total DNA synthesis by either hydroxyurea or cytosine arabinoside did not prevent vitellogenin mRNA synthesis and the demethylation of ^{m}CpG (19). This suggests that ^{m}CpG demethylation may not be passive. Limited DNA replication, such as gene amplification which might account for passive demethylation has been excluded by gene titration experiments carried out before and after estradiol treatment (22). Two plausible mechanisms of demethylation remain: active demethylation by excision repair and/or by activity of a demethylase (4). An enzymatic removal of a methyl group from 5 methyl cytosine in DNA is mechanistically improbable because the transfer of the methyl group from S Adenosine-L-Methionine to a carbon atom is irreversible at neutral pH. However, Razin et al (39) have recently demonstrated that in differentiating Friend erythroleukemia cells a replacement of 5 methylcytosine with cytosine occurs presumably via an enzymatic pathway.

The difference in the DNA footprints given by

estradiol receptor (ER) and NHP1 suggests a simultaneous binding of both proteins to the ERE. The homodimer of estrogen receptor could bind to the arms of the palindrome in the major grove of the double helix facing the same side, whereas NHP1 could possibly be tightly associated to the spacer GCG of the palindrome in the major grove facing the opposite side of the double helix. The binding of NHP1 to the ERE could play a role of tagging the DNA for the subsequent binding of the estrogen receptor (25) or it may contribute to the biology of estrogen action in general by contacting and stabilizing specific binding of the receptor with its enhancer element (24). Finally NHP1 could have a direct enzymatic function essential for the active demethylation of the ^{m}CpG present in the spacer of the palindrome. We have previously shown that upon stimulation of vitellogenin gene by estradiol there is a strand specific demethylation of the CpG present in the ERE and neighbouring sequenses (4). In our present case it is interesting to note that the sequence AGCG containing the methylation site plays a capital role for the binding of NHP1 to the ERE and in experiments where this sequence was deleted or replaced by CAG resulted in a very poor binding of NHP1 underlining the importance of the CpG for the binding to the ERE. The possible involvement of NHP1 in a general mechanism of active demethylation could account for its apparent ubiquity. Preliminary experiments (Jost unpublished results) carried out with affinity chromatografied NHP1 indicate clearly that this protein has specific nicking activity when tested with the ERE. This could represent one of the first steps in the process of active demethylation of 5 methyl cytosine. However the possible relationship between ER and NHP1 remains to be firmly established.

REFERENCES

1. Adams RL, Burdon RH (1985). "Molecular Biology of DNA Methylation." New York:Springer.
2. Razin A, Cedar H, Riggs AD, Conklin KF (1984). "DNA Methylation." New York:Springer.
3. Colgan V, Elbrecht A, Goldman P, Lazier CB, Deeley RJ (1982). J Biol Chem 257:14453-14460.
4. Saluz HP, Jiricny J, Jost JP (1986). Proc Natl Acad Sci USA 83:7176-7171.
5. Saluz HP, Feavers IM, Jiricny J, Jost JP (1988). Proc Natl Acad Sci USA 85:6697-6700.
6. Church GM, Gilbert W (1984). Proc Natl Acad Sci USA 81: 1991-1995.

7. Saluz HP, Jost JP (1986). Gene 42:151-157.
8. Saluz HP, Jost JP (1987). "A Laboratory Guide to Genomic Sequencing." Basel:Birkhäuser.
9. Maxam AM, Gilbert W (1980). Meth Enzymol 65:499-560.
10. Fritzsche E, Hayatsu H, Igloi GL, Iida S, Koessel H (1987). Nucl Acids Res 15:5517-5528.
11. Rubin CM, Schmid CW (1980). Nucl Acids Res 8:4613-4619.
12. Saluz HP, Jost JP (1989). Proc Natl Acad Sci USA, in press.
13. Burch JBE, Evans MI (1986). Mol Cell Biol 6:1886-1893.
14. Ephrussi A, Church GM, Tonegawa S, Gilbert W (1985). Science 227:134-140.
15. Gimble JM, Max EE (1987). Mol Cell Biol 7:15-25.
16. Pauli U, Chrysogelos S, Stein G, Stein J, Nick, H (1987). Science 236:1308-1311.
17. Becker PB, Ruppert S, Schuetz G (1987). Cell 51:435-443.
18. Wilks AF, Cozens PJ, Mattaj IW, Jost JP (1982). Proc Natl Acad Sci USA 79:4252-4255.
19. Wilks AF, Seldran M, Jost JP (1984). Nucl Acids Res 12:1163-1177.
20. Burch JBE, Weintraub H (1983). Cell 33:65-76.
21. Burch JBE (1984). Nucl Acids Res 12:1117-1135.
22. Jost JP, Schuerch AR, Walz A (1977). FEBS Lett 75:133-137.
23. Kumar V, Chambon P (1988). Cell 55:145-156.
24. Klein-Hitpass L, Tsai SY, Greene GL, Clark JH, Tsai MJ, O'Malley BW (1989). Mol Cell Biol 9:43-49.
25. Feavers IM, Jiricny J, Moncharmont B, Saluz HP, Jost JP (1987). Proc Natl Acad Sci USA 84:7453-7457.
26. Ben-Hattar J, Jiricny J (1988). Gene 65:219-227.
27. Keshet I, Yisraeli J, Cedar H (1985). Proc Natl Acad Sci USA 82:2560-2564.
28. Feenstra A, Fewell J, Lueders K, Kuff E (1986). Nucl Acids Res 14:4343-4352.
29. Kruczek I, Doerfler W (1986). Proc Natl Acad Sci USA 80:7586-7590.
30. Vardimon L, Kressmann A, Cedar H, Maechler M, Doerfler W (1982). Proc Natl Acad Sci USA 79:1073-1077.
31. Stein R, Razin A, Cedar H (1982). Proc Natl Acad Sci USA 79:3418-3422.
32. Fradin A, Manley JL, Prives CL (1982). Proc Natl Acad Sci USA 79:5142-5146.
33. Busslinger M, Hurst J, Flavell RA (1983). Cell 34:197-206.
34. McGeady ML, Jhappan C, Ascione R, Vande Woude GF (1983). Mol Cell Biol 3:305-314.
35. Langner KD, Vardimon L, Renz D, Doerfler W (1984). Proc Natl Acad Sci USA 81:2950-2954.
36. Watt F, Molloy PL (1988). Genes and Development 2:1136-1143.

37. Keshet I, Lieman-Hurwitz J, Cedar H (1986). Cell 44:535-543.
38. Hansen RS, Ellis NA, Gartler SM (1988). Mol Cell Biol 8:4692-4699.
39. Razin A, Szyf M, Kafri T, Roll M, Giloh H, Scarpa S, Carotti D, Cantoni GL (1986). Proc Natl Acad Sci USA 83:2827-2831.

ON THE MECHANISM OF PROMOTER INACTIVATION BY SEQUENCE-SPECIFIC METHYLATION

Walter Doerfler, Miklós Toth,[1] Ralf Hermann, Ursula Lichtenberg,[2] and Arnd Hoeveler[3]

Institute of Genetics, University of Cologne
Cologne, Germany

ABSTRACT We have used the adenovirus system as a model for studies on sequence-specific promoter methylation and gene inactivation. In experiments on the late E2A promoter of adenovirus type 2 (Ad2) DNA, it has been demonstrated that the sequence-specific (5'-CCGG-3') introduction of three 5-methyldeoxycytidine (5-mC) residues into this promoter causes inactivation in transient expression systems in Xenopus laevis or mammalian cells, after the genomic fixation of the methylated promoter in mammalian cells, or in a cell free transcription system which uses nuclear extracts of HeLa cells. In an effort to understand the biochemical mechanism of promoter inhibition or inactivation by sequence-specific methylation, the binding of cellular proteins to the methylated or unmethylated late E2A promoter of Ad2 DNA has been tested. With the isolated promoter, major binding differ-

---

[1] On leave from: Institute of Biochemistry, Biological Research Center, Hungarian Academy of Sciences, Szeged, Hungary.
[2] Present address: The George William Hooper Foundation, University of California, San Francisco, CA 94143
[3] Present address: Centre d'Immunologie, INSERM-CNRS, Marseille, France

ences between the 5'-CCGG-3' methylated and the unmethylated DNA have not been found. In vivo footprinting studies on the integrated late E2A promoter in the methylated or unmethylated promoter have not revealed differences either. On the other hand, late E2A promoter fragments of 50 or 73 nucleotides in lengths, which comprise two of the 5'-CCGG-3' sequences, exhibit differences between the unmethylated and methylated forms in the binding capacity to cellular proteins. It is conceivable that in the cell the differences in DNA-protein binding are functional rather than physical and/or that subtle physical binding differences are obscured by a large number of proteins whose DNA interactions are unaffected by promoter methylation. It will also be necessary to search for structural differences between the methylated and the unmethylated promoter sequence. We have also determined the pattern of methylation at all 13 5'-CG-3' sequences in the essential part of the late E2A promoter between nucleotides +24 and -160 relative to the promoter cap site. The genomic sequencing method has been used in these studies. In the Ad2-transformed hamster cell line HE1, the late E2A promoter is active and all 13 5'-CG-3' dinucleotides are unmethylated. Conversely, in cell lines HE2 and HE3 the same promoter is inactive and all 5'-CG-3' sites are methylated in both strands. Interestingly, when the in vitro 5'-CCGG-3' methylated late E2A promoter, which carries three 5-mC residues, is genomically fixed in the hamster cell genome, spreading of methylation can be observed with increasing passage of the transformed cells. From two preimposed 5-mC residues, as a point of crystallization as it were, methylation extends to neighboring, previously unmethylated 5'-CG-3' sequences. This spreading mechanism may, at least formally, account for the generation of complex patterns of DNA methylation. Since the inactivated late E2A promoter is markedly hypermethylated, although the introduction of a few 5-mC residues would suffice to effect its inactivation, the question arises whether a given pattern of methylation may functionally relate to different cellular mechanisms. It is conceivable that patterns found in certain regions of a genome are composed of several interdigitating patterns each of which has a separate functional significance.

INTRODUCTION

As demonstrated for the DNA of many mammalian cell lines and for the DNA of living organisms, very specific patterns of DNA methylation exist in the mammalian genome. In general, these patterns can be highly preserved in cell lines and seem to be inheritable in organisms. There is evidence that, under certain conditions, changes in these patterns can occur. The rules according to which these patterns are established or changed are unknown. During the analyses on the biochemical mechanisms underlying many biological phenomena, it has become apparent that specific DNA-protein interactions are at the core of regulatory processes. In the now classical studies on prokaryotic systems (for review, 1), methylated DNA sequence motifs have been recognized as efficient modulators of DNA-protein interactions. A methyl group in an established and genetically selected DNA sequence is the only possibility of variation on a genetic theme. In a functional sense, the extent and direction of these modulations cannot a priori be predicted for a given sequence but have to be experimentally determined in each case. From these general considerations, the concept can be deduced that probably several different cellular mechanisms, which are dependent on or influenced by DNA-protein interactions, can be affected by the methylation of specific DNA sequences. It is also conceivable that the functional effects of DNA methylation are mediated via complex structural alterations of DNA. The major, and probably the only, methylated nucleotide in mammalian DNA is 5-methyldeoxycytidine (5-mC) which is frequently, but not exclusively, found in the dinucleotide combination 5'-CG-3'.

For the past twelve years, our laboratory has worked on the analysis of the effects that sequence-specific methylations of eukaryotic (viral) promoters exert on gene activity (2-6). In what way, if any, is the regulation of gene expression dependent on specific patterns of promoter methylation? The main findings of past work have been reviewed recently (7, 8). These results have significantly contributed to the concept that sequence-specific promoter methylations lead to the long-term shut-off or the inhibition of gene activity. This notion is now supported by data from studies on many different eukaryotic promoters. The 5'-CG-3' sequences, whose methylation causes gene inactivation, are promoter specific and have to be determined for each eukaryotic promoter. Promoter methylation is now accepted as one major factor that contributes to eukaryotic gene regulation.

In this chapter, we are going to present a brief synopsis of previously published work on the effects of sequence-specific promoter methylations. The adenovirus system (for reviews, 9) has served as our model system. Subsequently, experiments will be described that bear on the mechanism by which promoter methylations lead to gene inactivation. Finally, we shall summarize results on the establishment and spreading of methylation patterns starting from pre-imposed methyl groups in adenovirus DNA sequences that had been genomically fixed in mammalian genomes by transfection and selection for a co-transfected selectible marker. In the same system, the patterns of active and of inactivated integrated adenoviral promoters have been assessed.

THE ADENOVIRUS SYSTEM - ADVANTAGES OF A VIRAL MODEL

The adenovirion, adenovirus type 2 (Ad2) as one of the model viruses, harbors a double-stranded DNA molecule of 35,937 nucleotide pairs

(10). With the chemical or restriction endonuclease techniques applied so far, methylated nucleotides have not been detected in the DNA of Ad2 or of adenovirus type 12 (Ad12) when the DNA was extracted from the virion (2, 11, 12). Similarly, there is no evidence that the free intracellular viral DNA in productively or abortively infected cells becomes methylated at early or late times after infection (12, 13). It is an unresolved question whether very few, possibly highly specific cytidine residues in the free viral DNA are methylated. This problem requires further analyses with very sensitive techniques. It will have to be investigated whether cytidine residues, e.g., in the major late promoter of adenovirus DNA are methylated early but not late in the viral infection cycle, and whether specific promoter methylation may thus play a role in the regulation of early versus late viral transcription. It is certain that there is not a major degree of methylation in free intracellular viral genomes.

In abortively and productively infected hamster cells, in adenovirus-transformed cells or in Ad12-induced hamster tumor cells, the viral genome or parts of it can be integrated into the host genome (14-17). The integrated viral genomes presumably become subject to a regime of methylation that is similar to the methylation of the host genome. Integrated Ad12 (2-4, 18) or Ad2 genomes (13) in established hamster cell lines exhibit characteristic methylation patterns of their genes. It has been with the DNA from these cell lines that inverse correlations between the extents of DNA methylation and gene expression have first been described. In general, the early adenovirus genes are expressed in the transformed and hamster tumor lines and are hypomethylated, whereas the late genes are shut off and hypermethylated. The differential patterns of integrated viral gene expression instigated our studies on DNA methylation (3, 4). A particularly striking example of such inverse correlations is presented by the E2A gene of Ad2 DNA in the integrated form in the Ad2-transformed hamster cell lines HE1, HE2, and HE3 (13, 19). In cell lines

HE2 and HE3, which do not express this gene, the fourteen 5'-CCGG-3' sequences in the E2A gene are all methylated. In cell line HE1, the same sequences are all unmethylated, and the E2A gene is expressed (13, 20). This gene encodes a single-strand DNA binding protein (21) which is essential for DNA replication.

In general, the patterns of Ad12 DNA integration in newly induced hamster tumors resemble those in established tumor or transformed cell lines that have been permanently kept in culture (22, 23). However, the levels of methylation in the integrated Ad12 genomes are much lower in tumors than in established cell lines. However, when cells from these tumors are explanted into culture and are continuously passaged for many cell generations, a shift over several passages is observed in the level of DNA methylation which eventually approaches the levels found in continuously passaged Ad12-transformed cell lines (24). It is unknown what factors determine these shifts in DNA methylation and, in particular, how the striking specificity in the patterns of methylation is attained. It is conceivable that the explantation of the cells and their transition from a tumor in a living animal to cells in culture may contribute to these changes in methylation patterns. The shifts in the extent of methylation cannot be due to the selection in culture of a subpopulation of tumor cells, since the cells from the tumor had been cloned after explantation and the cloned cells had been continuously passaged. It has been in these cloned cell populations that the increase in DNA methylation has been observed (22, 24). Even when the levels of DNA methylation of Ad12 DNA integrated in the genomes of hamster tumor cells are low, viral genes are not necessarily expressed. Obviously, a specific pattern of methylation is only one of the factors contributing to the long-term silencing of integrated viral genes (22). It has also been shown in the Ad2-transformed hamster cell line HE3 that prolonged animal passage (36 passages) of this transformed cell line does not alter the pattern of methylation of the integrated Ad12 genome as detectable by HpaII and MspI cleavage (25).

For detailed work on the inactivation or inhibition of eukaryotic promoters by sequence-specific methylation, we have chosen the E1A promoter of Ad12 DNA (26) and the late E2A promoter of Ad2 DNA (27-30).

The study of the molecular biology in mammalian systems has profited in more than one respect from the adenovirus and other viral systems. Obviously, resorting to the adenovirus system in studies on the functional significance of DNA methylation in mammalian cells has held many advantages not the least of which has been the availability of a wealth of information on these viral genomes and of a variety of interesting adenovirus-transformed cell lines (31). Adenoviruses can infect cells productively, abortively; they can transform cells and consequently insert their genomes into that of the host and thus generate virus-cell genome complexes of considerable stability. Lastly, Ad12 can induce rapidly growing tumors in rodents and with particularly high frequency in hamsters (32).

SEQUENCE-SPECIFIC PROMOTER METHYLATIONS CAUSE GENE INACTIVATION OR INHIBITION

Inhibition of the in vitro Methylated Late E2A Promoter of Ad2 DNA

In these experiments, we have mimicked the 5'-CCGG-3' methylation status in the late E2A promoter that we had previously observed to be inactive in hamster cell lines HE2 and HE3. When the entire E2A gene (33) or only its late promoter (27) is 5'-CCGG-3' methylated in vitro by using the HpaII DNA methyltransferase, and when these constructs are subsequently microinjected into the nuclei of Xenopus laevis oocytes, the gene is not expressed. Under the same conditions, the non-methylated constructs are transcribed into messenger RNAs. In co-injection experiments of the methylated E2A gene and a non-methylated histone gene from sea urchin, the E2A gene is not transcribed, the histone gene is transcribed. Thus, the inhibitory effect of sequence-specific DNA methylation is specific and not an artifact of the experimental procedure (33).

Furthermore, we have demonstrated that the late E2A promoter that has been 5'-CCGG-3' methylated at nucleotide positions -215, +6, and +24, relative to the +1 position at the site of transcriptional initiation (cap site), is also inhibited or inactivated after transfection into mammalian cells. In these experiments, a late E2A promoter-chloramphenicol acetyltransferase (CAT) gene construct (pAd2E2AL-CAT) has been used (28).

The same promoter is also inhibited by the methylation of the same 5'-CCGG-3' sequences when the activity of the late E2A promoter is assessed in an *in vitro* cell-free transcription system which employs nuclear extracts from HeLa cells (34). For the unequivocal demonstration of promoter inhibition in the cell-free transcription system, three preconditions have to be fulfilled. i) The extracts have to have high protein concentration. ii) The template has to be in a circular form. iii) Each extract has to be titrated for the optimal amount of template DNA which assures maximal inhibition due to promoter methylation. The structural requirement of circular DNA suggests that the DNA has to be in a certain conformation for the methylation signal to exert its genetic signal function.

Lastly, the unmethylated or the 5'-CCGG-3' methylated pAd2E2AL-CAT construct has been genomically fixed by co-transfection of BHK21 hamster cells with either construct plus the unmethylated pSV2-neo construct which carries the gene for neomycin phosphotransferase (neo) under the control of the early SV40 promoter (35). Transfected colonies have been selected for neomycin resistance and for the persistence of the pAd2E2AL-CAT construct in a non-rearranged configuration. Among 19 cell lines carrying the non-methylated construct, 18 cell lines express the CAT gene and contain the late E2A promoter in the non-methylated form. Of the 14 cell lines which have been established upon transfection of the *in* vitro 5'-CCGG-3' methylated pAd2E2AL-CAT construct, 7 cell lines do not express the CAT gene and maintain the preimposed methylation pattern (36). In the other 7 cell lines, the promoter has become partly or almost completely demethylated and consequently expresses the CAT gene to a lesser or a very pronounced extent, respectively. The results of these experiments have not only established the inactivation due to methylation of the late E2A promoter after genomic fixation in mammalian cells, but have also raised the question of why the genomically fixed 5'-CCGG-3' methylated late E2A promoter becomes demethylated in some cell lines, not in others. Could position effects of the integrated foreign DNA play a role in determining the stability of a pre-imposed methylation pattern (36)?

As an ensemble, these data show that under a variety of experimental conditions the late E2A promoter of Ad2 DNA is inactivated or strongly inhibited in mammalian cells when three 5'-CCGG-3' sequences have been methylated. Similar conclusions have been documented for the E1A promoter of Ad12 DNA. In this promoter, two 5'-CCGG-3' sequences have sufficed to be methylated in order to effect inactivation (26, 37). Surprisingly, for the E1A promoter of Ad12 DNA methylation of one deoxyadenosine sequence can also lead to inactivation (37). It is conceivable that depending on the function of certain promoter motifs, methylation of such sequences interferes or does not interfere with the binding of specific proteins and the function of the promoter.

Release of the Inhibitory Action of Promoter Methylation

One of the requirements for a genetic signal, such as DNA methylation, that has been shown to entail the long-term inactivation of a promoter, is its reversibility. Demethylation by DNA replication and concomitant suppression of the mechanisms responsible for maintenance methylation is one formal possibility in altering an existing pattern of methylation. This pathway of demethylation is contingent upon cell replication. Alternatively, pathways leading to transient demethylation, e.g., repair functions in which a methylated nucleotide would be excised and replaced by its non-methylated homologue, have been proposed (38) and remain to be proven.

For the reversal of the inactivation of the methylated E2AL promoter of Ad2 DNA, we have also considered more subtle processes that are capable of reactivating the promoter, although it remains methylated. The adenovirus E1A gene-encoded 289 amino acid protein (20, 28, 30), a classical transactivator (for review, ref. 39), and a strong enhancer from an immediate early gene in the human cytomegalovirus (29) have both been shown to reactivate the methylated E2AL promoter. It is likely that the E1A protein can reactivate only certain methylated promoters and that other promoters require different transactivators.

The finding that promoter methylation does not entail inactivation unconditionally underscores current concepts about the high level of complexity in the regulation of promoter activity. The presence of 5-mC even in decisive and sensitive promoter sequences does not eo ipso guarantee promoter inactivation. As outlined above, a series of other preconditions has to be fulfilled to render promoter methylations consistent with inactivation.

METHYLATION OF PROMOTER MOTIFS AND THE BINDING OF SPECIFIC PROTEINS

One possible explanation for the inhibitory effect of three 5-mC residues in the late E2A promoter sequence is the positive or negative modulation of the sequence-specific binding of proteins. A synthetic oligonucleotide of 50 or a restriction fragment of 73 basepairs (bp) in lengths, which both comprise the +24 and +6 5'-CCGG-3' sequences of the late E2A promoter, have been methylated at these two sites or have been left unmethylated and have been subsequently incubated with a partly purified nuclear extract from human HeLa cells. Protein binding has been monitored by electrophoretic migration delay of the [^{32}P]-labeled oligonucleotide or restriction fragment on polyacrylamide gels. The formation of one of the DNA-protein complexes is abrogated when the 5'-CCGG-3' methylated oligonucleotide or fragment has been used in the binding assays. The same complex can also be obliterated by adding the same (non-methylated) oligonucleotide as competitor to the reaction mixture. The results document that the *in vitro* methylation of sequences in the E2AL promoter not only inhibit promoter function, but also abrogate the binding of specific proteins (40). In contrast, when a 377 bp fragment of the late E2A promoter has been used for studies on the binding of specific nuclear proteins, no difference has been detectable between the unmethylated and the 5'-CCGG-3' methylated promoter (41). We have found very recently that the genomic footprinting technique does not reveal differences between the unmethylated and the methylated late E2A promoter in cell lines uc2 and mc23, respectively (M. Toth and W. Doerfler, in preparation). In these hamster cell lines, the unmethylated (uc2) or the 5'-CCGG-3' methylated (mc23) E2AL promoter-CAT gene construct have been genomically fixed (36). Thus, the results on protein binding with short fragments may not fully reflect the complexity of possible DNA-protein interactions in the E2AL promoter region. Alternatively, it is likely that on a longer DNA fragment from the late E2A promoter several possibilities for DNA-protein and DNA-protein-protein interactions exist that may obscure the slight differences in binding of one or a few methylation sensitive sites which can be better recognized on a short promoter fragment. Furthermore, it is conceivable that the functionally decisive differences between an unmethylated active and a methylated inactive promoter are structural and only secondarily affect the binding capacity of proteins.

From studies on a number of eukaryotic promoters, there is evidence for the effect of methylation in promoter motifs on the binding of proteins or for the lack thereof. In the globin promoter, interference with protein binding due to DNA methylation has not been observed (G. Felsenfeld, cited by ref. 42). In the early E2A promoter of Ad2 DNA, the methylation of one HhaI site has abolished protein binding and promoter function (43). In contrast, the methylation of an Sp1 binding site in several eukaryotic promoters does not seem to affect protein binding (44, 45). Methylation of the major late promoter of Ad2 DNA abrogates the binding of specific proteins (46). Extensive further work on well analyzed promoters will be required, before the code governing DNA-protein interactions and their specific interference by 5-mC can be deciphered.

METHYLATION PATTERNS IN ACTIVE AND INACTIVE PROMOTERS - THE SPREADING OF METHYLATION PATTERNS

It has become apparent that the notion of an inverse correlation between promoter methylation and gene activity has been particularly striking for certain viral promoter sequences (3, 4, 13, 18). We have performed a study to determine the methylation status of all 5'-CG-3' residues in the late E2A promoter of Ad2 DNA in cell line HE1, with an active late E2A promoter, and in cell lines HE2 and HE3 with inactive late E2A promoters (47). The genomic sequencing method (48) permits the detection of 5-mC residues also in 5'-CG-3' sequences other than the 5'-CCGG-3' sites. The patterns of methylation have been determined in all 5'-CG-3' sequences over a region of about 180 nucleotides required for gene activity in the late E2A promoter of integrated Ad2 DNA. In cell line HE1 or uc2, the late E2A promoter is active and all thirteen 5'-CG-3' sequences between positions +24 and -160 are unmethylated. In cell line HE2 (HE3), the same promoter is permanently shut off, and all 5'-CG-3' sequences are methylated in either strand. Thus, the inverse correlation is perfect in these cell lines over a region of about 180 nucleotides in the late E2A promoter (47). The same promoter segment has been analyzed in cell lines mc23 and mc40 in which a late E2A promoter-chloramphenicol acetyltransferase (CAT) gene construct has been genomically fixed after transfection and prior in vitro 5'-CCGG-3' methylation. In cell line mc23, the preimposed methylation pattern is stable and the cells contain an inactive CAT gene. Genomic sequencing has confirmed the presence of 5-mC at the 5'-CCGG-3' sequences and has revealed the spreading of methylation to neighboring 5'-CG-3' sequences along the entire promoter. Some of these sites are hemimethylated. In cell line mc40, several of the 120 integrated copies are demethylated in positions +24 and +5. The cell line expresses the CAT gene. Thus it appears possible that a certain pattern of DNA methylation can be established by the gradual spreading of methylation from one or a few points of crystallization, as it were. It is unknown whether this spreading effect can also originate in 5-mC residues in flanking sequences or has to be initiated inside the sequence that is to become de novo methylated as in the example of cell line mc23. These results bear on the puzzling problem of how de novo patterns of DNA methylation can be initiated. Is it conceivable that the gradual spreading of the inactivation of large parts of the genome, e.g., in the inactivation of one of the X chromosomes, may be related to the spreading of a methylation pattern?

CONCLUSIONS AND UNSOLVED PROBLEMS

The conclusions presented here are based on the results of a detailed analysis of adenovirus promoters integrated into host DNA and of in vitro studies with isolated adenovirus promoters. The major findings are the following.

i) In cell lines, which carry genomically fixed adenovirus DNA, methylation patterns of viral genes are stable over many cell generations. The establishment of these stable patterns can require time and several cell divisions. At early times after the fixation of the foreign genes, the level of methylation of the previously unmethylated viral genome can be very low. The factors determining patterns of de novo methylation are not well understood.

ii) Inverse correlations between adenovirus gene methylation and the levels of expression of these integrated viral genes have been described a decade ago.

iii) There is evidence from hamster cell lines carrying an *in vitro* specifically 5'-CCGG-3' methylated, recently integrated adenovirus promoter-CAT gene (pAd2E2AL-CAT) construct that a pre-established pattern of methylation can spread in the genome. This conclusion is based on the results of genomic sequencing experiments in the late E2A promoter. Perhaps such spreading effects are responsible for the complete methylation of all 5'-CG-3' sequences in the inactive E2A promoter in hamster cell line HE2, and in a more general sense, for the establishment of *de novo* methylation patterns. In cell line HE1 which expresses the E2A gene, the same 5'-CG-3' sequences are all unmethylated.

iv) Sequence-specific promoter methylations cause gene inactivation. The sequences in a promoter, which upon methylation lead to promoter inactivation, seem to be different for each promoter and have to be determined experimentally. For the late E2A promoter of Ad2 DNA, sequence-specific methylation at nucleotide positions -215, +6 and +24 relative to the +1 cap site has been shown to inactivate this promoter. Data supporting this conclusion have been derived from microinjection studies of *Xenopus laevis* oocytes, from transfection experiments of mammalian cells, from the *in vitro* transcription of the methylated or the unmethylated promoter construct in a cell-free system using nuclear extracts of HeLa cells, and from cell lines in which the 5'-CCGG-3' methylated or the unmethylated pAd2E2AL-CAT construct has been genomically fixed.

v) It has been discussed whether promoter methylation is cause or consequence of gene inactivation. As will be apparent from previous reasoning, this may be a pseudo argument. The methylation of one or very few 5'-CG-3' promoter sequences suffices to inactivate a promoter or seriously interfere with promoter function. As a consequence of the introduction of one or of several 5-mC residues, which cause gene inactivation, the extent of methylation then spreads starting from the very few sequences that have been initially methylated. Hence, cause and consequence are not mutually exclusive, since the limited extent of methylation, that is sufficient to cause inactivation, can subsequently give rise to the establishment of a complex pattern of de novo DNA methylation.

vi) We have shown for the late E2A promoter of Ad2 DNA that the methylation of the 5'-CCGG-3' sequences at positions +6 and +24 prohibits the binding of cellular proteins. This effect can only be demonstrated with short promoter fragments, and the band shift technique has proven a useful and sensitive method for these studies. When longer fragments or the entire late E2A promoter are investigated in vitro or in vivo for protein binding differences between the unmethylated and the 5'-CCGG-3' methylated forms, such differences fail to be detectable. In studies on longer promoter fragments, subtle binding differences may be obscured because many different proteins will probably bind to a longer promoter fragment. It has also to be considered that structural changes in the promoter are caused by sequence-specific methylations and that alterations of promoter structure may be important in gene inactivation.

Sequence-specific promoter methylation occupies an important place among the mechanisms that can lead to long-term gene inactivation. In that sense, DNA methylation might influence also more complex biological phenomena that depend on a specifically programmed progression of the expression of families of genes, such as differentiation, the development of the immune response, oncogenesis, paternal and maternal imprinting of gene activity patterns and others. We have also adduced evidence that the functional effects of 5-mC are pliable and, at least partly, reversible by cis- or trans-acting factors.

With respect to the biological function(s) of DNA methylation, more questions have been raised than answers have been forthcoming. Many basic biological mechanisms other than promoter function may depend on specific methylation patterns. When looking at larger and larger segments of cellular genomes in cell lines or, probably more pertinently, in living organisms, questions on the establishment, the inheritance, and the biological functions of complex methylation patterns must be posed and formulated in integrals accessible to experimental analyses. It has been recently hypothesized (8) that a given methylation pattern in a segment of a mammalian genome may be the sum total of several patterns, each with a specific meaning. These individual patterns may interdigitate and give rise to a complicated array of methyl groups. It will be the challenge of the future to decipher yet another genetic code that may hold the answer to many unresolved biological problems.

ACKNOWLEDGMENTS

M.T. acknowledges receipt of a fellowship by the Alexander-von-Humboldt-Foundation, Bonn, Germany. We are indebted to Petra Böhm for editorial assistance. Work performed in the authors' laboratory has been supported over many years by the Deutsche Forschungsgemeinschaft through SFB74-C1 and SFB274-TP2.

REFERENCES

1. Arber W (1974). DNA modification and restriction. Progr Nucleic Acids Res Mol Biol 14:1.
2. Sutter D, Westphal M, Doerfler W (1978). Patterns of integration of viral DNA sequences in the genomes of adenovirus type 12-transformed hamster cells. Cell 14:569.
3. Sutter D, Doerfler W (1979). Methylation of integrated viral DNA sequences in hamster cells transformed by adenovirus 12. Cold Spring Harbor Symp Quant Biol 44:565.
4. Sutter D, Doerfler W (1980). Methylation of integrated adenovirus type 12 DNA sequences in transformed cells is inversely correlated with viral gene expression. Proc Natl Acad Sci USA 77:253.

5. Doerfler W (1981). DNA methylation - a regulatory signal in eukaryotic gene expression. J Gen Virol 57:1.
6. Doerfler W (1983). DNA methylation and gene activity. Ann Rev Biochem 52:93.
7. Doerfler W (1989a). Complexities in gene regulation by promoter methylation. In Eckstein F, Lilley DM (eds): "Nucleic Acids and Molecular Biology," vol 3, Berlin, Heidelberg, New York, London, Paris, Tokyo: Springer, p. 92.
8. Doerfler W (1989b). The significance of DNA methylation patterns: promoter inhibition by sequence-specific methylation is one functional consequence. Transact. Royal Soc., London, in press.
9. Doerfler W, ed. (1983, 1984). "The Molecular Biology of Adenoviruses." Curr Top Microbiol Immunol, vol 109, 110, 111, Berlin, Heidelberg, New York, London, Paris, Tokyo: Springer.
10. Roberts RJ, Akusjärvi G, Aleström P, Gelinas RE, Gingeras TR, Sciaky D, Pettersson U (1986). A consensus sequence for the adenovirus-2 genome. In Doerfler W (ed): "Adenovirus DNA: The Viral Genome and its Expression." Developments in Molecular Virology, vol 8, Boston: Martinus Nijhoff, p. 1.
11. Günthert U, Schweiger M, Stupp M, Doerfler W (1976). DNA methylation in adenovirus, adenovirus-transformed cells, and host cells. Proc Natl Acad Sci USA 73:3923.
12. Wienhues U, Doerfler W (1985). Lack of evidence for methylation of parental and newly synthesized adenovirus type 2 DNA in productive infections. J Virol 56:320.
13. Vardimon L, Neumann R, Kuhlmann I, Sutter D, Doerfler W (1980). DNA methylation and viral gene expression in adenovirus-transformed and -infected cells. Nucleic Acids Res 8:2461.
14. Doerfler W (1968). The fate of the DNA of adenovirus type 12 in baby hamster kidney cells. Proc Natl Acad Sci USA 60:636.
15. Doerfler W, Gahlmann R, Stabel S, Deuring R, Lichtenberg U, Schulz M, Eick D, Leisten R (1983). On the mechanism of recombination

between adenoviral and cellular DNAs: The structure of junction sites. Curr Top Microbiol Immunol 109:193.
16. Doerfler W, Spies A, Jessberger R, Lichtenberg U, Zock C, Rosahl T (1987). Recombination of foreign (viral) DNA with the host genome. Studies in vivo and in a cell free system. In Rott R, Goebel W (eds): "38. Colloquium Mosbach 1987, Molecular Basis of Viral and Microbial Pathogenesis," Berlin, Heidelberg, New York: Springer, p. 60.
17. Deuring R, Doerfler W (1983). Proof of recombination between viral and cellular genomes in human KB cells productively infected by adenovirus type 12: structure of the junction site in a symmetric recombinant (SYREC). Gene 26:283.
18. Kruczek I, Doerfler W (1982). Unmethylated state of the promoter/leader and 5'-regions of integrated adenovirus genes correlates with gene expression. EMBO J 1:409.
19. Vardimon L, Doerfler W (1981). Patterns of integration of viral DNA in adenovirus type 2-transformed hamster cells. J Mol Biol 147:227.
20. Knust B, Brüggemann U, Doerfler W (1989) Reactivation of a methylation-silenced gene in adenovirus-transformed cells by 5-azacytidine or by E1A transactivation. J Virol 63:0000.
21. Van der Vliet PC, Levine AJ (1973). DNA-binding proteins specific for cells infected by adenovirus. Nature New Biol 246:170.
22. Kuhlmann I, Doerfler W (1982). Shifts in the extent and patterns of DNA methylation upon explantation and subcultivation of adenovirus type 12-induced hamster tumor cells. Virology 118:169.
23. Kuhlmann I, Achten S, Rudolph R, Doerfler W (1982). Tumor induction by human adenovirus type 12 in hamsters: loss of the viral genome from adenovirus type 12-induced tumor cells is compatible with tumor formation. EMBO J 1:79.

24. Kuhlmann I, Doerfler W (1983). Loss of viral genomes from hamster tumor cells and non-random alterations in patterns of methylation of integrated adenovirus type 12 DNA. J. Virol 47:631.
25. Cook JL, Lewis AM, Jr, Klimkait T, Knust B, Doerfler W, Walker TA (1988). In vivo evolution of adenovirus 2-transformed cell virulence associated with altered E1A gene function. Virology 163:347.
26. Kruczek I, Doerfler W (1983). Expression of the chloramphenicol acetyltransferase gene in mammalian cells under the control of adenovirus type 12 promoters: effect of promoter methylation. Proc Natl Acad Sci USA 80:7586.
27. Langner K-D, Vardimon L, Renz D, Doerfler, W (1984). DNA methylations of three 5' C-C-G-G 3' sites in the promoter and 5' region inactivate the E2a gene of adenovirus type 2. Proc Natl Acad Sci USA 81:2950.
28. Langner K-D, Weyer U, Doerfler W (1986). Trans effect of the E1 region of adenoviruses on the expression of a prokaryotic gene in mammalian cells: resistance to 5'-CCGG-3' methylation. Proc Natl Acad Sci USA 83:1598.
29. Knebel-Mörsdorf D, Achten S, Langner K-D, Rüger R, Fleckenstein B, Doerfler W (1988). Reactivation of the methylation-inhibited late E2A promoter of adenovirus type 2 DNA by a strong enhancer of human cytomegalovirus. Virology 166:166.
30. Weisshaar B, Langner K-D, Jüttermann R, Müller U, Zock C, Klimkait T, Doerfler W (1988). Reactivation of the methylation-inactivated late E2A promoter of adenovirus type 2 by E1A (13S) functions. J Mol Biol 202:255.
31. Doerfler W (1982). Uptake, fixation, and expression of foreign DNA in mammalian cells: the organization of integrated adenovirus DNA sequences. Curr Top Microbiol Immunol 101:127.
32. Trentin JJ, Yabe Y, Taylor G (1962). The quest for human cancer viruses. Science 137:835.

33. Vardimon L, Kressmann A, Cedar H, Maechler M, Doerfler W (1982). Expression of a cloned adenovirus gene is inhibited by in vitro methylation. Proc Natl Acad Sci USA 79:1073.
34. Dobrzanski P, Hoeveler A, Doerfler W (1988). Inactivation by sequence-specific methylations of adenovirus promoters in a cell-free transcription system. J Virol 62:3941.
35. Southern PJ, Berg P (1982). Transformation of mammalian cells to antibiotic resistance with a bacterial gene under control of the SV40 early region promoter. J Mol Appl Gen 1:327.
36. Müller U, Doerfler W (1987). Fixation of unmethylated or 5'-CCGG-3' methylated foreign DNA in the genome of hamster cells: gene expression and stability of methylation patterns. J Virol 61:3710.
37. Knebel D, Doerfler W (1986). N^6-Methyldeoxyadenosine residues at specific sites decrease the activity of the E1A promoter of adenovirus type 12 DNA: J Mol Biol 189:371.
38. Razin A, Szyf M, Kafri T, Roll M, Giloh H, Scarpa S, Carotti D, Cantoni GL (1986). Replacement of 5-methylcytosine: a possible mechanism for transient DNA demethylation during differentiation. Proc Natl Acad Sci USA 83:2827.
39. Nevins JR (1987). Regulation of early adenovirus gene expression. Microbiol Rev 51:419.
40. Hermann R, Hoeveler A, Doerfler W (1989). Sequence-specific methylation in a downstream region of the late E2A promoter of adenovirus type 2 DNA interferes with protein binding. Submitted.
41. Hoeveler A, Doerfler W (1987). Specific factors binding to the E2A late promoter region of adenovirus type 2 DNA: No apparent effects of 5'-CCGG-3' methylation. DNA 6:449.
42. Keshet I, Lieman-Hurwitz J, Cedar H (1986). DNA methylation affects the formation of active chromatin. Cell 44:535.
43. Kovesdi I, Reichel R, Nevins JR (1987). Role of an adenovirus E2 promoter binding factor in E1A-mediated coordinate gene control. Proc Natl Acad Sci USA 84:2180.

44. Harrington MA, Jones PA, Imagawa M, Karin M (1988). Cytosine methylation does not affect binding of transcription factor Sp1. Proc Natl Acad Sci USA 85:2066.
45. Höller M, Westin G, Jiricny J, Schaffner J (1988). Sp1 transcription factor binds DNA and activates transcription even when the binding site is CpG methylated. Genes Devel 2:1127.
46. Watt F, Molloy PL (1988). Cytosine methylation prevents binding to DNA of a HeLa cell transcription factor required for optimal expression of the adenovirus late promoter. Genes Devel 2:1136.
47. Toth M, Lichtenberg U, Doerfler W (1989). Genomic sequencing reveals 5-methyldeoxycytosine-free domain in active promoters and the spreading of preimposed methylation patterns. Proc Natl Acad Sci USA 86:0000.
48. Church GM, Gilbert W (1984). Genomic sequencing. Proc Natl Acad Sci USA 81:1991.

METHYLATED DNA-BINDING PROTEIN FROM MAMMALIAN CELLS

Melanie Ehrlich, Xian-Yang Zhang, Clement K. Asiedu,
Rana Khan, and Prakash C. Supakar

Dept. of Biochemistry, Tulane Medical School,
New Orleans, LA 70112

ABSTRACT Methylated DNA-binding protein (MDBP) is the first eukaryotic protein shown to bind specifically to certain DNA sequences only when they are methylated at their CpG dinucleotides. This DNA-binding protein which is widely distributed in mammalian cells recognizes a 14 base-pair DNA sequence, which contains considerable degeneracy including the acceptability of substitution of T residues for m^5C residues. Evidence is presented for MDBP being a transcription regulatory protein. A model is proposed in which DNA replication-associated hemimethylation of methylation-dependent MDBP sites could help direct differentiation.

INTRODUCTION

Much evidence indicates that methylation of CpG dinucleotide sequences is involved in controlling expression of genes in vertebrate cells (1-8). CpG sites are symmetrical (palindromic) and a majority appear to be methylated on both strands in mammalian cells (9). A minor fraction of CpG sites is unmethylated and some of these unmethylated sites probably have important roles to play in transcription control often in a manner reflecting tissue-specific differences in methylation and gene expression (1). However, many questions remain to be answered about mammalian DNA methylation. Some of these are as follows.

1. Why do some normal tissues in humans or other mammals have as much as 15% more m^5C residues (~10^7 per haploid genome) than others in the same organism (Table 1; 10-12)?

TABLE 1
5-METHYLCYTOSINE CONTENTS OF VARIOUS HUMAN TISSUES
AND CELL POPULATIONS

Source of DNA	m^5C content[a] (millions of residues per haploid genome)
Values for normal samples	
thymus	70
brain	69
lymphocytes	67
spleen	65
lungs	64
liver	62
heart	61
sperm	59
placenta	53
Values for cultured cells	
undifferentiated embryonal carcinoma cells	74
differentiated embryonal carcinoma cells	66
Raji cell line	60
skin fibroblast cell strain	50
HeLa cells	49
RPMI 1788 cell line	40
Range of values for different malignancies	
20 metastases	46-72
62 primary malignancies	25-69
Mean value for different malignancies	
20 metastases	54
62 primary malignancies	58

[a] These values were calculated from the average of multiple HPLC determinations of the mol% m^5C in DNA digested to deoxynucleosides (percentage of bases as m^5C) assuming that there are 7×10^9 bases per haploid human genome (11, 13, 20). The Raji and RPMI 1788 cells are lymphoblasted cell lines. The relative standard deviations for these analyses were 1-3%.

2. Why is tumor formation or tumor progression so often associated with massive genomic hypomethylation (Table 1; 13-15)?
3. Mammalian DNA methyltransferases catalyze de novo methylation in vitro (methylation of 5'-CpG-3'/3'-GpC-5' sites) although not quite as well as maintenance methylation (methylation of 5'-CpG-3'/3'-Gpm^{5}C-5' sites). How are they prevented from indiscriminately methylating all CpG sites over the course of an organism's lifetime and how does de novo methylation occur during differentiation?
4. How is selective "demethylation" accomplished in a cell-specific manner and how is it limited to only certain sites (1)?
5. How is gamete-specific hypomethylation established (16-20)? Do single-copy or repetitive DNA sequences that are hypomethylated only in germ cells and very early embryos play a role in gametogenesis, the beginning of embryogenesis, or genomic imprinting?
6. Does the DNA hypomethylation specifically associated with extraembryonic tissues in mammals function in their differentiation (11, 21)?
7. Does DNA methylation play a role in regulating DNA replication, transposition, and influencing DNA repair in mammalian cells as it does in bacteria (22) and, in the case of transposition, probably also in higher plants (23-24)?
8. How does methylation of certain DNA sequences lead to the establishment of repression or aid in its maintenance?

To answer the last question, we undertook a search for a DNA methylation-specific DNA-binding protein in extracts of human placental nuclei. We found such a protein initially by assaying protein fractions for better binding to fragmented human DNA enriched in m^{5}C than to the exactly analogous, m^{5}C-depleted human DNA (25). We showed that such a fraction also binds much better to M13mp8 double-stranded DNA methylated at every C residue of one strand than to the analogous but unmethylated M13mp8 DNA (25). In addition, this protein fraction bound to a naturally m^{5}C-rich phage DNA, XP12 DNA, much better than to various unmethylated DNAs (25). The binding of this protein, methylated DNA-binding protein (MDBP), to various DNAs of high sequence complexity depends much more on the concentration of m^{5}CpG sites than on the level of total m^{5}C residues (26).

MDBP is a Sequence-Specific DNA-Binding Protein

We found that MDBP is sequence-specific in its DNA binding. It recognizes only six different sites in pBR322 or M13mp8 RF DNA and only if they are first methylated in vitro at their CpG sites (26, 27). The first high-affinity MDBP site that was identified by footprinting experiments is a sequence in pBR322 DNA, pB site 1 (Fig. 1), which consists of 14 base-pairs (bp). MDBP binds specifically to this site only when the three CpG dinucleotide pairs in this region are methylated with human DNA methyltransferase (27). Site-specific methylation by incorporation of m^5C-containing oligonucleotides in this region indicated that methylation of two of these CpG dinucleotides on the upper strand (Fig. 1) sufficed for specific binding by MDBP; however, methylation of all three gave a higher affinity site (28). That MDBP binding is highly sequence-specific was demonstrated with a number of pB site 1 mutants containing 1-bp changes that abolished detectable binding by MDBP (28). However, some mutations at pB site 1 increase binding up to several fold such as an A→G transition at the first position of pB site 1 (Fig. 1; 29).

Subsequent studies have uncovered 21 additional binding sites for MDBP in mammalian or viral DNAs (Table 2 and unpubl. results). All of these are related to a common 14-bp consensus sequence and many have T residues substituting for m^5C residues of pB site 1 as described below (Fig. 1). This consensus recognition sequence displays hyphenated dyad symmetry as do many DNA regulatory sequences. Consistent with the studies of site-directed mutagenesis of pB site 1, one or two base differences from the consensus sequence often suffice to abrogate MDBP binding in naturally occurring DNA sites (Table 3). However, several high-affinity MDBP binding sites deviate from the consensus sequence at two positions and yet bind well (Table 2). The effect of such deviations depends on the nature of the base change and the exact sequence of the rest of the bases in the 14-bp region. The consensus sequence itself contains much degeneracy with either purine or either C or T allowed at 7 of the 14 positions in the upper strand of the recognition sequence (Fig. 1). In contrast, at two positions in this strand, A is almost always present and, at a third position, G is almost invariably found (Fig. 1). Mutagenesis and dimethyl sulfate interference studies are consistent with this consensus sequence in indicating that every base-pair in this 14-bp region is important for MDBP recognition (29).

pB site 1 of pBR322 tet gene

5'- A T m⁵C G T C A • m⁵C G G m⁵C G A T -3'

Cytomegalovirus site 2 in the IE enhancer

5'- A T <u>T</u> <u>A</u> C C A • <u>T</u> G G <u>T</u> G A T -3'

MDBP Consensus Sequence
_____

5'- R T m⁵Y R Y Y A • m⁵Y R G m⁵Y R A Y -3'

Matches
at 27   25 17 19 23 27 27 24   24 26 26 21 25 24 27
sites

FIGURE 1. The sequences of two high-affinity MDBP sites and a consensus sequence deduced from these sites plus 25 others are shown (27, 33, Table 2 and unpubl. results). The sequence of only one strand is indicated although MDBP recognizes only double-stranded DNA (29). Y indicates the common pyrimidines of DNA, C or T; R is A or G; m⁵Y is m⁵C or T (5-methyluracil). The pBR322 site, pB site 1 (positions 1141-1162), binds to MDBP only when methylated in vitro at its CpG dinucleotides (26). The human cytomegalovirus site CMV site 2 is the second of two MDBP recognition sites in the enhancer of the immediate-early (IE) genes. Both MDBP sites in this enhancer do not require cytosine methylation for binding because they contain T residues replacing several of the m⁵C residues of pB site 1. At the three critical 5'-m⁵YR-3' dinucleotide positions in pB site 1 and CMV site 2, the m⁵Y and the purine which has an m⁵Y opposite it in the complementary strand are underlined.

TABLE 2
SOME MAMMALIAN OR VIRAL SITES
WHICH ARE SPECIFICALLY BOUND BY MDBP

Site	Sequence[a]
CMV site 1 (-443, IE enhancer)	G T **T** C(x) C C A **T** A G **T** A A C
CMV site 2 (-226, IE enhancer)	A T **T A** C C A **T** G G **T** G A T
HSV site 1 (∼-2, UL17)	A T **M G** C C A **M G** G **M G** A T
rSA site 1 (∼+840)	G C(x) **T G** T C A **M G** G **M G** A C
h HLA site 1 (∼+30)	G C(x) **M G** T C A **T** G G **M G** C(x) C
h actin site 1 (intron 4)	G C(x) **T G** C C A **T** G G **T** G A C
hu site 1	A T **T** C(x) C C A **T** G G **T** A A C
hu site 2	G G(x) **T G** C C A **T** G G **T** G A C
hu site 3	G T A(x) G T C A **T** A G A(x) G A T

[a] The sequences are written 5' to 3'. The 5-methylcytosine (M) positions and the corresponding T residues as well as the adjacent purine residues 3' to them are underlined. The bases which deviate from the consensus sequence are indicated by an "x" underneath. The sites that bind to MDBP with a high affinity and contain m^5C residues show much less or negligible binding when unmethylated. The different sites tested were as follows in the order given in the table: human cytomegalovirus immediate early (IE) enhancer sites 1 and 2, herpes simplex virus site 1 in the promoter region of a gene of unknown function (UL17), a rat serum albumin gene site, the human HLA-A2 gene site, a site in human smooth muscle (aortic) actin gene, and three anonymous human sites cloned on the basis of their recognition by MDBP (33 and unpubl. results).

TABLE 3
SOME MAMMALIAN OR VIRAL SITES THAT WERE TESTED FOR BINDING TO MDBP AND FOUND NOT TO BIND

Site	Sequence[a]
hTCR	A T A(x) G C C A M̲ G̲ G M̲ G̲ A T
Ad 2	A T M̲ G̲ C C G(x) M̲ G̲ G M̲ G̲ A T
CMV site 4	G T T̲ A̲ T T A^A T(x) A̲ G T̲ A̲ A T
CMV site 5	A C(x) C(x) G C C A T G̲ T̲(x) T̲ G A C
EBV	C(x) C(x) M̲ G̲ T C A M̲ G̲ G T̲ G̲ A C
h oxytocin	G C̲(x) T̲ G̲ C C A M̲ G̲ C̲(x) M̲ G̲ A C
r Bn Ca prot. kinase	A C(x) C(x) A C C A M̲ G̲ G T̲ G̲ A C
h c-<u>fos</u>	A T T̲(x) C̲ C C A M̲ G̲ G T̲ C̲(x) A C
m c-<u>abl</u>	G T G̲(x) G C C A M̲ G̲ G M̲ G̲ A A(x)
m his tRNA	A M G̲(x) A T C A M̲ G̲ G M̲ G̲ A G(x)
r transthyretin	A C̲(x) T̲ T̲(x) T C A M̲ G̲ G M̲ G̲ A C
h 45 S	G G̲(x) C̲(x) C̲(x) T C A M̲ G̲ G T̲ G̲ A C
h int-1	G M G̲(x) G C C A M̲ G̲ G M̲ G̲ G G

[a]Sequences, written as in Table 2, were from the human T-cell receptor gene, human adenovirus 2 DNA, as well as sequences in two positions near the IE CMV enhancer positions -580 and -605 of the IE transcription unit, and sequences in the Epstein Barr virus DNA, the rat brain calcium-dependent protein kinase gene, the human c-<u>fos</u> and mouse c-<u>abl</u> genes, a mouse histidyl tRNA gene, rat transthyretin gene, human 45 S rRNA gene, and human int-1 gene. These sequences were tested by standard band shift assays (29) as part of 22-bp duplexes containing the naturally occurring sequences.

MDBP Binds to Certain DNA Sites in the Absence of DNA Cytosine Methylation

Included in the degeneracy of the consensus sequence is the acceptability of either m^5CpG, TpG, or TpA dinucleotides at positions 3 and 4, 8 and 9, and 11 and 12 (Fig. 1). 5-Methylcytosine and T residues share the same 5-methyl group. Therefore, recognition at these residues probably principally involves contact of the protein with this methyl group and, in one case, with the N7 group of the purine residue opposite it on the other strand, as previously indicated in dimethyl sulfate interference assays (29). All MDBP binding sites detected thus far have at least two and usually more of the six m^5YpR sequences in both strands of the consensus sequence (Fig. 1, Table 2 and data not shown). When TpR occupies several of these positions and the rest of the 14-bp region matches the consensus sequence well, then the site will be recognized by MDBP in the absence of C methylation. Sequences with some replacement of m^5C residues with T residues may still require methylation of a remaining CpG dinucleotide for binding when there are deviations from the consensus sequence as in pB site 2 in pBR322 DNA, 5'-GM̄GATCATGGM̄GAC-3', and M13 sites 2 and 3 in M13mp8 DNA, 5'-GT̄T̄ATTATTGM̄GTT-3' and 5'-CT̄GATTAM̄GGTGCT-3' (26) or when there are an inadequate number of m^5YpR sites without C methylation, as in rSA site 1 (Table 2).

MDBP is a Ubiquitous DNA-Binding Protein

By band-shift assays (also known as gel retardation assays) to identify specific MDBP·DNA complexes, we showed that MDBP activity is present in nuclear extracts from a variety of mammalian sources (30). These included various cultured human cells (HeLa, embryonal carcinoma, and H-9 lymphoblastoid cells), rodent cells and tissues (murine LTK⁻ cells, rat lung, and spleen, and liver), and calf thymus. MDBP activity in each of the studied mammalian nuclear extracts forms a family of different complexes with methylated but not with unmethylated duplexes containing pB site 1 (30). These migrate as closely spaced bands or one broad band (30). Formation of all of these complexes is competed by oligonucleotide duplexes containing methylated pB site 1 or by naturally m^5C-rich phage XP12 DNA (34 mol% m^5C; 31, 32) but not by nonspecific DNAs (29, 30). The electrophoretic mobilities of these complexes of pB site 1 and MDBP from mouse cells and various human cells is similar. The analogous complexes from rat tissue and calf thymus samples have a higher electrophoretic mobility apparently due to partial proteolysis which did not inhibit binding (30). Cultured mosquito cell extracts did not display an activity similar to that of the mammalian extracts (30).

Binding of Hemimethylated vs. Bifilarly Methylated DNA Sites

When only one strand of a methylation-dependent MDBP site such as pB site 1 or HSV site 1 (Table 2) is methylated, MDBP will bind to it but no binding is seen in the absence of methylation at these sites (Fig. 2; 29, 33). The more of the three CpG sites per strand that are methylated, the more binding is obtained. Furthermore, methylation of both strands (bifilar methylation) results in much better binding than methylation of only one strand (hemimethylation) as seen in Fig. 2. Although MDBP sites have dyad symmetry, they are not completely symmetrical (Table 2, Fig. 1). Similarly, methylation of only one strand of pB site 1 gave very different levels of binding depending on which of the two strands is methylated (Fig. 2). A preference for one hemimethylated form over the other was also seen at the MDBP sites in several mammalian gene regions (unpubl. results).

The weaker binding of a methylation-dependent, hemimethylated MDBP site than of the bifilarly methylated site could have important physiological consequences at MDBP sites methylated in mammalian cells. Immediately after DNA replication in mammalian cells, previously bifilarly methylated sites, which should be the predominant type of methylated sites, will be made hemimethylated (34-37). If, for example, a methylation-dependent mammalian MDBP site were normally bifilarly methylated and MDBP functioned as a repressor at that site, then DNA replication could transiently decrease repression due to localized denaturation of this site at the replication fork and its hemimethylated state immediately after replication. Furthermore, the extent of repression in the daughter cells could be unequal depending on which strand was methylated at the resulting hemimethylated sites in the two progeny cells. Such differential expression of a given gene in two progeny cells could be important during differentiation. In E. coli, conversion of bifilarly methylated dam sites to the hemimethylated state has been linked to increased expression of a number of genes

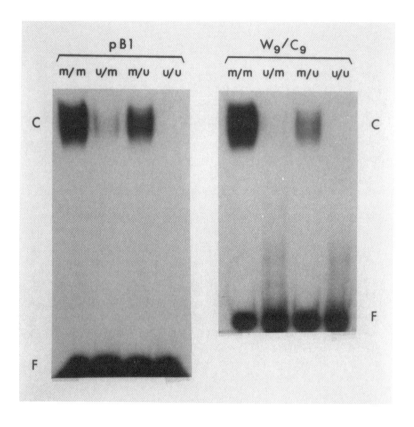

FIGURE 2. Comparison of binding of variously methylated forms of pB site 1 to MDBP. pB site 1 was present in a 14-bp oligonucleotide duplex (pB1) or a 35-bp duplex (W9/C9) as previously described (29). These duplexes were synthesized with both strands methylated at the three CpG site (m/m), neither strand methylated (u/u), only the upper strand (the one depicted in Fig. 1) methylated (m/u), or only the lower strand methylated. These ^{32}P-labeled ligands were used in a standard band shift assay with the hydroxyapatite fraction of human placental MDBP as previously described (29). The positions of the bands of the complex of MDBP and oligonucleotide duplex (C) and of the free ligand (F) in the autoradiogram are indicated. Binding to the u/m hemimethylated form of W9/C9 as well as W17/C17 was clearly seen in longer exposures whereas no binding to the completely unmethylated (u/u) forms of these duplexes was observed.

(38, 39). Evidence for different effects of hemimethylation on two progeny DNA strands depending on which of the strands is methylated has been reported for E. coli (39).

MDBP as a Transcription Regulatory Protein

Evidence suggests that MDBP plays a role in transcription control. First, MDBP recognizes 14-bp binding sites that contain hyphenated dyad symmetry and preliminary studies suggest that MDBP is a dimeric protein (unpubl. results). Such symmetry in DNA-binding sites and in proteins that bind to DNA is often found associated with transcription regulatory proteins. The rather long, albeit degenerate, consensus sequence for MDBP binding sites provides the sort of high degree of specificity and infrequency required for regulation of transcription. Secondly, the specificity of MDBP for certain DNA sequences only when they are methylated is consistent with a role in methylation-associated transcription control. Indeed, MDBP is the first mammalian protein shown to have specificity for methylated DNA sequences. Evidence for another methylation-specific mammalian DNA-binding protein has been recently reported (6). Thirdly, the binding of MDBP to certain sites independent of the methylation status of the DNA and the ubiquitous distribution of this protein in mammalian cells is consistent with the hypothesis that it is involved in transcription of certain constitutively expressed genes as well as in that of other inducible, differentiation-related genes. It is possible that methylation-independent MDBP sites, which contain T residues replacing m^5C residues, evolved from m^5C-containing sites by spontaneous deamination of m^5C residues, hotspots for such transition mutations during the course of evolution (40-42).

Furthermore, we have found two high-affinity MDBP sites in the enhancer of the immediate early genes of human cytomegalovirus (CMV sites 1 and 2, Table 2). Both of these are methylation-independent, high-affinity MDBP sites and so could function in vivo regardless of the state of methylation of the viral genome. This strong enhancer has been shown to have its in vitro activity competed by an oligonucleotide duplex containing CMV site 1 (43). This suggests that MDBP acts as a positive transcription factor at the enhancer. We are currently analyzing the role of MDBP recognition of these two sites in enhancer function in collaboration with L. Hennighausen. Lastly, we have found another methylation-independent MDBP site in the first half of the first intron on the human c-myc proto-oncogene (unpublished results). This region has been implicated in negative regulation of transcription of the c-myc gene (44, 45). We propose that MDBP, like certain other sequence-specific DNA-binding proteins acts as a positive transcription factor at certain DNA sites and a negative regulatory factor at others depending on the DNA elements surrounding the MDBP site. In the latter capacity, MDBP could be an important link between negative control of transcription and DNA methylation in certain mammalian genes.

ACKNOWLEDGEMENTS

This research was supported in part by Public Health Service Grant GM33999.

REFERENCES

1. Cooper, DN (1983) Eukaryotic gene methylation. Hum Genet 64: 315-333.
2. Yisraeli, J, Adelstein, RS, Melloul, D, Nudel, U, Yaffe, D Cedar, H (1986) Muscle-specific activation of a methylated chimeric actin gene. Cell 46:409-416.
3. Guntaka, RV, Gowda, S, Wagner, H, Simon, D (1987) Methylation of the enhancer region of avian sarcoma virus long terminal repeat suppresses transcription. FEBS Let 221:332-336.
4. Muller, U, Doerfler, W (1987) Fixation of the unmethylated or the 5'-CCGG-3' methylated Adenovirus late E2A promoter-cat gene construct in the genome of hamster cells: Gene expression and stability of methylation patterns. J Virol 61:3710-3720.
5. Carbone, AM, Marrack, P, Kappler, JW (1988) Remethylation at sites 5' of the murine Lyt-2 gene in association with shutdown of Lyt-2 expression. J Immunol 141:1369-1375.
6. Saluz, HP, Feavers, IM, Jiricny, J, Jost, JP (1988) Genomic sequencing and in vivo footprinting of an expression-specific DNase I-hypersensitive site of avian vitellogenin II promoter reveal a demethylation of a mCpG and a change in specific interactions of proteins with DNA. Proc Natl Acad Sci USA 85:6697-6700.

7. Alberti, S, Herzenberg, LA (1988) DNA methylation prevents transfection of genes for specific surface antigens, Proc Natl Acad Sci USA 85:8391-8394.
8. Razin, A, Levine, A, Kafri, T, Agostini, S, Gomi, T, Cantoni, GL (1988) Relationship between transient DNA hypomethylation and erythroid differentiation of murine erythroleukemia cells. Proc Natl Acad Sci USA 85:9003-9006.
9. Ehrlich, M, Wang, RY-H (1981) 5-Methylcytosine in eukaryotic DNA. Science 212:1350-1357
10. Vanyushin, BF, Mazin, AL, Nasilyev, VK, Belozensky, AN (1973) The content of 5-methylcytosine in animal DNA: The species and tissue specificity. Biochim Biophys Acta 299:397-403.
11. Ehrlich, M, Gama-Sosa, MA, Huang, LH, Midgett, R, Kuo, KC, McCune, RA, Gehrke, C (1982) Amount and distribution of 5-methylcytosine in human DNA from different types of tissues or cells. Nucleic Acids Res 10:2709-2721.
12. Gama-Sosa, MA, Midgett, R, Slagel, VA, Githens, S, Kuo, KC, Gehrke, CW, Ehrlich M (1983a) Tissue-specific differences in DNA methylation in various mammals. Biochem Biophys Acta 740:212-219.
13. Gama-Sosa, MA, Slagel, VA, Trewyn, RW, Oxenhandler, R, Kuo, KC, Gehrke, CW, Ehrlich, M (1983b) The 5-methylcytosine content of DNA from human tumors. Nucleic Acids Res 11:6883-6894.
14. Lapeyre, JN, Becker, FF (1979) 5-Methylcytosine content of nuclear DNA during chemical hepatocarcinogenesis and in carcinomas which result. Biochem Biophys Res Comm 87:698-705.
15. Feinberg, AP, Gehrke, CW, Kuo KC, Ehrlich, M (1988) Reduced genomic 5-methylcytosine content in human colonic neoplasia. Cancer Res 48:1159-1161.
16. Sturm, KS, Taylor, JH (1981) Distribution of 5-methylcytosine in the DNA of somatic and germline cells from bovine tissues. Nucleic Acids Res 9:4537-4546.
17. Sanford, J, Forrester, L, Chapman, V (1984) Methylation patterns of repetitive DNA sequences in germ cells of Mus musculus Nucleic Acids Res 12:2823-2836.
18. Ponzetto-Zimmerman, C, Wolgemuth, DJ (1984) Methylation of satellite sequences in mouse spermatogenic and somatic DNAs Nucleic Acids Res 12:2807-2822.
19. Zhang, X-Y, Wang, RY-H, Ehrlich, M (1985) Human DNA sequences exhibiting gamete-specific hypomethylation. Nucleic Acids Res 13:4837-4851.

20. Zhang, X-Y, Loflin, PT, Gehrke, CW, Andrews, PA, Ehrlich, M (1987) Hypermethylation of human DNA sequences in embryonal carcinoma cells and somatic tissues but not in sperm. Nucleic Acids Res 15:9429-9449.
21. Sanford, J, Chapman, VM, Rossant, J (1985) DNA methylation in extraembryonic lineages of mammals. Trends Gen 1:89-93.
22. Messer, W, Noyer-Weidner, M (1988) Timing and targeting: The biological functions of dam methylation in E coli. Cell 54:735-737.
23. Schwartz, D, Dennis, E (1986) Transposase activity of the Ac controlling element in maize is regulated by its degree of methylation. Mol Gen Genet 205:476-482.
24. Chandler, VL, Walbot V (1986) DNA modification of a maize transposable element correlates with loss of activity, Proc Natl Acad Sci USA 83:1767-1771.
25. Huang, L-H, Wang, RY-H, Gama-Sosa, MA, Shenoy, S, Ehrlich, M (1984) A protein from human placental nuclei binds preferentially to 5-methylcytosine-rich DNA. Nature 308:293-295.
26. Wang, RY-H, Zhang, X-Y, Khan, R, Zhou, Y-W, Huang, L-H, Ehrlich, M (1986) Methylated DNA-binding protein from human placenta recognizes specific methylated sites on several prokaryotic DNAs. Nucleic Acids Res 14:9843-9860.
27. Wang, RY-H, Zhang, X-Y, Ehrlich, M (1986) A human DNA-binding protein is methylation-specific and sequence-specific, Nucleic Acids Res 14:1599-1614.
28. Zhang, X-Y, Ehrlich, KC, Wang, RY-H, Ehrlich, M (1986) Effect of site-specific DNA methylation and mutagenesis on recognition by methylated DNA-binding protein from human placenta. Nucleic Acids Res 14:8387-8398.
29. Khan, R, Zhang, X-Y, Supakar, PC, Ehrlich, KC, Ehrlich, M (1988) Human methylated DNA-binding protein. J Biol Chem 263:14374-14383.
30. Supakar, PC, Weist, D, Zhang, D, Inamdar, N, Zhang, X-Y, Khan, R, Ehrlich, KC, Ehrlich, M (1988) Methylated DNA-binding protein is present in various mammalian cell types. Nucleic Acids Res 16:8029-8044.
31. Kuo, TT, Huang, T-C, Teng, M-H (1968) 5-Methylcytosine replacing cytosine in the deoxyribonucleic acid of a bacteriophage for Xanthomonas oryzae. J Mol Biol 34:373-375.
32. Kuo, KC, McCune, RA, Gehrke, CW, Midgett, R, Ehrlich, M (1980) Quantitative reversed-phase high performance

liquid chromatographic determination of major and modified deoxyribonucleosides in DNA. Nucleic Acids Res 8:4763-4776.
33. Zhang, XY, Supakar, PC, Khan, R, Ehrlich, KC, Ehrlich, M (1989) Related sites in human and herpesvirus DNA recognized by methylated DNA-binding protein from human placenta. Nucleic Acids Res 17:1459-1474.
34. Wookcock, DM, Adams, JK, Cooper, IA (1982) Characteristics of enzymatic DNA methylation in cultured cells of human and hamster origin the effect of DNA replication inhibition. Biochem Biophys Acta 696:15-22.
35. Woodcock, DM, Simmons, DI, Crowther, PJ, Cooper, IA, Trainor, KJ, Morley, AA (1986) Delayed DNA methylation is an integral feature of DNA replication in mammalian cells. Exp Cell Res 166:103-112.
36. Gruenbaum, Y, Szyf, M, Cedar, H, Razin, A (1983) Methylation of replicating and post-replicated mouse L-cell DNA. Proc Natl Acad Sci USA 80:4919-4921.
37. Katianianen, TL, Jones, PA (1985) DNA methylation in mammalian nuclei. Biochemistry 24:5575-5581.
38. Marinus, MG (1987) DNA methylation in Escherichia coli. Ann Rev Genet 21:113-131.
39. Messer, W, Noyer-Weidner, M (1988) Timing and targeting: The biological functions of dam methylation in E coli. Cell 54: 735-737.
40. Wang, RY-H , Kuo, KC, Gehrke, CW, Huang, LH, Ehrlich, M (1982) Heat- and alkali-induced deamination of 5-methylcytosine and cytosine residues in DNA. Biochim Biophys Acta 697:371-377.
41. Ehrlich, M, Norris, KF, Wang, RY-H, Kuo, KC, Gehrke, CW (1986) DNA cytosine methylation and heat-induced deamination, Bioscience Rep 6:387-393.
42. Cooper, DN, Youssoufian, H (1988) The CpG dinucleotide and human genetic disease. Hum Genet 78:151-155.
43. Ghazal, P, Lubon, H, Hennighausen, L (1988) Multiple sequence-specific transcription factors modulate cytomegalovirus enhancer activity in vitro. Molec Cell Biol 8:1809-1811.
44. Zajac-Kaye, M, Gelmann, EP, Levens, D (1988) A point mutation in the c-myc locus of a Burkitt lymphoma abolishes binding of a nuclear protein. Science 240:1776-1780.
45. Chung, J, Sinn, E, Reed, RR, Leder, P (1986) Trans-acting elements modulate expression of the human c-myc gene in Burkitt lymphoma cells. Proc Natl Acad Sci USA 83:7918-7922.

TRANSGENE METHYLATION IMPRINTS ARE ESTABLISHED POST-FERTILIZATION

Ross McGowan
Thu-Hang Tran
Jean Paquette
Carmen Sapienza

Ludwig Institute for Cancer Research
687 Pine Avenue West
Montreal, Quebec
Canada H3A 1A1

ABSTRACT: In some lines of transgenic mice, the methylation of MspI sites within or adjacent to the transgene locus is affected by the sex of the parent from which the transgene was inherited. These differences are consistent with a role for DNA methylation in the process of genome imprinting. We have examined two transgenic lines in which the methylation state of a quail troponin I transgene is affected by gamete-of-origin and surprisingly, our analyses indicate that the somatic transgene methylation patterns are not established during gametogenesis, but after fertilization. Although it has generally been assumed that imprinting is accomplished during gametogenesis, a model in which cells of the early embryo are mosaic with respect to transgene methylation phenotype may accomodate all of the available data and requires only that the process be irreversibly initiated before pronuclear fusion in the mouse zygote.

INTRODUCTION

The methylation state of some transgene loci changes in a predictable manner depending upon their maternal or paternal inheritance (1,2,3,4). These observations are consistent with a role for differential DNA methylation in the process of genome imprinting. However, a direct relationship between this type of "methylation imprinting" and imprinting as defined by pronuclear transplantation (5,6,7,8,9) or genetic experiments (10,11,9) has not been established.

We have previously demonstrated (2) that particular Msp I sites within a quail troponin I transgene always became methylated in the somatic tissues of offspring which inherited the locus from a female. However, when the same locus was inherited from a male, some but not all offspring exhibited transgene hypomethylation in their somatic tissues (2). We therefore attempted to determine whether such variability was due to post-fertilization modification or differences in the transgene methylation state of male gametes.

In this study, we present two lines of evidence which indicate that the gamete-of-origin-dependent transgene methylation imprint observed in somatic tissues is established after fertilization and is therefore subject to modification in the early embryo: 1)A founder male transgenic mouse (in which the transgene was established from unmethylated DNA injected into the male pronucleus) contained a transgene locus which was methylated. Moreover the methylation phenotype observed in the founder was indistinguishable from that of his offspring; 2) The transgene methylation pattern found in male gametes differs from the transgene methylation pattern found in both embryonic and extraembryonic tissue of offspring. Therefore, the transgene methylation pattern of male germ cells is not reproduced in any somatic lineage.

RESULTS

Transgene Methylation in Founder Animals

Fig. 1a shows the gamete-of-origin-dependent methylation changes exhibited by quail troponin I transgene locus 354. In this pedigree, MspI sites

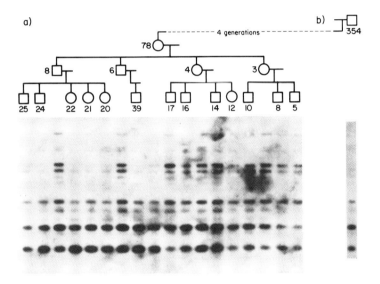

Figure 1: Tail DNAs from the 354 quail troponin I transgenic line, digested with BamH1 and HpaII restriction endonucleases. The pedigree of individuals hemizygous for the transgene locus is shown across the top. Fig. 1a shows the autoradiographic hybridization pattern obtained using an internal BamH1-Kpn1 fragment of the quail troponin I gene (described in 2) as a probe. An increase in methylation is evident after passage of the transgene through females (offspring of female #78) whereas passage through the male produces a decrease in methylation (offspring of males # 6 and 8). Fig 1b shows the hybridization pattern of the founder male of this line, which is four generations removed from female #78. All methods are as described in (2).

within the transgene become unmethylated on passage through the male germline and methylated on passage through the female germline. Fig. 1b shows that somatic tissue from the founder of this line, male 354, contains a transgene locus which is methylated in the same way as the offspring (individuals nos. 20-25, 39) of transgenic males. Hybridization of the same probe used in Figs. 1a and 1b to MspI digests of these same DNA's reveals only the smallest (870bp) hybridizing DNA fragment (data not shown). Because the transgene locus in male 354 was not derived through gametogenesis, but by microinjection of unmethylated DNA into the male pronucleus of the zygote, these sequences must have been _de novo_ methylated during embryogenesis. The faithful reproduction of this pattern in the offspring of males nos.6 and 8 indicates that the transgene locus is treated in the same way despite the experience of male gametogenesis.

Transgene Methylation in the Soma and Germline

In quail troponin I transgene line 379, we observed a distinct transgene methylation pattern in the testes of males in which the transgene was hypermethylated in somatic tissue. Taken in isolation, these results are consistent with hypothesized germ-line "switching" of genome imprints (5,7,12). However, if the level of transgene methylation is established in male gametes, and no post-fertilization changes take place, then all somatic tissue should show a methylation pattern identical to that found in the gamete.

Fig. 2 shows that the methylation phenotype exhibited by testes of males of the 354 line (and mature sperm, data not shown) is distinct from that found in any other adult tissue (Fig. 1 and data not shown). Moreover, the testes-specific pattern is similar in all adult males examined (Fig. 2 and data not shown), regardless of which somatic transgene methylation phenotype they exhibit. Among more than 100 offspring of transgenic males, we have never observed a methylation phenotype like that found in testes (for example, compare offspring of male 8, Fig. 1, with testes of male 8, Fig. 2). These data imply that the patterns observed in somatic tissues must be established after fertilization.

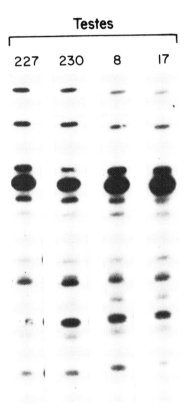

Figure 2: Hybridization of BamHI plus HpaII cleaved DNAs from testes of several individuals from the 354 troponin I transgenic line (probed as in previous figure). The methylation phenotype of these DNAs is the same in all cases even though males 227 and 230 inherited the transgene from a female and males 8 and 17 inherited the locus from a male.

Transgene Methylation in Embryonic and Extraembryonic Tissues

In the 354 lines, as in the 379 line (2) somatic tissues derived from different primary germ layers display the same methylation phenotype (data not shown). From these data we infer that primitive ectoderm formation (the last common progenitor of these tissues) is the latest stage in development by which these patterns must be established. For the phenotype to be established at a later stage, multiple, independent methylation or demethylation events must occur to exactly the same extent in each tissue.

Because all derivatives of primitive ectoderm from any individual display the same transgene methylation phenotype, we wished to determine whether the phenotype might be established at a stage earlier than primitive ectoderm formation. The first morphologically distinguishable differentiation event which occurs during mouse development is the segregation of the trophectoderm lineage, which contributes only to extraembryonic tissues, from the inner cell mass, which gives rise to primitive ectoderm and primitive endoderm (13). Therefore, the transgene methylation phenotype of placenta was compared to that found in skin from the same day 14 embryos of the 379 line. Fig. 3 shows that the transgene methylation phenotype of placenta is similar in all cases. Moreover, it is not affected by gamete-of-origin. Qualitatively similar results were obtained for the 354 line (data not shown).

To examine the possibility that the methylation pattern observed in day 14 placenta might be different from that of pure trophectoderm, we analyzed the methylation pattern of ectoplacental cone from day 8.5 embryos, which serves as an uncontaminated source of trophectoderm (14). Fig. 3 shows that the methylation phenotype of this tissue is not substantially different from day 14 placenta. The lack of hybridization to placental DNA of non-transgenic embryos derived from transgenic females (offspring of females 95 and 439 in Fig. 3) indicates that the level of contaminating maternal tissue in our dissections of day 14 placenta is less than 5%.

These observations indicate that a somatically stable transgene methylation phenotype is not established before the divergence of trophectoderm and inner cell mass lineages.

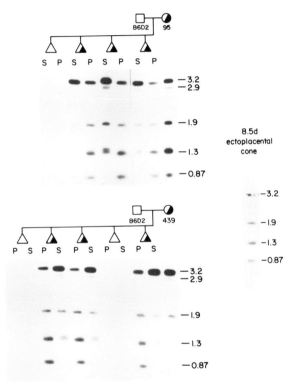

Figure 3: Hybridization of BamH1 plus HpaII cleaved DNAs from transgenic (1/2 filled symbols) or non-transgenic (open symbols) embryos. Both embryonic tissue (skin from the dorsal side of embryos, lanes S) and extra-embryonic tissue (predominantly placenta but not excluding extra-embryonic membranes, lanes P) were dissected from 14 day embryos and DNA was extracted. Dissections of extra embryonic tissue yielded samples which were almost completely free of contaminanting maternal decidual tissue as demonstrated by the

lack of hybridization to DNAs from non-transgenic individuals derived from transgenic females (offspring of female 95 and 439, above and data not shown. The autoradiographic hybridization pattern of BamHI and HpaII cleaved DNAs from day 8.5 ectoplacental cone, an uncontaminated source of trophectoderm, is not substantially different from the pattern seen for day 14 placenta.

DISCUSSION

Our results give rise to two, seemingly disparate, conclusions: 1) both the 354 and 379 transgene loci display somatic methylation phenotypes which reflect gamete-of-origin; and 2) these phenotypes are not established in the gamete, but are established after fertilization. Because derivatives of all primary germ layers of any individual display the same transgene methylation phenotype, it is most likely that the final establishment of this phenotype occurs prior to primitive ectoderm formation. Moreover the embryo has established lineages with different methylation phenotypes prior to the emergence of primitive ectodermal and trophectodermal derivatives.

Two explanations may be proposed for these data. All cells of the early embryo could have the same transgene methylation phenotype, reflecting only the gametogenic origin of the transgene. In this case the phenotype must be specifically altered in both primitive ectoderm and trophectoderm cells. Alternatively, preimplantation embryos could be mosaics composed of cells with at least two different methylation phenotypes, and cells bearing these different phenotypes are ultimately differentially distributed between the trophectoderm and inner cell mass lineages.

The latter model requires a mechanism by which to generate methylation mosaicism and some basis upon which to discriminate between maternally and paternally derived transgenes. One solution to the problem of generating mosaicism is to invoke the activity of a *de novo* methylase in only some cells of the early embryo (Fig. 4). Evidence that such an activity exists in early embryos has been demonstrated for both exogenous (15) and endogenous (16) sequences, but neither of these studies was designed to distinguish whether the activity was cell-specific. We have no direct evidence that such methylation mosaics are generated at either the 379 or 354 quail troponin I transgene loci but the observed differences between the methylation phenotype of embryonic and extraembryonic tissues are consistent with this interpretation.

The mechanism by which paternally and maternally derived transgenes might be distinguished is unknown, but the presence or absence of methylated sequences themselves could in some way serve as a primary post-fertilization imprinting signal. For example, the presence of unmethylated DNA in the male pronucleus, or the presence of methylated DNA in the female pronucleus will result in an "unbalanced" mosaic if both daughter cells are not equivalent in de novo methylating activity (Fig. 4). The transgene will therefore be "hypomethylated" when inherited from males and "hypermethylated" when inherited from females.

The global operation of an imprinting mechanism which resulted in somatic mosaicism (i.e. the apparent erasure of the imprint in some but not all cells) would predict unusual genetic behavior of alleles at imprinted loci. If imprinting sometimes resulted in allele inactivation, then mosaic individuals would be composed of cells having only one active allele at an imprinted locus, as well as cells having two active alleles at the same imprinted locus (Sapienza, in press). That fraction of cells which contained an epigenetically inactivated allele would be functionally hemizygous at imprinted loci. Inactivation of the remaining allele by a somatic mutation (sequence alteration, nondisjunction or mitotic recombination) would then result in loss of function at this locus and the expression of a cell-specific mutant phenotype, such as is observed in some types of human tumors (18,19).

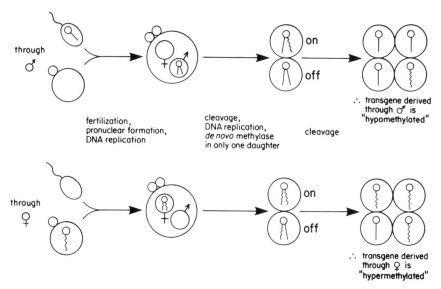

Figure 4; Model for the generation of methylation mosaicism at a hemizygous transgene locus. The transgene is depicted as being hypermethylated in female gametes (bottom of figure) and hypomethylated in male gametes (top of figure). If daughter cells are different with respect to the expression of a de novo methylase, subsequent divisions will result in the creation of a mosaic embryo which is skewed in the direction of the initial (gamete-of-origin) methylation state of the transgene.

Similarly, the inheritance of an allele which had been previously inactivated by sequence alteration (i.e. a "classical" mutation) would result in the expression of a mutant phenotype only in that fraction of cells which retained an epigenetically inactivated corresponding allele. If individuals differ in the degree to which they are mosaic, then such mutant alleles would display a dominant mode of inheritance, but would exhibit variable expressivity or incomplete penetrance (20,21,22,23,17). An interesting prediction of this model is that hemizygous transgene loci should sometimes display a mosaic pattern of expression, when examined at the level of individual cells. Experiments designed to test this prediction are currently in progress.

ACKNOWLEDGEMENTS

We are grateful to Pat Hallauer for the troponin I transgenic mice, Alan Peterson for critical discussion, Catherine Italiano, Mira Puri, Irene Tretjakoff and Susan Gauthier for technical assistance, Terri Genio for typing the manuscript and Linda Sapienza and Robert Derval for artwork.

REFERENCES

1. Reik, W., Collick, A., Norris, M.L., Barton, S.C., and Surani, M.A.H. (1987). Genomic imprinting determines methylation of parental alleles in transgenic mice. Nature 328, 248-251.
2. Sapienza, C., Peterson, A.C., Rossant, J., and Balling, R. (1987). Degree of methylation of transgenes is dependent on gamete of origin. Nature 328, 251-254.
3. Swain, J.L, Stewart, T.A., and Leder, P. (1987). Parental legacy determines methylation and expression of an autosomal transgene: a molecular mechanism for parental imprinting. Cell 50: 719-727.
4. Hadchouel, M., Farza, H., Simon, D., Tiollais, P., and Pourcel, C. (1987). Maternal inhibition of hepatitis B surface antigen gene expression in transgenic mice correlates with de novo methylation. Nature 329, 454-456.
5. McGrath, J., and Solter, D. (1984). Completion of mouse embryogenesis requires both the

maternal and paternal genomes. Cell 37, 179-183.
6. McGrath, J., and Solter, D. (1984). Inability of mouse blastomere nuclei transferred to enucleated zygotes to support development in vitro. Science 226, 1317-1319.
7. Surani, M.A.H., Barton, S.C., and Norris, M.L. (1984). Development of reconstituted mouse eggs suggests imprinting of the genome during gametogenesis. Nature 308, 548-551.
8. Surani, M.A.H., Barton, S.C., and Norris, M.L. (1986). Nuclear transplantation in the mouse: heritable differences between parental genomes after activation of the embryonic genome. Cell 45, 127-136.
9. Solter, D. (1988). Differential imprinting and expression of maternal and paternal genomes. Ann. Rev. Genet. In press.
10. Cattanach, B.M., and Kirk, M. (1985). Differential activity of maternally and paternally derived chromosome regions in mice. Nature 315, 496-498.
11. Beechey, C.V., and Searle, A.G. (1987) Mouse News Letter 77, 126-127.
12. Monk, M. (1988). Genomic imprinting. Genes Dev. 2, 921-925.
13. Hogan, B., Costantini, F., and Lacy, E. (1986). Manipulating the Mouse Embryo: A Laboratory Manual. (Cold Spring Harbor, New York: Cold Spring Harbor Laboratory).
14. Rossant, J., and Croy, B.A. (1985). Genetic identification of tissue of origin of cellular populations within the mouse placenta. J. Embryol. exp. Morphol. 86, 177-189.
15. Jähner, D., Stuhlmann, H., Stewart, C.L., Harbers, K., Löhler, J., Simon, I., and Jaenisch, R. (1982). De novo methylation and expression of retroviral genomes during mouse embryogenesis. Nature 298, 623-628.
16. Sanford, J.P., Clark, H.J., Chapman, V.M., and Rossant, J. (1987). Differences in DNA methylation during oogenesis and spermatogenesis and their persistence during early embryogenesis in the mouse. Genes Dev. 1, 1039-1046.

17. Sapienza, C. (1989). Genomic Imprinting and Dominance Modification. Ann. N.Y. Acad. Sci., in press.
18. Koufos, A., Hansen, M.F., Copeland, N.G., Jenkins, N.A., Lampkin, B.C., and Cavenee, W.K. (1985). Loss of heterozygosity in three embryonal tumours suggests a common pathogenic mechanism. Nature 316, 330-334.
19. Hansen, M.F., and Cavenee, W.K. (1988). Retinoblastoma and the progression of tumor genetics. Trends Genet. 4, 125-128.
20. Herrmann, J. (1977). Delayed mutation model: carotid body tumors and retinoblastoma. Mulvihill, J.J., Miller, R.W. and Fraumeni, Jr., J.F., (eds.): In Genetics of Human Cancer, (New York: Raven Press), pp. 417-438.
21. Matsunaga, E.(1978). Hereditary retinoblastoma: delayed mutation or host resistance? Am. J. Hum. Genet. 30, 406-424.
22. Glanz, A.,and Fraser, F.C.(1984). Risk estimates for neonatal myotonic dystrophy. J. Med. Genet. 21, 186-188.
23. Farrer, L.A., and Conneally, P.M. (1985). A genetic model for age at onset in Huntington disease. Am. J. Hum. Genet. 37, 350-357.

ALLELE-SPECIFIC METHYLATION IN HUMAN CELLS[1]

Hamid Ghazi, Anne E. Erwin, Robin J. Leach, and Peter A. Jones

Department of Biochemistry and
Kenneth Norris Jr. Comprehensive Cancer Center
University of Southern California Medical School
Los Angeles, California 90033

ABSTRACT The methylation status of individual alleles of the c-Ha-*ras*-1 and insulin genes in human fetal tissues and sperm cells was determined. These genes, with the exception of the CpG island of the c-Ha-*ras*-1 gene, were heavily methylated in sperm. One of the alleles of c-Ha-*ras*-1 and insulin genes was sometimes more methylated than the other allele in cultured fetal fibroblasts. DNA methylation is inherently mutagenic; this allele-specific methylation might predispose the more methylated copy to a higher mutation rate.

INTRODUCTION

Approximately 3-4% of cytosine residues in the mammalian genome are methylated in the 5 position (1). This methylation may play an important role in the multilevel control of eukaryotic gene expression (2). Strong correlations between the methylation status and chromatin conformation of certain genes exist. Generally, in normal mammalian tissue, transcriptionally active or potentially active genes are hypomethylated (3). The inactivation of one of the two copies of the X

---

[1] This work was supported by OIG grant R35CA49758 from the National Cancer Inst.

chromosome in the somatic cells of mammals is thought to be reinforced by methylation of CpG islands of housekeeping genes (4,5). Recent studies using transgenic mice have implicated DNA methylation as a potential regulatory mechanism for genomic imprinting (6-8). In this article we will present evidence which shows that some autosomal genes can be differentially methylated in human cells. We will also present ideas on how differential DNA methylation might influence the rate of formation of new mutations and their possible role in the genetics of certain childhood cancers.

Methylation Status of c-Ha-*ras*-1 in Human Cells

We were interested in studying the methylation status of individual alleles of autosomal genes in different human cell types (9). We took advantage of a class of restriction fragment length polymorphisms known as variable number of tandem repeats (VNTR) polymorphisms, which are associated with some genes, to distinguish between the methylation status of the two alleles. The c-Ha-*ras*-1 gene is located on the short arm of chromosome 11 and contains a VNTR region, made up of a 28 base pair tandem repeat unit, located in the 3' end of the gene (Fig 1A). The c-Ha-*ras*-1 gene has a high G+C content, and a CpG island is located in its promoter and 5' coding regions. While there are no MspI sites (CCGG) present within the tandem repeat region, the rest of the gene has many MspI sites which are not resolved on standard Southern blots (Fig 1A). The methylation of CCGG sites flanking the VNTR regions was studied in cultured epithelial cells and fibroblasts by differential cutting with MspI and its methylation-sensitive isoschizomer, HpaII. Our previous studies had shown that in fibroblast cells cultured from human fetuses the c-Ha-*ras*-1 alleles were differentially methylated (9). In samples that were heterozygous for the c-Ha-*ras*-1, two bands corresponding to the two alleles were present in the MspI digests. When the same samples were digested with HpaII, one of the bands shifted to a higher molecular weight, indicating that one or both of the CCGG sites flanking the VNTR region were methylated. There was little or no change in the intensity of the other band, indicating that in this allele the CCGG sites flanking the VNTR region were not methylated. These

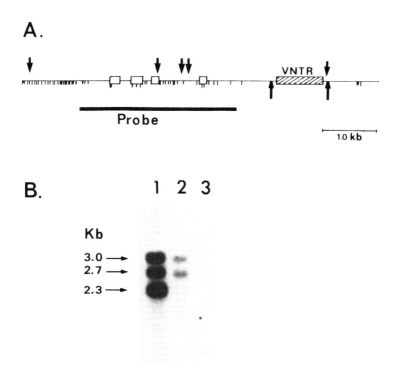

FIGURE 1. Methylation of the c-Ha-<u>ras</u>-1 gene in human sperm. (A) A schematic of the c-Ha-<u>ras</u>-1 gene is shown. The open boxes are the exons and the tick marks represent MspI sites. The two bottom arrows point to the MspI sites flanking the VNTR region. The TaqI sites are represented by the top arrows. The dark line marks the region that was used in (B) as a probe. (B) Five ug of sperm DNA were digested with TaqI (lane 1), TaqI and HpaII (lane 2), and TaqI and MspI (lane 3). Digested DNAs were electrophoresed, blotted and hybridized to the probe shown in (A).

observations clearly showed that the two alleles of a gene could be differentially methylated at the sites flanking the VNTR region.

We also examined the methylation status of the c-Ha-<u>ras</u>-1 gene in human sperm samples and found that both copies of the gene were heavily methylated in the regions flanking the VNTR (9). However, despite the heavy methylation of the VNTR flanking sites, the c-Ha-<u>ras</u>-1 associated CpG island remained unmethylated (Fig 1B). We have demonstrated this by cutting the gene into four fragments with TaqI and determining the methylation of the CCGG sites within the two larger fragments, one containing the CpG island and the first three exons, the other corresponding to the last exon and the VNTR region (Fig 1A). In the TaqI digest of DNA extracted from an individual who was heterozygous for the VNTR region, three bands were visible: a 2.3 Kb band which contains the CpG island, a 2.7 Kb band and a 3.0 Kb band corresponding to the VNTR containing regions of the two c-Ha-<u>ras</u>-1 alleles (Fig 1B, lane 1). MspI cut all three bands into very small fragments due to the presence of multiple recognition sites (Fig 1B, lane 3). However, HpaII cut only the 2.3 Kb fragment (Fig 1B, lane 2), which indicated that, despite the heavy methylation of the 3' region of c-Ha-<u>ras</u>-1 gene in sperm, the CpG island remained unmethylated. This finding is consistent with Bird's hypothesis that the CpG island of genes located on autosomes are always unmethylated in the germ line (10).

Since methylation patterns are known to be variable in culture (11,12), we examined cultured fetal urothelial cells and uncultured muscle tissue to see if the allele-specific methylation patterns were derived as a consequence of culturing. No significant differences in the methylation patterns of the c-Ha-<u>ras</u>-1 gene in the cultured and uncultured tissues were observed (Fig 2). The HpaII digest of the two samples gave rise to similar band patterns with the higher molecular weight allele being less methylated than the lower.

Methylation Status of Insulin Gene in Human Cells

The insulin gene, located on the short arm of chromosome 11, also has a high G+C content, but lacks a CpG island (13). The VNTR region, flanked by PvuII

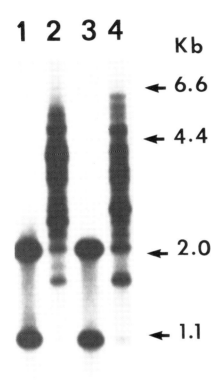

FIGURE 2. Methylation of the c-Ha-ras-1 gene in cultured fetal epithelial cells and uncultured fetal muscle tissue. Five ug of DNA from cultured fetal epithelial cells (lanes 1,2) and fetal muscle tissues (lanes 3,4) were digested with MspI (lanes 1,3) and HpaII (lanes 2,4), electrophoresed, blotted and probed with the VNTR region of the ras gene.

sites, is located in the 5' region of the gene and is made up of non-identical repeat units (Fig 3A). The methylation status of the individual insulin alleles was examined by performing a double digest with PvuII, which leaves the VNTR region intact, and the methylation sensitive HpaII or AvaI, which cut within the VNTR region (Fig 3A). In the PvuII digest of DNA from a fetal fibroblast sample heterozygous for insulin gene, two bands corresponding to the two VNTR regions were present (Fig 3B, lane 4). Both HpaII and AvaII cut the top band into smaller fragments, indicating that sequences in their recognition sites were not methylated in the VNTR region (lanes 5, 6). However, the same enzymes did not cut the bottom VNTR band, suggesting that the recognition sites within this region were methylated. We have made similar observations with other fetal fibroblast samples that were heterozygous for the insulin gene, and found the presence of allele-specific methylation within the VNTR region (results not shown).

In human sperm samples, the CCGG sites within and around the VNTR region as well as in the body of both copies of the insulin gene were heavily methylated. In the PvuII digest of sperm DNA, two bands corresponding to the VNTR region for the two insulin alleles were present (Fig 3B, lane 1). HpaII and AvaI did not cut either of the VNTR regions, indicating that their recognition sites were methylated within this region (Fig 3B, lanes 2, 3). MspI cuts the insulin gene into very small fragments (Fig 3C, Lane M). However, in the HpaII digest, only a single high-molecular-weight band around 20 Kb is present, suggesting heavy methylation throughout the gene (Fig 3C, lane H).

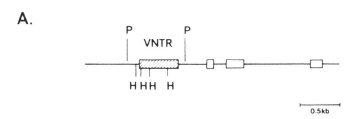

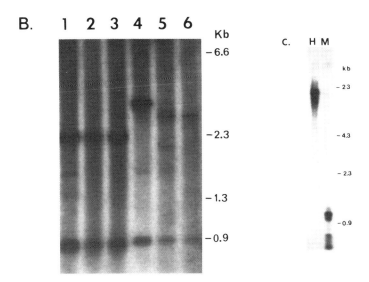

FIGURE 3. Methylation of the insulin gene in human fetal and sperm cells. (A) A schematic of the gene is shown. PvuII sites are represented by P and MspI/HpaII sites in the VNTR region are represented by H. Open boxes represent the exons. (B) Five ug of DNA from the sperm cells (lanes 1-3) and fetal fibroblast cells (lanes 4-6) were digested with PvuII (lanes 1,4), PvuII and AvaI (lanes 2,5), and PvuII and HpaII (lanes 3,6). Digested DNAs were electrophoresed, blotted, and probed with the insulin gene. (C) Same as part B. MspI is represented by M and HpaII is represented by H.

Methylation Status of Parental Alleles

The allele-specific methylation patterns of c-Ha-ras-1 and insulin genes in fetal cells show that the two copies can be differentially methylated. These genes, with the exception of the CpG island in c-Ha-ras-1, are heavily methylated in sperm. This suggests that perhaps the more methylated allele in the fetal cells came from the sperm, implying that the methylation levels of the oocyte genome are lower than in the sperm genome. The methylation levels of human oocytes are not known, but they have been studied in mice.

Studies of mouse gametes have shown that in general the mouse genome is more methylated in sperm relative to the oocyte. Monk's group examined the patterns generated by end- labelling HpaII-restricted fragments of DNA from sperm, oocyte, and cells from different stages of embyrogenesis. The egg genome was noticeably undermethylated relative to the sperm genome (14). Rossant and her colleagues examined several repetitive and low copy number sequences and also found that these sequences are less methylated at the CCGG sites in the oocyte relative to the sperm. An intermediate pattern of methylation was observed in embryonic tissues (15). These studies show that the mouse genome is more methylated in sperm relative to oocytes and that the patterns of methylation of the gametes persist in the early stages of embryogenesis.

Recent studies with transgenic mice have also demonstrated differential methylation patterns of parental genomes (6-8). The methylation status of a transgene located on an autosome was determined as it was passed through the maternal or paternal germ lines. In general, it was found that these loci were less methylated in the somatic tissues of the offspring when inherited from the father. Through successive passage of the transgenes in alternative crosses between male and female transgenic mice, the pattern of methylation of the transgene was changed. It was shown that the pattern of methylation of the maternally derived transgenes in the testis was different from that in the somatic tissues so that the patterns of methylation must be removed and reestablished during gametogenesis. These studies clearly show that the maternal and the paternal chromosomes are differentially methylated; it is possible that the differential methylation is a regulatory mechanism for genomic imprinting.

Genomic imprinting is a process by which the two parental genomes are marked and utilized differentially. Nuclear transplantation and genetic studies in mice suggest that the maternal and paternal genomes have different roles in development (16,17). Specific regions of functionally equivalent chromosomes were shown to be utilized differentially based on the parental origin (18). Embryos with disomy for maternal or paternal chromosomes 1,4,5,9,13,14 and 15 developed normally. However, paternal or maternal duplication in combination with maternal or paternal deficiency of chromosome 6 or certain regions of chromosomes 2,8, and 17 leads to abnormal development and is often lethal. It is thought that the imprinted maternal and paternal alleles functionally complement each other in a developing embryo (19). The reciprocal nature of imprinting suggests that certain regions of the sperm genome may be more methylated relative to the oocyte genome and vice versa.

Role of DNA Methylation in Cancer Genetics

Studies of certain childhood cancers have shown that specific chromosomes in tumor cells undergo a reduction to homozygosity when compared to the normal cells of the patient (20). Recent studies of sporadic tumors show that there is a strong preference for the retention of the paternal alleles within some tumors. In Wilms' tumor, 14 out of 15 studied cases preferentially retained the paternal copy of chromosome 11 (21-24). Similar studies of osteosarcoma, where it is thought that mutation in the retinoblastoma locus on chromosome 13 is involved in tumorigenesis, have shown that in 12 out of 13 cases the paternal chromosome 13 is retained in the tumor cells (25). These studies suggest a preference for mutation of the paternal alleles in these chromosomal regions.

All the studied cases where a preferential reduction to homozygosity of the paternal chromosomes have been observed are sporadic. Based on Knudson's two hit-model for the development of cancer, the first event is a mutation in one of the copies of the gene, often followed by a loss of the remaining wild-type copy (26). In this model, both copies of the gene should have an equal chance of being mutated in the sporadic, non-hereditary, form of the disease. The preferential mutations in the paternal chromosome in the sporadic cases implies that the paternal and maternal chromosomes, although homologous, are not identical. One possible explanation is that imprinting predisposes certain regions of the paternal chromosome to be more susceptible to mutation than the maternal chromosomes.

Earlier, we showed that in human fetal cells the two alleles are differentially methylated and suggested that the paternal allele is the more methylated copy. The differential methylation of the parental alleles might influence their individual mutation rates (27). CpG dinucleotides are "hot spots" for mutation in the mammalian genome (28) and as a result are underrepresented in the genome (10). Cytosine and 5-methylcytosine undergo frequent deamination in the cell and give rise to uracil and thymine. The DNA repair mechanism of the cell can readily recognize the U-G mismatch, excise and replace the uridine with a cytosine to generate the original base pair. However, in the case of 5-methylcytosine, the deamination product is thymine, which is a normal base in the DNA, and the repair enzymes must now distinguish between replacement of the T or the G in the mismatch. It has been shown that there is a 90%

preference to excise and replace the T, to get back the original C-G base pair (29). However, in some cases the C is replaced by an A, resulting in the development of a point mutation.

Some of these mutations are silent, but others could affect the function of the gene products. A good example is the blood clotting factor VIII, which has been shown to be nonfunctional in hemophilia A. In 25% of mutations examined by direct sequencing, it was shown that point mutations in the CpG sequences located in the body of the gene were involved (30). Thus, if the two alleles of a gene are differentially methylated, it would be predicted that they would have different mutation rates. In our laboratory, we are currently investigating the methylation levels and the nature of the mutations in the retained paternal retinoblastoma locus in sporadic cases of retinoblastoma.

In this paper we have presented evidence showing that the two alleles in a human cell can be differentially methylated and that heavy methylation of non-CpG island regions in genes can occur during spermatogenesis. Evidence from certain childhood cancers suggests preferential mutation of the paternal allele, which might be influenced by differences in methylation levels between the maternal and paternal alleles.

ACKNOWLEDGMENTS

We thank C.M. Marziasz and J. Chen for critical reading of the manuscript, G. Bell for providing insulin probes and Cheryl Ouellette for secretarial assistance.

REFERENCES

1. Riggs AD, Jones PA (1983).
   5-Methylcytosine, gene regulation and cancer. Adv Cancer Res 40:1-30.
2. Doerfler W. (1983). DNA methylation and gene

activity. Ann Rev Biochem 52:93-124.
3. Michalowsky LA, Jones PA (1989). Gene structure and transcription in mouse cells with extensively demethylated DNA. Mol Cell Biol 9:885-892.
4. Keith DH, Singer-Sam J, Riggs AD (1986). Active X chromosome DNA is unmethylated at eight CCGG sites clustered in a guanine-plus-cytosine-rich island at the 5' end of the genes for phosphoglycerate kinase. Mol and Cell Biol 6:4122-4125.
5. Hansen RS, Ellis NA, Gartler SM (1988). Demethylation of specific sites in the 5' region of the inactive X-linked human phosphoglycerate kinase gene correlates with the appearance of nuclease sensitivity and gene expression. Mol Cell Biol. 8:4692-4699.
6. Sapienza C, Peterson AC, Rossant J, Balling R (1987). Degree of methylation of transgenes is dependent on gamete of origin. Nature 328:251-254.
7. Reik W, Collick A, Norris ML, Barton SC, Surani MAH (1987). Genomic imprinting determines methylation of parental alleles in transgenic mice. Nature 328:248-251.
8. Swain JL, Stewart TA, Leder P (1987). Parental legacy determines methylation and expression of an autosomal transgene: A molecular mechanism for parental imprinting. Cell 50:719-727.
9. Chandler LA, Ghazi H, Jones PA, Boukamp P, Fusenig NE (1987). Allele-specific methylation of the human c-Ha-ras-1 gene. Cell 50:711-717.
10. Bird AP (1986). CpG-rich islands and the function of DNA methylation. Nature 321:209.
11. Wilson VL, Jones PA (1983). DNA methylation decreases in aging but not in immortal cells. Science 220:1055-1057.
12. Shmooler-Reis RJ, Goldstein S (1982). Interclonal variation in methylation patterns for expressed and non-expressed genes. NAR 10:4293-4304.
13. Gardiner-Garden M, Frommer M (1987). CpG island in vertebrate genomes. J Mol Biol 196:261-282.
14. Monk M, Boubelik M, Lehnert S (1987). Temporal and regional changes in DNA methylation in the embryonic, extraembryonic and germ cell lineages during mouse embryo development. Development 99:371-382.
15. Sanford JP, Clark HJ, Chapman VM, Rossant J (1987). Differences in DNA methylation during oogenesis and spermatogenesis and their persistence

during early embryogenesis in the mouse. Genes and Development 1:1039-1046.
16. Surani MAH, Barton SC, Norris ML (1984). Development of reconstituted mouse eggs suggests imprinting of the genome during gametogenesis. Nature 308:548-550.
17. McGrath J, Solter D (1984). Completion of mouse embryogenesis requires both the maternal and paternal genomes. Cell 37:179-183.
18. Cattanach BM, Kirk M (1985). Differential activity of maternally and paternally derived chromosome regions in mice. Nature 315:496-498.
19. Surani MAH, Barton SC, Norris ML (1987). Experimental reconstruction of mouse eggs and embryos: an analysis of mammalian development. Biol. of Reproduction 36:1-16.
20. Hansen MF, Cavenee WK (1987). Genetics of cancer redisposition. Cancer Res 47:5518-5517.
21. Reeve AE, Housiax PJ, Gardner RJM, Chewings WE, Grindley RM, Millow LJ (1984). Loss of a Harvey ras allele in sporadic Wilms' tumor. Nature 309:174-176.
22. Schroeder WT, Chao LY, Dao DD, Strong LC, Pathnak S, Riccardi V, Lewis WH, Saunders GF (1987). Nonrandom loss of maternal chromosome 11 alleles in Wilms tumors. Am J Genet 40:413-420.
23. Mannens M, Slater RM, Heyting C, Bliek J, de Kraker J, Coad N, de Pagter-Holthuizen R,, Pearson RL (1988). Molecular nature of genetic changes resulting in loss of heterozygosity of chromosome 11 in Wilms' tumor. Hum Genet 81:41-48.
24. Williams JC, Brown KW, Mott MG, Maitland NJ (1989). Maternal allele loss in Wilms' tumour. The Lancet I:283-284.
25. Toguchida J, Ishizaki K, Sasaki MS, Nakamura Y, Ikenaga M, Kato M, Sugimot M, Kotoura Y, Yamamuro T (1989). Preferential mutation of paternally derived RB gene as the initial event in sporadic osteosarcoma. Nature 338:156-158.
26. Knudson AG (1971) Mutation and Cancer: statistical study of retinoblastoma. PNAS 68:820-823.
27. Jones PA, Buckley JD (1989). The role of DNA methylation in cancer. In Adv Cancer Res (Klein, G & Vande Woude G eds) Academic Press, N.Y.,V. 54.
28. Barker D, Schafer M, White R (1984). Restriction sites containing CpG show a higher frequency of

polymorphism in human DNA. Cell 36:131-138.
29. Brown TC, Jiricny J (1987). A specific mismatch repair event protects mammalian cells from loss of 5-methyecytosine. Cell 50:945.
30. Antonarakis SE, Kazazian HH (1988). The molecular basis of hemophilia A in man. Trends in Genet 4:233.

THE GENETICS OF MOSAIC METHYLATION PATTERNS IN HUMANS

Alcino J. Silva, Kenneth Ward,[*] Raymond White

Department of Human Genetics and Howard Hughes Medical Institute,
*Department of Obstetrics and Gynecology, University of Utah, Salt Lake City, Utah 84132

ABSTRACT We have developed a genetic approach to the study of human methylation; we found variation among the methylation of alleles, and we demonstrated that this variation is inherited in pedigrees. The strategy developed for these studies can also be used to study the relationship between imprinting and methylation in native mammalian loci. Furthermore, several human tissues were shown to be mosaic for methylation, including leiomyomas which are clonal tumors of myometrium. Studies of leiomyomas showed that single cells can generate independent lineages with determined methylation patterns. We have also established that the methylation of neighboring sites can be an independent and probabilistic event.

INTRODUCTION

Mammalian methylation is involved in the regulation of gene expression (for a review see 1); in parental imprinting (for a review see 2); and in X-inactivation (for a review see 3). However, it is not known if methylation is a primary signal for phenomena such as gene expression (4), or if it simply maintains patterns established by other mechanisms (5, 6). Perhaps, understanding of the mechanisms that control methylation may shed new light on its role. In this manuscript we report that despite their diversity, human methylation patterns are not random, and that a probabilistic mechanism can account for the specificity and reproducebility of these patterns.

A GENETIC APPROACH TO THE STUDY OF METHYLATION.

We developed a novel genetic approach (7) that allows us to discriminate the methylation status of identical sites on homologous chromosomes (henceforth referred to as allelic sites), and to follow the inheritance of methylation patterns in human pedigrees. Our approach depends on the use of methylation-sensitive enzymes (8). HpaII was the methylation-sensitive enzyme most often used because it has an isoschizomer' MspI, unaffected by methylation at the inner cytosine of the recognition sequence·

The contribution of each allele to the methylation pattern observed with methylation-sensitive enzymes was defined with the help of polymorphisms. The type of polymorphism that we used in these studies reflected variable numbers of small tandem repeats (VNTRs) (9, 10). Heterozygous individuals have a different number of repeats in each of the two alleles of a VNTR locus, resulting in different electrophoretic mobility. The methylation sites studied in these VNTR loci flanked the region with the tandem repeats. This allowed us to distinguish between the methylation patterns that were maternally-derived and those that were paternally-derived. Chandler et al. (11) also developed a similar method that takes advantage of polymorphism to separate allelic methylation patterns.

The resolution of the methylation patterns was improved by a second digestion with a methylation-insensitive enzyme(s) whose restriction site(s) flank closely both the methylation site(s) analyzed and the VNTR.

An allele's methylation pattern is defined by the number of bands present, and by the relative intensity of those bands measured by a densitometer. The heterozygous individuals in Figure 1 include five bands: four polymorphic bands with the repeats, and a band of constant length that does not include the repeats. The number of bands from each allele is related to the number of methylation-sensitive sites analyzed. Polymorphism allows the unequivocal separation and identification of the bands in the methylation pattern that belong to each of the alleles (figure 1).

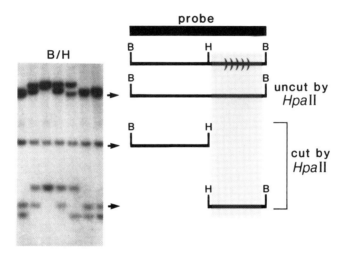

Tissue: Blood
Locus: Immunoglobulin heavy chain

FIGURE 1. The methylation of the 3.4B2 locus. Genomic DNA isolated from the peripheral blood of seven individuals was digested with BamHI (B) and HpaII (H), electrophoresed in a 1% agarose gel, and hybridized with radiolabeled p3.4B2 (Silva et al. 1987). The restriction map of the HpaII and BamHI sites digested, the fragments expected from the digestion, and the position of the probe are shown.

HUMAN TISSUES ARE MOSAIC FOR METHYLATION

Mammalian tissues differ in their levels of methylation. However, with the exception of sperm, the pattern of methylation is not uniform throughout all cells within the differentiated tissues we studied (blood, colon, serosal membrane, ovary, myometrium, mucosal membrane, various carcinomas, metastic growths, and benign tumors).

Figure 2 is an example of mosaic methylation of the JCZ67 locus in blood. Analysis of the bands in the MspI and BamHI lanes of the autoradiograph in figure 2 indicate that this individual is heterozygous for this VNTR system.

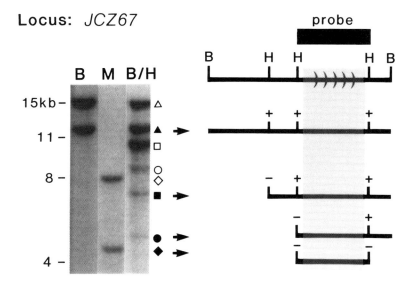

FIGURE 2. Allelic variation between the methylation of a site in the JCZ67 locus. Genomic DNA isolated from peripheral blood of an individual was digested with MspI (M), BamHI (B), and with BamHI/HpaII (B/H). The samples were electrophoresed in a 0.6% agarose gel and hybridized with pJCZ67.

The methylation pattern obtained for this locus after digestion with BamHI and with HpaII includes three allelic pairs of bands: two marked with triangles identical in size to the two bands seen in the BamHI lane, indicating that some of the cells in peripheral blood of this individual have all three sites methylated (diagram in figure 2); two marked with squares, and two others marked with diamonds. Each one of these allelic pairs of bands represents a different arrangement of methylated and unmethylated sites within the JCZ67 locus; the presence of multiple arrangements for each allele demonstrates that cells within blood are mosaic for methylation. Since the cell population in blood is heterogeneous, we asked if each cell type was associated with an homogeneous methylation pattern. Analysis of the methylation patterns of polymorphonuclear neutrocytes for JCZ67, showed that even though the methylation pattern is different from the one obtained for whole blood, it is nevertheless mosaic. Similar results were obtained for the two other main cell populations in blood: lymphocytes, and monocytes.

THE MOSAIC PATTERNS ARE ALLELE-SPECIFIC.

The approach developed permitted us to study the methylation patterns of each allele in diploid loci, and we found that methylation of homologous chromosomes can differ. Figure 2 shows an example of variation of methylation between alleles of the JCZ67 locus. The pair of bands marked with squares in figure 2 have a striking difference in intensity: the smaller band is less intense than the larger one. This difference in intensity is not due to the lesser number of repeats in the small allelic band, because the other two allelic pairs of bands (figure 2), which also contain the VNTR, have similar intensities. The same is true for the bands in the BamHI and in the MspI control lanes. Densitometry of the bands in the methylation pattern of the small allele confirmed the difference in intensity between the allelic bands marked with squares. Similar results were obtained by Chandler et al. (11) for the methylation of the H-ras gene in cells in culture.

THE TRANSMISSION OF METHYLATION PATTERNS IN HUMAN PEDIGREES.

The individual analyzed in figure 2 is the father in Utah kindred #1456. Thus, we followed the methylation of his two alleles of the JCZ67 locus in this kindred, and figure 3 shows an example of this analysis done for the larger allele. Densitometric analysis of the methylation patterns at this locus in pedigrees showed that the methylation of the two allelic variants was stably inherited. The relative intensities of the bands in the larger allele in the three generations studied are 34%+7%, 55%+2%, and 11%+6% respectively for the bands marked with triangles, squares and circles ;the numbers are presented with one standard deviation. Similar measurements for the smaller allele yielded the following relative intensities: 71%+8%, 21%+6%, and 8%+2% for the larger, the middle and the smaller bands respectively. A T distribution analysis of these results demonstrated that the intensities of the large and middle bands in the methylation patterns of the two alleles yielded a significant difference at the 98% confidence level. We have repeated this type of study in other pedigrees and for other loci with always the same result: the methylation (partially methylated, fully methylated or completely unmethylated) of the sites studied remained unchanged in the generations of the pedigrees studied (table 1).

Methylation must be determined in cis because we demonstrated that sites in homologous chromosomes can be differentially methylated, and that the methylation patterns are not random since they are inherited in pedigrees. These cis-acting elements could be, for example, DNA sequences or DNA structures (1), or even proteins that are associated with the site. Weissbach et al. (12) and Bolden et al. (13) demonstrated that GC-rich oligonucleotides are preferentially methylated de novo by HeLa DNA methylases, and Bird (5) has suggested that a permanent association with transcriptional factors obstructs the activity of methylases and is responsible for the hypomethylation of HTF islands. Moreover, Wang et al. (14) identified a sequence- and methylation-specific human protein.

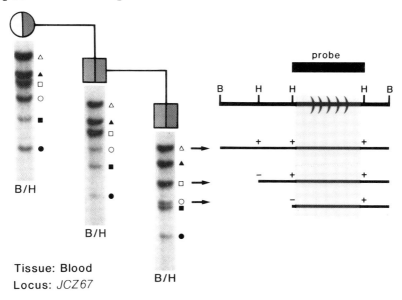

FIGURE 3. The mendelian inheritance of methylation in the JCZ67 locus. DNA samples isolated from peripheral blood of three members of kindred 1456 (paternal grand-father, father, and a sibling all sharing one allele) were digested with BamHI/HpaII (see diagram of kindred above autoradiograph); the samples were electrophoresed in a 0.6% gel, transferred to a nylon membrane, and probed with pJCZ67. Each class of fragments in the diagram has a symbol that identifies it in the autoradiograph. The allele present in the three generations is marked with dark symbols.

It is likely that most pairs of homologous chromosomes will have polymorphic methylation patterns because we found allelic variations in methylation patterns in three out of 10 loci studied. The inheritance of allelic variations in methylation shown here demonstrates that each chromosome carries a unique program or blue-print for methylation. However, the frequency of differences in methylation between homologues suggests that most are not likely to affect gene expression. Perhaps these variant methylation sites have another role in the human genome; for example, they could serve as reference points for phenomena that involve interactions between homologues.

TABLE 1
SUMMARY OF METHYLATION DATA FOR LOCI IN PERIPHERAL BLOOD

Locus	Chromosome localization	#Pedigrees tested	#variant sites and frequencies*1
YNZ22	17p	6	1, 21/22
YNH37	17p	3	0, 0/16
YNH24	2q	3	1, 3/21
CMM65	16p	2	0, 0/9
JCZ67	7q	3	1, 3/9
RMU3	17q	2	0, 0/5
MCOC12	14	2	0, 0/5
c-Ha-ras1	11p	3	0, 0/12
3.4B2	14q	3	0, 0/15
CMM101	14q	4	0, 0/8
YNZ32	4	2	0, 0/6

*1-ratios of unrelated heterozygous individuals with variant methylation patterns versus total number of heterozygous individuals tested.

We compared the methylation pattern of sperm and blood and we found that sites methylated in sperm can become unmethylated in blood, and that sites unmethylated in sperm may be methylated in blood. This demonstrates that the entity inherited in pedigrees is not the methylation status of the locus but the elements that determine that methylation.

Comparisons among the methylation patterns of several tissues from the same individual showed that allelic sites equally methylated in one tissue, could be methylated differently in other tissues (colon versus blood for example). Since we have demonstrated that the methylation of peripheral blood cells is inherited in pedigrees, it is unlikely that the differences observed between tissues are random. This indicates that the cis-elements interact with cell-specific factors in the regulation of methylation. These trans-elements could be proteins that modify the activity of the methylases. For example, methinin is an inhibitor for DNA methylation (15).

METHYLATION OF CELL LINEAGES GENERATED IN VIVO.

Leiomyomas are an appropriate model system for studies of the control of methylation-patterns in cell lineages, because they are clonal. Leiomyomas enabled us to compare methylation-patterns in phenotypically and genotypically similar cell lineages that had been generated in the uterus.

Despite their clonal origin and their homogeneous histology (figure 4), we found that leiomyomas are mosaic for the methylation of seven loci studied, and that the generation of these mosaic methylation-patterns were not random. We compared the mosaic patterns of several independent leiomyomas from the same individual, and we often found similar patterns. These results suggest that single cells can form independent lineages with similar methylation patterns. Interestingly, Rubin et al. (16) have shown mosaic patterns of gene expression in histologically homogeneous rat fetal epithelial cells.

The cells in the periphery of leiomyomas have been shown to make the greatest contribution to the growth of these tumors, while the cells in the core of the tumor reveal very little growth. Since we found that the methylation of the core and periphery of leiomyomas is similar, and since there is little cell mixing in the formation of these solid tumors, than this shows that in the formation of these tumors there is not a clear separation between an initial de novo methylation stage and a maintenance methylation stage that replicates the patterns. In the absence of cell mixing, maintenance methylation would eventually result in regions in the periphery of the tumors less heterogeneous than regions within the core of the tumors.

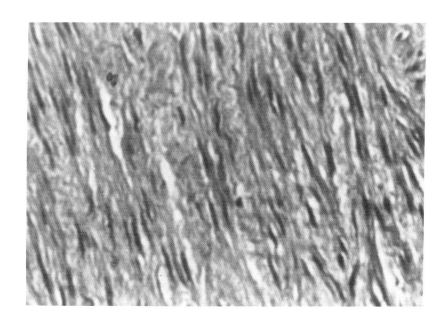

FIGURE 4. Leiomyomas can be histologically homogeneous. A light microscopy view (400X) of a section of a small leiomyoma (approximately 0.5-cm radius). The sections were obtained immediately after hysterectomy. Tumor tissue was embedded in paraffin, sectioned, and stained with hematoxylin and eosin.

METHYLATION CAN BE PROBABILISTIC.

The mechanism that determines methylation is unknown. We propose that methylation can be a probabilistic event, and that the methylation of one site can be independent of the methylation of neighboring sites. This hypothesis can be tested by determining whether the probability of finding two sites methylated within the same allele equals the products of the probabilities of finding each of the sites methylated. Using the method described above we are able to measure the probabilities of methylating single sites for polymorphic alleles (figure 5).

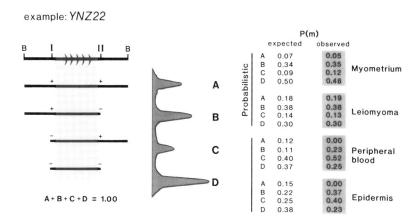

FIGURE 5. Probabilistic methylation at the YNZ22 Locus. Densitometry of the bands in the methylation patterns of the YNZ22 locus yielded the relative abundance of all the bands, which were converted into the probabilities of finding those bands in the populations studied: myometrium, leiomyomas, blood, and epidermis. The left column in the table shows the probabilities expected if methylation were probabilistic for the four possible methylation choices at sites I, and II: methylation at both sites (A), methylation at site I but not II (B), methylation at site II but not I (C), and no methylation at both sites (D). The right column shows the probabilities of the four combinations A, B, C, ,and D measured with densitometry in myometrium, leiomyomas, peripheral blood, and epidermis. The results obtained for

myometrium and leiomyomas are different from the results obtained for blood and epidermis; the methylation of the two sites studied is not independent for blood and epidermis.

We tested two loci so far, and we found in both evidence for our model (figure 5). Interestingly, nucleosome phasing is though to be associated with DNA methylation (17), and Fedor et al. (18) have suggested that the positioning of nucleosomes can be a non-random probabilistic event restricted by the presence of other proteins contacting the DNA. This is consistent with a model that claims that the probabilistic phasing of nucleosomes might be responsible for the probabilistic nature of methylation.

CONCLUSION

The inheritance of allele-specific methylation patterns demonstrates the existence of a specific mechanism that establishes the relative abundance of particular cells with specific methylation patterns in tissues. This process is determined because these site-, tissue- and allele-specific patterns are reproducible in different generations. We called this process blueprinting to distinguish it from another allele-specific process referred to as imprinting. Contrary to imprinting, blueprints for methylation are preserved between generations.

The probabilistic model suggests that the complex nature of methylation could be controlled by groups of probabilities that are specific to cell lineages, or cell types during development. The inheritance of methylation patterns in pedigrees could be due to the inheritance of the elements that condition the probabilities.

REFERENCES

1. Cedar, H (1988). DNA methylation and gene activity. Cell 53: 3-4.
2. Solter, D (1988). Differential imprinting and expression of maternal and paternal

genomes. Annu Rev Genet 22: 127-46
3. Lyon, M F (1988). X-chromosome inactivation and the location and expression of X-linked genes. Am J Hum Genet 42: 008-016
4. Razin, A, Riggs, A D (1980). DNA methylation and gene function. Science 210: 604-610
5. Bird, A (1986). CpG islands and the function of DNA methylation. Nature 321, 209-213
6. Lock, L F, Takagi, N, Martin, G R (1987). Methylation of the HPRT gene on the inactive X occurs after chromosome inactivation. Cell, 48: 39-46
7. Silva, A, White, R (1988). Inheritance of allelic blueprints for methylation patterns. Cell 54: 145-52
8. Bird, A, Southern, E M (1978). Use of restriction enzymes to study eukaryotic DNA methylation: I. The methylation pattern in ribosomal DNA from Xenopus laevis. J Mol Biol 118: 27-47
9. Bell, G I, Selby, M J, Rutter, W (1982). The highly polymorphic region near the human insulin gene is composed of simple tandemly repeating sequences. Nature 295: 31-35
10. Nakamura, Y, Leppert, M, O'Connell, P, Wolff, R, Holm, T, Culver, M, Martin, C, Fujimoto, E, Hoff, M, Kumlin, E, and White, R (1987). Variable number of tandem repeat (VNTR) markers for human gene mapping. Science 235: 1616-1622
11. Chandler, L A, Ghazi, H, Jones, PB, Fusening, N E (1987). Allele-specific methylation of the human c-Ha-ras-1 gene. Cell 50: 711-717.
12. Weisbach, A, Nalin, C M, Ward, C A, Bolden, A H (1985). The effect of flanking sequences on de novo methylation of C-G pairs by the human DNA methylase. Biochemistry and Biology of DNA Methylation, Cantoni, G L, Razin, A (New York: Alan R Liss, Inc) pp 79-94
13. Bolden, A H, Nalin, C M, Ward, C A,

Poonian, M S, Weissbach, A (1986). Primary DNA sequence determines sites of maintenance and de novo methylation by mammalian DNA methyltransferases. Mol Cell Biol 6 (4): 1135-1140

14. Wang, R Y, Zhang, X Y, Ehrlich, M (1986). A human DNA-binding protein is methylation-specific and sequence-specific. Nucleic Acids Res 14(4): 1599-1614

15. Lyon, S B, Buonocore, L, Miller, M (1987). Naturally occurring methylation inhibitor: DNA hypomethylation and hemoglobin synthesis in human K562 cells. Mol Cell Biol 7(5): 1759-1763

16. Rubin, D C, Ong, D E, Gordon, J I (1989). Cellular differentiation in the emerging fetal rat small intestinal epithelium: mosaic patterns of gene expression. Proc Natl Acad Sci USA 86: 1278-1282

17. Pages, M, Roizes, G (1988). Structural organization of satellite I chromatin of calf liver. Eur J Biochem 174: 391-398

18. Fedor, M J, Lue, N F, Kornberg, R D (1988). Statistical positioning of nucleosomes by specific protein-binding to an upstream activating sequence in yeast. J Mol Biol 204: 109-127.

Index

Ad12–transformed cell line, 334
A2058, 23S rRNA, 34
Adeno
   E2A promoter, 275
   gene, 281
Adenoma, premalignant, 212
Adenosine
   analog, 104
   dimethylation, ribosomal RNA, 34
   dimethyltransferase, 20
   methylation, 118
Adenovirus, 154, 164
   message, 84
   model, 283
   promoter
      –CAT gene (pAd2E2AL–CAT), 342
      integrated, 341
   system, 155, 165, 329, 332
Ado–Met hydrolase, 115
AGACU sequence, 88
AIDS
   disease, 153
   –related complex (ARC), 154
Aldosterone, sodium transport, 247
Alkylation, EMS, 233
Allele, 397
   autosomal gene, 382
   human fetal cell, 390
   inheritance, 376
   insulin, 386
   methylation, 395
      pattern, 396
      status, 381
   mutant, 377
   pattern, 388, 406
   –specific methylation, 381
Allelic
   site, 403
   variation, 398
Alpha subunit sequence, 54
Amino acid
   ksgA methylase, 24
   sequence comparison, 25
Analog AdoHcy, 253
Ancestral
   cryopreserved cell, 217
   ksgA methylase, 19
Antibiotic groups, 126
Antigenic activation, 154
APRT gene, 271
Assembly role 50S, 2
A/T pairs 117
5–aza–C, 178

409

treated, demethylated Mov13 fibroblast, 175
treatment, 155, 187
  Mov 13 fibroblasts, 179
5-azaCdR
  treated cell growth, 219
  treatment, 214
5-azacytidine
  (azaC^r), 140
  mutant, 142
  virus, 141
  (5-azaCR), 155, 211, 212, 230, 258, 286
  DNA methylation inhibitor, 229
  and hypomethylation, 205
  induced activation, T-DNA gene, 187
  inhibition, 199
  in plant tissue, 201
  potent inhibition, 188
  and transfection, 172
  treatment, 157, 232
Azacytidine
  incorporation, 205
  treatment, 205
5-aza-2'-deoxycytidine (5-azaCdR), 211, 214

ß-galactosidase gene, 51
Bifilarly methylated DNA site, 359
Binding site Sp1, 277

CA methylation, 242
Cancer
  colorectal, 225
  gentics, DNA methylation role, 389
  Knudson model, 390
5' cap
  eukaryotic mRNA, 68
  related phenomena, 79

mRNA, 253
Cap
  analog
    commercial, 63
    inhibition, 71
  formation sequence reaction, 46
  structure, 57
    mRNA stability, 68
  5'-capped RNA transcripts, 69
Capping
  enzyme
    subunit, 48
    yeast, 50
  guanine nucleotide, 68
Carcinogenesis, 211, 212
  chemical, 214
  human, 213
  colorectal, 224
Carcinoma cell P19, embryonal, 173
CAT
  coding region, 145
  gene, 147, 336
  protein synthesis induction, 160
CCGG site, 388
  and VNTR region, 382
CDNA
  clone, SV 26S mRNA (pTS22), 75
  library, 211, 214, 221
  probe, 222
Cell
  capture, 223
  line
    lineage, 231
    mouse C3H 10T1/2, 214
    treated with 5-azaCdR, 217
CEM T-cell line, 160
5'-CG-3' residue, 340
C-Ha-ras-1 gene, 384
Chloramphenicol
  acetylatransferase
    assay (*CAT*), 156
  gene, 239, 340
  to acetylchloramphenicol, 160
Chromatin structure of COL1A1 gene, 177

Chromatography, 6
Chromosomal transposon Tn554, 125
   ermA in, 127
Chromosome
   11 c–Ha–ras–1 gene location, 382
   homologour, 400
   imprinting, 390
   tumor suppressor gene on, 213
Cis
   acting regulatory elements, 181
   level, 160
   model, 240
   suppression, 180
Clonal tumor myometrium, 395
Clone encoded polypeptide, 48
Collagen I gene (COL1A1)
   human, 181
   –producing cell, 178
   promoter, 175, 176
      activity, 177
      Mov13, 175
Colonic mucosa, human, 212
Colorectal cancer, human, 213
Concanavalin–A (Con–A), 103, 104
   presence, 111
Consensus sequence, 88
Core regions, ribosome function, 39
CpA abundance in dinucleotide DNA, 242
CpG
   demethylated, 314
   diad–symmetrical sequence, 40
   dinucleotide, 188, 354
      hot spot, 390
   5'island, 266
   island, 257, 258
      autosomal, 286
      c–Ha–ras–1, 381
      methylated, 154
      mouse, 286
      –rich, 154, 285, 286, 289
      sequence, 268
      unmethylated, 384
   and non–CpG sites, 158

present, avian vitellogenin
   gene, 314
residue, 270
–rich, region, 241
site, 277, 351
   EMS modified, 230, 232
   ethylated, 239
   in prolactin gene, 229
   upstream, 302, 303
CpNpG trinucleotide, 188
Crosslinking, 2
Crown gall disease, 189
Cryopreservation, 217, 219
CX2 cell derivatives, isolated, 192
Cytomegalovirus (CMV)
   enhancer, 281
   site, human, 361
Cytoplasmic
   demethylation, 101
   rRNA, 97
Cytosine
   methylation, 140
   to 5–methylcytosine, 154
   residue, 381
   –sensitive, 193

Dcm MTase, 115
DEAE–cellulose paper, 86
3–deazaadenosine (DZA), 104, 247, 248
   and concanavalin–A, 108
   cytotoxicity, 248
   methylation inhibitor, 103
   presence, 111
   stimulating effect, 107
Deazaadenosine, 103
3–deaza–Ari
   effect on AdoMet and AdoHcy, 250
   potent antiviral, 249
3–deaza–($\pm$) aristeromycin, 247, 248
3–deaza nucleoside, 247

inhibition, 253
Demethylase activity, 99
Demethylation, 99, 148
   α–actin, 264
   CpG, 321
   Hpa II site, 321
   insulin I gene, 267
   mCpG, 317
   ribosomal RNA, 98
   T–DNA gene, 191
   studies, 96
Deoxyribonucleotide
   hybridization, 8
   synthesis, 200
Deoxyoligomer
   958–977, 9
   1506–1525 and rRNA, 9
Differential screening identification, 222
Dimethylation
   A2058 in 23S rRNA, 31
   adenosine, 19
Dimethylsulfate (DMS), 318
   footprinting, 322
Dinucleotide sequence, dCpdG, 139
DNA
   Ad2 E2A promoter, 340
   bank in YEp24 yeast, genomic, 51
   binding protein, 188
      human extract, 353
   double–stranded unmethylated, 141
   extracted, 384
   footprints by estradiol, 325–326
   FV3
      –encoded, 140
      probed with oligonucleotide, 144
   hemi–methylated or single–stranded, 141
   host cell hypomethylation, 155
   human Hpa II site, 290
   hypermethylation, 154, 187, 353
      global, 212
   linear amplification, 308
   male female difference analysis, 293
   MeTase, 140, 148
      activity, 141

   methylase, 199, 205, 230, 242
      activity, copolymer, 234
      affinity, ethylated CpG, 235
      enzymes, 159
      HeLa, 400
      and hemimethylated DNA, 208
      in maize, 200
      weight, 200
   methylation, 174, 179, 187, 258, 271, 299, 353
      altered, 211, 212, 213
      antagonist, 157
      assay, 285
      of COL1A1 gene, 177
      complete eukaryotic mapping, 280
      correlation, 313
      at CpG dinucleotides, 171
      FV3, 148
      in gene regulation, 172
      in genome imprinting, 368
      HIV role, latent, 164
      inactivates, 155
      influence, 343
      inhibition, 140, 205
      latency control, 165
      mapping SV40 gene, 278
      measure, 295
      de novo pattern, 341, 343
      partial eukaryotic mapping, 279
      pattern, 188, 194, 331
      perturbation in, 214
      postsynthetic, 114
      in plant, 188, 200
      in plant and mammal cell, 207
      regulatory, 382
      role, 199
      role, embryo gene activity, 259
      site–specific, 144
      vertebrate, 40
      and virus avoidance, 116
      in wheat difference, 205
   methyltransferase (MeTase), 8, 140, 199, 275, 353
      viral–encoded, 139
   modification–restriction enzyme, 113

molecule comparison, 116
and oligonucleotide hybridization, 129
protein binding, 330
  interaction, 331, 339
  single–strand, 334
repetitive mCyt residue, 204
replication, 120, 200
  and nascent, 118
  and recombination, 148
sequence, 331
strand upper (US), 319
synthesis effect, vaccinia virus, 253
tail quail troponin I, 369
thymidine incorporation, 251
transfection, 258
unmethylated, 116
vaccinia treated with 3–deaza–Ari, 252
DNAse I
  conformation sensitive, 259
  digested chromatin, pea, barley, corn, 195
  footprinting, 240
  hypersensitive site, 313, 316, 319, 325
  sensitive
    chromatin conformation, 196
    DNA plant, 193, 195
DNAase sensitivity, methylated, 260
Double helix, 326

E1A promoter, 145
  Ad12 DNA, 335
    methylation, 337
E2A
  gene, 342
    methylated, 334
  promoter, 340
    Ad2 DNA, 338, 342
    Ad2 DNA, late, 337
EcoP 15
  coding, 118
  encoding cells, 114
  endonuclease, 120

EcoRII
  behavior, 117
  and EcoP15 enzymes, 120
  hsd enzyme, 116
  target sites, 116
Ectoderm primitive, 374
Ehrlich ascites cell, 57, 58
  enzyme activity, 63
Electrotransfer to nylon membrane, 306
Embryogenesis, 258, 353
Embryo, mouse, Hpa II site #7 interest, 292
5′–end analyses of artificial mRNA, 71
Endonucleolytic action, 117
Enzymatic methylation, EMS–induced, 240
Enzyme
  GMP intermediate covalent, 46
  Hpa II/Msp I, 321
  ribosome reaction, 12–13
  stimulation test, 236
  yeast capping, 45
Epigenetic variant, 229
Erm
  family, 126
  gene
    encoding, 20
    inducible by erythromycin, 29
  methylase, 19, 20
    contents, 26
    descendants, 34
    evolutionary descendance, 27
    and ksgA methylase, 26
Erythromycin
  antibiotics, 20
  and kasugamycin, 34
  resistance, 31
Escherichia coli
  cell families, 113
  extract, 3
Estradiol, 313, 316
  injected, 323
  result, 321
  treatment, 325
Estrogen
  dependent binding activity 321

response element (ERE), 239, 313, 321
contact points, 324
Ethionine
treated mice, 58, 60
liver, 63
treatment, 98
Ethyl methanesulfonate (EMS), 230
hypothesis, 232
Eukaryotic
rRNA modification, 37
vertebrate, 37–38
virus FV3, 148

Fertilization, 367
Frog virus 3 (FV3) DNA
activation, 147
DNA, 140
purified, 141, 146
factor, 148
infection, 141, 145, 146
mutant, 147
classification, 143
replication, 145
wild type, 141

G + C
content, 382
rich, 154–155
Galactosamine (GAL) treatment, 91, 96, 97
animals, 99
Gamete-of-origin, 367, 368
Gametogenesis, 353, 367
GCG site placement, 237
Gel
analysis
L-cell transfection, 266
PRL+ EMS, 240
electrophoresis, 305
Gene II, avian vitellogenin, 302
Gene
α
actin, 260
CAT chimeric, 260
activation, 205, 313
CAT α actin promoter link, 261
c–Ha–ras–1 and insulin, 381
class I–IV, 21
cloning, bacterial host, 299
coding RNA adenosine dimethyl-transferase, 21
demethylation, 187
E2A of Ad2 DNA, 333
early SV40, 281
encode, 334
expression, 207
alteration, 212
and COL1A1 promoter, 173
differential, 258
regulation, 187
housekeeping and tissue–specific, 296
inactivation, 342
methylation, eukaryotic organism, 257
prolactin, 238
EMS inactivate, 239
tissue specific, 257, 259
Generation, pIM13 from pNE131, 127
Genetic
code, 344
instability, 231
parasitism, 189
structure of HIV, 155
Genome
DNA double–stranded, 139
FV3, 140, 144
imprinting, 367
non–modified, T7 and T3, 119
RSV, 88
Genomic
DNA, 52, 233, 315
isolated, 397
methylation, 140
expression screening, 45
footprinting, 305, 339
in vivo, 317
hypomethylation and tumor formation, 353
imprinting, 353, 389

and DNA methylation, 382
library of mRNA clone, 47
sequence, 313
   avian vitellogenin gene, 304
   classic scheme, 301
   Taq polymerase, 302
sequencing, 242, 263, 300, 314, 321
   detect 5–mC residue, 340
   HIV, 165
   scheme, 315
   Southern blot, 53
   template, DNA, 286
GMP
   comparison, 71
   and lysine residue, 46
Growth hormone
   (GH/3)
     cell PRL–deficient, 239
     cell tumor, cell line 230
   (GH), 229, 230
   –deficient line, 231
GTP and guanyltransferase, 63
Guanine–7–
   methylation, 58, 60
   methyltransferase, 61, 64
   isolated, 58
Guanosine residue, 317, 318, 323
   DMS reaction, 319
Guanylyltransferase gene, 51, 52, 54

HeLa
   cell, 329
   lysate system, 325
Helix distortion, 236
Hemiethylated TG/CA site, 242
Hemimethylated
   CpG site, 242
     comparison, 241
   DNA, 263
     molecules, 120
     site, 233, 359
Hepatocyte, 94
   metabolic status, 99
Hepatotoxins, 94, 100
   and liver damage, 93
   unrelated, 99
Herpes
   Simplex, 155
   viruses, 154
Heterogeneity, morphological character, 213
HhaI and HpaII sites, 241
3′ HhaI site S194, 266
High performance liquid chromatography (HPLC), 4, 7
HIV–1 infection incubation, 163
HIV
   infection, 153, 155
     T– cells, 154
   latency, 154
     reactivation, 155
   LTR, 162
     probe, DNA, 163
     silence transcription, 153
     target region, 158
     transfected, 157
   LTRCAT DNA (pU3RIIICAT), 160
   trans–acting factor, 159
   transactivator *tat*, 153
Homocysteine thiolactone (HCY), 104
Homologous regions, 33
Homology, 19
   between eukaryote and prokaryote, 39
   between sites, 23S and 16S rRNA, 33
Hpa II, 396
   –PCR
     assay site 7, 294
     method, 295
HPLC. *See* High performance liquid chromatography
H–ras gene in culture, 399
Hybridization
   autoradiographic, 374
   of BamHI plus HpaII, 371
   differential, 221
   to DNA, 374
   embryo, 373
   membrane, 307
   probe, 306

results, 130
Hydrazine, 315
Hydroxylapatite chromatography, 202
Hypermethylated cap
  analog, 70
  normal cell mRNA, 78
  structure, 68, 74
    guanine, 67
  on SV 26S mRNA translation, 77
Hypermethylation
  inactivation, 157
  of HIV LTR CpG sequence, 165
Hypomethylated
  colorectal carcinoma, 212
  gene, 172
Hypomethylation, 97, 98, 99, 108, 155, 224, 313
  carcinogenesis change, 213
  DNA, 212, 213
  HTF island, 401
  gamete–specific, 353
  rRNA, 94, 95, 99

IE promoter, 148
Immunoscreening of λgt11 yeast, 47
Imprint
  mechanism, global, 375
  parental, 172
Inhibition, ß–globin synthesis, 70
Initiation factor, 154
In vitro
  5′–CCGG–3′ methylated, 342
  DNAse I footprint, 320
  enzymatic methylation, 158
  HpaII site, 262
  HSV TK DNA, 281
  methylated
    DNAs, 188
    late E2A promoter, 335
    SV40 and HSVI TK DNA, 276
  methylation, 97, 160, 162, 204
    adenovirus, 161
    plasmid DNA, 156
    /transfection, 154
  model system, 213
  protein

in vivo binding, 343
  synthesis, 97
  synthetic and rRNA methylation, 100
  synthesis cloning 9
In vivo
  cell treatment, 5–azaC, 189
  footprinting, 313, 330
    DMS, 318
    ERE, 323
  interaction protein, 322
  methylated rRNA, 99
  methyl–labeled rRNA, 99
  protein–DNA interaction, 319
    and *in vitro*, 314, 317
  recognition sequence, 299
  RNA transcription, 258
Isotopic equilibrium, 251

Junctional sequence, Tn554, 134
  inserts, 132
Junction complement, GATGTC, 134

Kasugamycin (ksgA mutants)
  antibiotic, 20
  mode of action, 31
Kinetic property, 63
KsgA
  cistron, 29
  and erm
    amino acid sequence, compared, 23
    system correspondence, 30
  gene, 20
    assumed ancestral methylase, 27
  methylase
    ancestral, 34
    cross–reaction, antibody, 28
    homology, 19
  mRNA regulating elements, 29

L5 cell line, 111
  myoblast, 103, 104
    fusion effect, Con–A, 108
    in vitro, 110

Latency
    and 5–azaCR treatment, 214
    HIV
        activation, 157
        LTR, 156
λC3 structure, restriction site, 49
Leiomyomas, 395, 405
    clonal, 403
    histologically homgeneous, 404
Leukocyte, female, 286
Linear amplification, 315
5' LTR promoter, 180
Lytic replication, latency *tat* role, 154

M5C
    depleted, human DNA, 353
    formation, 4
    methyltransferase purification 5
    and m2G
        activity preference, 5
        formation, 8
    rich phage, 353
M5CpG site, 353
M6
    2A formation in 23S RNA, 12
    A synthesis and formation, 3
Macrolide lincosamide streptogramin (MLS), 20, 126
Maloney retrovirus, 155
Mammotroph, PRL–producing, 230
MDBP. *See* Methylated DNA binding protein
5'–Me7G cap structure, eukaryotic mRNA, 68
Methionine, 108, 250
Methylase
    encoding, 126
    HpaII or MspI, 165
    ksgA and erm, 30
    non–CpG, 164
Methylated
    bases, 2
        RSV virion RNA, 85
        cap, mRNA, 84
        DNA, 140

binding P consensus sequence, 355
binding protein (MDBP), 351, 353; *see also* Protein, methylated DNA binding
binding protein pattern, 358–359
E2A promoter Ad2 DNA and unmethylated, 329
nucleotide, 2, 3, 60, 84
    formed, 4
    number, 2
pB site 1 to MDBP binding comparison, 360
phase ϕ DNA, 204
plasmid, 148
    DNA, 160
23S RNA nucleoside analysis, 12
SV40 DNA, 281
Methylate T7 DNA, Dam, Dcm MTases, 115
Methylation, 2
    activities, 3
    analysis of Igκ, 265
    3.4B2 locus, 397
    c–Ha–*ras*–1, 385
    comparison, sperm and blood, 402
    CpG island housekeeping gene, 382
    data loci, blood, 402
    EMS induced, 229
    eukaryotic, 283
    finishing–up, 101
    gene, human sperm, 383
    heterogenous at RSV RNA site, 88
    HIV
        DNA, 153
        LTR, 156
        LTR CpG sites, 161
        profile, 165
    Hpa II, 283
    hybridization–protection, 9
    ICR169 promoter, 147
    inhibitor, 89, 247
    insulin gene, fetal and sperm cell, 387
    internal universal substrate, 84
    JCZ67 locus, 401

levels, protein–free DNA, plant, 195
LTK fibroblasts, 261
mammal, gene expression regulation, 395
mc6A site, 87
methyltransferase, rat liver, 281
mosaic model, 376
MspI site, 367
mRNA expression, 58
native DNA, 203
pAd12E1A–*cat* template, 146
partial SV40, late gene, 279
pattern, 399
    Hpa II site, 285
    inheritance, 406
    mosaic, 403
    parental genome, 388
    promoters, 340
    and T–DNA gene expression, 190
    transgene, 367
phenotype, 368, 372
    gamete–of–origin, 374
probabilistic event, 404
promoter inhibition, 276
sensitive
    enzyme, 396
    isoschizomer, 382
    site, 397
sex–specific mouse DNA, 290
site, 353
30S ribosome sequence deduction, 13
state APRT gene, 269
TK gene, 281
and transcription, COL1A1 gene, 174
5 methylcytosine (m^5C), 164, 325, 356
    content, 212
        human tissue, 352
    and CpG dinucleotide, 187
    deamination, 390
    free DNA fragment, 159
    pattern, 286
Methylcytosine (mCyt), 199
    content, 200

DNA
    plant, 201
    substituted phage, 204
    level, 205
    regulatory, 207
    in wheat, 199
        DNA distribution, 204
        formation, 200
5–methyldeoxycytidine (5–mC), 332, 338
    residue, 342
7–methyl–guanosine, 60
Methyl sensitive site, 258
Methyltransferase (MTases), 114, 115, 148
    activity, 142
    assay, 59
    distribution, 4
    Hpa II, 280
    in nuclear extracts, 57
    rRNA isolation, 12
Mg^{++} concentration, 4
MLS antibiotic action, 30
Model test, 234
Modified cap structure, 68
Molecule, chimeric, 262
Moloney murine leukemia virus (M-MuLV), 172
Monolayer morphology normal, 217
Morphology, 5–azaCdR–treated cell, 218
Mosaic
    degree, 377
    methylation, 395
        JCZ67 locus blood, 398
        leiomyomas, 403
        model, 367, 374
        unbalanced, 375
Mov13
    cell, 173
    mice, 172
mRNA
    analysis, 145
    ß-globin with capping nucleotide, 74
    5′ cap, 70
        terminal structure, 58

vaccinia, 253
viral, 247
viral, methylation inhibition, 248
capped, 77
capping, enzyme yeast, 54
eukaryotic, 46
    and cap structure, 58
    higher, 84
    lower, 84
globin
    and histone, 84
    with nonstandard caps, 73
guanylyltransferase, 45, 46, 54
    expression, 49, 50
    gene disruption, 53
    in situ, 47
IE synthesis, 147
influenza number of m^6A residues, 84
level, tissue–specificity, PGK–1, 291
Me^{7}G capped, 78
methylase, 29
PGK comparison, 292
prolactin methylated, 84
protection against nuclease, 46
splicing
    complex in trypanosomes, 68
    transport stability inhibition, 89
and src protein, 88
SV
    26S translation and cap structure, 77
    terminated with hypermethylated cap, 75
synthesis, 154
yeast
    capping enzyme, 53, 54
    Northern blot analysis, 52
MspI
    control lane, 399
    site, 368
Murine alpha 1 type I procollagen (COL1A1), 172
Mutagenesis
    site–directed, RNA, 33

studies, sites 7414 and 7424, 88
Mutation
    and childhood cancer, 382
    rate, 381
Myoblast differentiation, 104

N7–
    ethyldGTP, synthesized, 236
    methylation, transcription initiation, 77
Neoplastic transformation, 211
N^6–methyladenosine (m^6A), 83
    distribution, 84
    residue localization, 88
    RNA role, 89
    in RSV RNA, 86
Nonmethylated sites, restricted, 120
Nontransformed
    ancestral cell, 217
    cell, treated, 221
Northern blot analysis, 291
De novo
    methylase, 374
    methylation, 178, 341, 342
    purine synthesis, 103
NPI binding, 165
Nuclear
    methyltransferase, 60
    protein, fractionated, 319
Nucleoside
    analysis, 10
    m^5C, m^5U, m^2G, 12
Nucleotide
    activity, ribosomal RNA, 14
    altered, 133
    capping, α–phosphate, 68
    modified
        distribution, ribosome, 42
        in human ribosome, 38
        in human rRNA, 38, 41
Nylon membrane, 315

Ocr
    protein
        inhibitor, Type II enzymes, 115
        phage T7 and T3, 115

Oligonucleotide, 86, 113
  label, 319
    synthetic, 320
    m^{5}C–containing, 354
    unmethylated and methylated, 320
Oncogene
  expression, DNA methylation in, 188
  ipt, 193
  mutation, 213
Onset of clinical manifestation HIV–1, 163
Oocyte, 176, 179
  DNA, 258
  frog, 281
  human, 388
  Xenopus laevis, 180, 276, 335
ORF–1, 50, 51
  and –2 open reading frames, 54

Particle 30S as substrate, 2
Peptidyl
  transferase center of A2058, 33
  tRNA and mRNA interaction, 30
Phenotype mutant, 376
7–phenyl cap analog derivative, 71
Phorbol esters, 154
Phosphoglycerate kinase (PGK–1), 286
  activity specific, 292
  enzyme, 292
Phosphorylation in wheat, 205
Phytohemagglutinin, 154
Piperidine, 315
Pituitary differentiation, 230
Plant
  DNA
    methylation, 188
    methylcytosine, 207
  gene expression, 188
  genome
    methylated, 188
    response to 5–azaC treatment, 188
  methylated DNA region, 207
  reversibility, 206
Plaque formation inhibition, vaccinia virus, 249
Plasmid
  adeno–virus 12 E1A promoter, 144–145
  circular, 180
  construct, 263
  DNA, 156
    encoding, 159
    unmethylated and hemimethylated, 201
  pGEM–3Z, 63
  pNE131, 125, 126
  sequences, 145
PNE131 as vector for ermC and Tn554, 135
Polyacrylamide gel, 162
Polyadenylation, 84
Polymerase
  chain reaction (PCR), 156, 159, 309
    assay 286, 295
    Hpa II site, 286–287
  cloned, 3
Polymerization, 30
Polyoma early gene and SV40, 275
Polypeptide, 142
  clone–specified, 48
  hormone, 230
  pattern, 200
  prolactin EMS–induced, 237–238
Pretransformation, 225
Pretransformed cell, 211, 221, 223
  ancestral, 220
Probabilistic
  methylation YNZ22 locus, 405
  phasing, 406
Prokaryote cell production, 114
Prolactin (PRL), 229, 230
  deficient
    GH/3 cell, 239
    line property, 231
  gene
    content, 242
    EMS target, 240
    promoter, 239
    rat, 238
  promoter, 240

synthesis, 230
Promoter
    α actin, 263
    element distal, 238
    of ICR169, 147
    methylation role, 342
    region SV40 and TK, 277
Pronuclear
    fusion, mouse zygote, 367
    transplantation imprinting, 368
Protein
    binding, 338
        globin promoter, 339
        nuclear to HIV, 163
        short fragments, 339
    DNA
        free, 196
        interaction,
            313, 324
        gene storage in maize, correlated, 188
        interaction, direct, 115
        kinase and phosphatase code, 148
        methylated DNA binding (MDBP), 208
    NHP1, 323
    synthesis, 93, 94
        absence, 147
        early defects, 99
        in vitro, 2, 30
    in vitro, 30
Proto–oncogene c–myc, human, 362
Proviral genome, 176
Provirus
    induced interference, Mov13 mice, 179
    insertion, 171, 178, 180, 181
        and COL1A1, 172–173

RAC sequence, 84, 88
rDNA 23S analysis, 9–10
5' region, mouse PGK–1 gene, 289
Reporter genes (CAT), 153
Residue
    bacterial 6mA, 263
    m^5C, 351

Restriction
    digestion, 307
    endonucleases, 115, 117
        (and MTases), 121
    enzyme, 113, 193
        site, 194
    pattern
        Tn554, 128, 131
        site, chicken liver, 314
Reticulocyte lysate, 70
Retroviral
    genome inserted, 171
    LTRs, 154
    murine, 164
Reverse transcriptase release, 154
RGACU
    consensus sequence, methylation, 85
    sequence, 88
Ribonuclease and pronase treatment, 202
Ribosomal
    function, 2
    gene methylation plant development, 188
    subunits, 20
Ribosome
    protein synthesis *in vitro*, 93
    resistant to MLS, 126
    16S RNA, 23S RNA, and 30S, 14
    30S synthetic, 7
    substrates, rRNA methyltransferase, 14
    triple mutant, 31
RNA (guanine–7–) methyltransferase, 57
    biosynthesis rate, 111
    chains nascent, 46
    genome, 85
    hypermethylation, 111
    messenger, 109
    methylation, 111
        affected by DZA and Con-A, 108
        Con–A increase, 108
        inhibition, DZA, 111
        internal, 85
        level changes, 103

422 Index

rate, 110
ribosomal, 93, 100
16S, 10
23S, 10–11, 14
in T–lymphocyte, 103, 104–105
variation, 109
methyl–deficient, 61
methyltransferase 16S
m⁵C substrates, 7
molecular weight, 6
natural, 6
23S methylated *in vivo*, 10
polymerase II, 46, 54, 140, 148
polymerase, 118, 144
T7, 10
purification, 62
16S, 2, 3, 4
enzyme fraction, 8
synthetic, 7
slicing and cap structure, 68
splicing, 54
16S and 23S ribosomal of *Escherichia coli*, 1–4
23S unmethylated synthetic, 6
synthesis, 110
effect, vaccinia virus, 253
transcript *in vitro*, 64
treated and untreated cell, 222
5′–triphosphatase, 45, 54
triphosphatase *in situ*, 47
Rous sarcoma virus (RSV), 84, 253
infection, 85, 88
mutant, 89
RNA, 86, 87
rRNA
eukaryote
complexity, 37
location, 43
human
2′–0–methylation frequency, 40
modified, nucleotide, 39, 41
LSU, 40–42
methylation, 3, 98, 99
and protein synthesis, 101
methyltransferase
methyl–deficient, 11

specific for, 13
polymerase T7, 3
protection, 8
23S, 9–10, 31
adenosine residue, 34
16S and 23S structure models, 32
SSU, 39–42

S–adenosylhomocysteine (AdoHcy)
hydrolase, 104, 247
reaction, 249
inhibitors, 247
S–adenosylmethionine, 204
Sensitivity
azacytidine, 206
DNA DNAse I, 187
Sequence
alteration, 376
cellular, 173
5′–CG–3′, 342
C–N, 201
dinucleotide, 208
cis acting, 257
CpG, 391
island, 268
dCpdG, 147, 148
DNA, 195
EMS target, 230
gel, 310
genomic, 258, 300
GGG, 165
HIV LTR, methylation sensitive, 164
hypermethylated, 155
LTR, 157
primer labeling, 308
procedure, 310
–specific
DNA–binding protein, 354
methylation, 343
promoter methylation, 329, 342
Sequencing gel, 315
Sexual dimorphism effect, ethionine, 97
snRNA maturation, 77
snRNP assembly, 77
Somatotrophs GH–producing, 230
Southern blot, 290

analysis, 129, 130
Splicing, mRNA, 84
Src
  gene, 88, 89
  mRNA, 89
Stability, mRNA, 84
Staphylococci, encode MLS resistance, 125
Stoichiometry, m^{5}C, 8
Streptomycin resistance mutations, 31
S–tubercidinylhomocysteine, 253
Subunit 30S, 3
  early assembly, 5
SV40
  enhancer, 283
  gene, 283
  and HSV TK, regulatory region, 277

T–antigen, 277
Taq
  I
    digest, 384
    site, 383
  polymerase, 308, 315
Target site, number increase, 188
Tat, 165
T7 DNA, 115, 116
  resistance, 113
  in vitro, Dam MTase, 116
T–DNA
  copy, silent, 194
  in crown gall tumor cell, 188
  gene expression, 187, 189
    analysis, 190
    re–initiation, 191
  methylation level, 192
  in plant cell, 187
  oncogene expression, 189
5′–terminal cap structure, 84
5′–triphosphatase activity, 47
5′–triphosphate termini, 46
Thymidylate synthase inhibition, 206
Thymine and EcoRII recognition sequence, 117
Tissue specificity, PGK mRNA level, 291

TK gene
  early and late SV40 enhancer, 282
  inactivation, 283
  methylation, 280
TLC analysis, 60
Tobacco T–DNA gene, 189
Trans–acting
  agent, 160
  factor, 155, 164, 165, 181
    and 5–mC, 343
  protein role, 147
Transactivation
  expression *cis*–level, 165
  HIV LTR inhibition, 159
Transcription
  and DNA methylation control, 362
  early and late, 277
  factor–binding assays, 159
  frog virus 3, 147
  inhibition, 276, 283
  initiation point, 286
  genes inhibited, 154
  and MDBP, 361
  PRL, 230
  promoted, 145
  TK gene, 276
  vitellogenin gene, 322
  in vitro cell–free, 336
Transcriptional
  activation, 178
  activity, 172, 175, 187
    COL1A1 promoters, 176
  block, 165
  control, 181
  inhibition, 155
    frog oocyte, 281
  initiation, 176
    cis–level, 164
  interference and downstream gene, 180
  repression, 171
    COL1A1, 181
  suppression, 175
Transfection
  /electroporation, 156
  transient, 262

Transformation
  cell destined, 213
  –committed cell, 214–215
  gene
    expression, 211
    isolation, 220
  neoplastic, 211
  phenotypically normal, 218
  role, 212
  in soft agar, 219
Transformed cell, 217
Transgene
  hypermethylated, female, 375
  hypomethylated, male, 375
  hypomethylation, 368
  locus, 367
  methylation
    found animal, 369
    phenotype, 367, 373
    soma and germline, 370
Transgenic mice, 367
Translation, mRNA, 84
Trimethylated cap, snRNA, 77
tRNA molecule, nucleotide in, 38
Trophectoderm cell, 374
Troponin I
  quail, 367
  transgene, 367
Tryptophan conversion, 189
TS22Δ34 translation, 78
Tumor
  cell, 212
  formation, 403
  human, 212
  sporadic, 389
Tumorous growth induction, 5–azaC, 190
Type III recognition sites, 119

Ultraviolet light, 157
Undermethylation, 89
Unmethylated
  DNA, 172, 263
  pAd2E2AL–CAT construct, 342

Variable number of tandem repeats (VNTR), 382

polymorphism, reflected, 396
region, 384, 386
system, heterozygous individual, 398
5′–vector sequence, 75
Viral
  DNA, 140
  LTR, 154
  mRNA
    synthesis, 154
    transcription, 68
  site MDBP
    nonbinding, 357
    specific, 356
Virion–associated protein 147
Virus
  –induced proteins, 147
  latency, preservation mechanism, 157
  replication, 154
Vitellogenin
  gene, 316, 317
    avian, 313
    embryonic, 314
    response element, 321
  mRNA, 316
    synthesis, 313, 321, 325
    transcription, 322
VNTR. *See* Variable number of tandem repeats

Wheat
  DNA, 201, 202
    fraction methylation, 204
    methylase, 200, 201, 203
    methylase, *in vitro*, 208
    methylation, azacytidine effect, 206
  embryo, 200, 201
  enzyme, 200
  germination, 200
Wild type
  cell, collagen–producing, 175
  copy, 390
  gene, collagen–expressing fibroblast, 178
  line, 231
  and PRL–deficient cell, 240

reversion, 232

X chromosome, 257, 381–382
  active, 296
  housekeeping gene, 270
  inactive, 286
  male
    leukocyte, 286
    single, 293

  methylation, 270
X inactivation, 286, 395
X–linked housekeeping gene HPRT, 286

Yeast
  capping enzyme
    α and β chains, 54
    gene, 48
  genomic DNA expression library, 48